Lecture Notes in Physics

Vol. 100: Einstein Symposion Berlin. Proceedings 1979. Edited by H. Nelkowski et al. VIII, 550 pages. 1979.

Vol. 101: A. Martin-Löf, Statistical Mechanics and the Foundations of Thermodynamics. V, 120 pages. 1979.

Vol. 102: H. Hora, Nonlinear Plasma Dynamics at Laser Irradiation. VIII, 242 pages. 1979.

Vol. 103: P. A. Martin, Modèles en Mécanique Statistique des Processus Irréversibles. IV, 134 pages. 1979.

Vol. 104: Dynamical Critical Phenomena and Related Topics. Proceedings, 1979. Edited by Ch. P. Enz. XII, 390 pages. 1979.

Vol. 105: Dynamics and Instability of Fluid Interfaces. Proceedings, 1978. Edited by T. S. Sørensen. V, 315 pages. 1979.

Vol. 106: Feynman Path Integrals, Proceedings, 1978. Edited by S. Albeverio et al. XI, 451 pages. 1979.

Vol. 107: J. Kijowski, W. M. Tulczyjew, A Symplectic Framework for Field Theories. IV, 257 pages. 1979.

Vol. 108: Nuclear Physics with Electromagnetic Interactions. Proceedings, 1979. Edited by H. Arenhövel and D. Drechsel. IX, 509 pages. 1979.

Vol. 109: Physics of the Expanding Universe. Proceedings, 1978. Edited by M. Demiański. V, 210 pages. 1979.

Vol. 110: D. A. Park, Classical Dynamics and Its Quantum Analogues. VIII, 339 pages. 1979.

Vol. 111: H.-J. Schmidt, Axiomatic Characterization of Physical Geometry. V, 163 pages. 1979.

Vol. 112: Imaging Processes and Coherence in Physics. Proceedings, 1979. Edited by M. Schlenker et al. XIX, 577 pages. 1980.

Vol. 113: Recent Advances in the Quantum Theory of Polymers. Proceedings 1979. Edited by J.-M. André et al. V, 306 pages. 1980.

Vol. 114: Stellar Turbulence. Proceedings, 1979. Edited by D. F. Gray and J. L. Linsky. IX, 308 pages. 1980.

Vol. 115: Modern Trends in the Theory of Condensed Matter. Proceedings, 1979. Edited by A. Pekalski and J. A. Przystawa. IX, 597 pages. 1980.

Vol. 116: Mathematical Problems in Theoretical Physics. Proceedings, 1979. Edited by K. Osterwalder. VIII, 412 pages. 1980.

Vol. 117: Deep-Inelastic and Fusion Reactions with Heavy Ions. Proceedings, 1979. Edited by W. von Oertzen. XIII, 394 pages. 1980.

Vol. 118: Quantum Chromodynamics. Proceedings, 1979. Edited by J. L. Alonso and R. Tarrach. IX, 424 pages. 1980.

Vol. 119: Nuclear Spectroscopy. Proceedings, 1979. Edited by G. F. Bertsch and D. Kurath. VII, 250 pages. 1980.

Vol. 120: Nonlinear Evolution Equations and Dynamical Systems. Proceedings, 1979. Edited by M. Boiti, F. Pempinelli and G. Soliani. VI, 368 pages. 1980.

Vol. 121: F. W. Wiegel, Fluid Flow Through Porous Macromolecular Systems. V, 102 pages. 1980.

Vol. 122: New Developments in Semiconductor Physics. Proceedings, 1979. Edited by F. Beleznay et al. V, 276 pages. 1980.

Vol. 123: D. H. Mayer, The Ruelle-Araki Transfer Operator in Classical Statistical Mechanics. VIII, 154 pages. 1980.

Vol. 124: Gravitational Radiation, Collapsed Objects and Exact Solutions. Proceedings, 1979. Edited by C. Edwards. VI, 487 pages. 1980.

Vol. 125: Nonradial and Nonlinear Stellar Pulsation. Proceedings, 1980. Edited by H. A. Hill and W. A. Dziembowski. VIII, 497 pages. 1980.

Vol. 126: Complex Analysis, Microlocal Calculus and Relativistic Quantum Theory. Proceedings, 1979. Edited by D. Iagolnitzer. VIII, 502 pages. 1980.

Vol. 127: E. Sanchez-Palencia, Non-Homogeneous Media and Vibration Theory. IX, 398 pages. 1980.

Vol. 128: Neutron Spin Echo. Proceedings, 1979. Edited by F. Mezei. VI, 253 pages. 1980.

Vol. 129: Geometrical and Topological Methods in Gauge Theories. Proceedings, 1979. Edited by J. Harnad and S. Shnider. VIII, 155 pages. 1980.

Vol. 130: Mathematical Methods and Applications of Scattering Theory. Proceedings, 1979. Edited by J. A. DeSanto, A. W. Sáenz and W. W. Zachary. XIII, 331 pages. 1980.

Vol. 131: H. C. Fogedby, Theoretical Aspects of Mainly Low Dimensional Magnetic Systems. XI, 163 pages. 1980.

Vol. 132: Systems Far from Equilibrium. Proceedings, 1980. Edited by L. Garrido. XV, 403 pages. 1980.

Vol. 133: Narrow Gap Semiconductors Physics and Applications. Proceedings, 1979. Edited by W. Zawadzki. X, 572 pages. 1980.

Vol. 134: $\gamma\gamma$ Collisions. Proceedings, 1980. Edited by G. Cochard and P. Kessler. XIII, 400 pages. 1980.

Vol. 135: Group Theoretical Methods in Physics. Proceedings, 1980. Edited by K. B. Wolf. XXVI, 629 pages. 1980.

Vol. 136: The Role of Coherent Structures in Modelling Turbulence and Mixing. Proceedings 1980. Edited by J. Jimenez. XIII, 393 pages. 1981.

Vol. 137: From Collective States to Quarks in Nuclei. Edited by H. Arenhövel and A. M. Saruis. VII, 414 pages. 1981.

Vol. 138: The Many-Body Problem. Proceedings 1980. Edited by R. Guardiola and J. Ros. V, 374 pages. 1981.

Vol. 139: H. D. Doebner, Differential Geometric Methods in Mathematical Physics. Proceedings 1981. VII, 329 pages. 1981.

Vol. 140: P. Kramer, M. Saraceno, Geometry of the Time-Dependent Variational Principle in Quantum Mechanics. IV, 98 pages. 1981.

Vol. 141: Seventh International Conference on Numerical Methods in Fluid Dynamics. Proceedings. Edited by W. C. Reynolds and R. W. MacCormack. VIII, 485 pages. 1981.

Vol. 142: Recent Progress in Many-Body Theories. Proceedings. Edited by J. G. Zabolitzky, M. de Llano, M. Fortes and J. W. Clark. VIII, 479 pages. 1981.

Vol. 143: Present Status and Aims of Quantum Electrodynamics. Proceedings, 1980. Edited by G. Gräff, E. Klempt and G. Werth. VI, 302 pages. 1981.

Lecture Notes in Physics

179

Dynamical Systems and Chaos

Proceedings of the Sitges Conference
on Statistical Mechanics
Sitges, Barcelona/Spain
September 5 – 11, 1982

Edited by L. Garrido

Springer-Verlag Berlin Heidelberg GmbH 1983

Editor

Luis Garrido
Departamento de Física Teórica, Universidad de Barcelona
Diagonal 647, Barcelona-28, Spain

ISBN 978-3-540-12276-0 ISBN 978-3-540-39594-2 (eBook)
DOI 10.1007/978-3-540-39594-2

2153/3140-543210

ACKNOWLEDGEMENT

I would like to take this opportunity to thank all those who collaborated in the organization of this Conference. In particular, my deep appreciation goes to Professor D. Jou from the Department of Thermology of the Autonomous University of Barcelona for his great enthusiasm in solving the numerous problems involved in preparing and running the Conference, and to Professor M. San Miguel and J.M. Sancho from the Department of Theoretical Physics of the University of Barcelona for their efforts and cooperation.

I also extend my warmest thanks to the City of Sitges for allowing us to use the museum "Maricel" as a lecture hall, and to the International University "Menendez Pelayo" for its economic support.

My final thanks go to my wife for her patience and unremitting cooperation.

L. Garrido

Participants

Dr. AGUIRRE, E., Universidad Complutense, Facultad de Matemáticas, Ciudad Universitaria, Madrid-3, Spain

Mr. ALSEDA I SOLER, Ll., Universitat Autónoma de Barcelona, Facultat de Ciencies Economiques, Bellaterra, Barcelona, Spain

Prof. ATTEN, P., C.N.R.S., Laboratoire d'Electrostatique, Grenoble, France

Prof. AUBANELL POU, A., Universitat de Barcelona, Facultat de Matemátiques, Barcelona-7, Spain

Prof. AUGE, J.; Universitat de Barcelona, Facultat de Matemátiques Gran Via 585, Barcelona-7, Spain

Prof. d'AURIAC, A., C.N.R.S., C.R.T.B.T., Grenoble, France

Miss BAESENS, C., Université Libre de Bruxelles, Faculté des Sciences, Campus Plaine, Blvd. du Triomphe, 1050 Bruxelles, Belgium

Dr. BARCONS, F., Universidad de Santander, Facultad de Ciencias, Santander, Spain

Dr. BEN-JACOB, E., Institute for Theoretical Physics, Santa Barbara CA 93106, USA

Mr. BENSENY, A., Facultat de Matemátiques, Universitat de Barcelona, Barcelona, Spain

Prof. BOUNTIS, T.C., Clarkson College of Technology, Potsdam, NY 13676, USA

Prof. BRANNER-JØRGENSEN, B., Mathematical Institute, Lyngby, Denmark

Prof. BREITENECKER, M., Institute for Theoretical Physics, University of Vienna, Vienna, Austria

Dr. van den BROECK, C. Vrye Universiteit Brussel, 1050 Brussel, Belgium

Mr. CALSINA, A., Secció de Matemátiques, Facultat de Ciencies, Universitat Autónoma de Barcelona, Bellaterra, Barcelona, Spain

Mr. CALVO HERNANDEZ, A., Facultad de Ciencias, Dpto. Termología Salamanca, Spain

Mr. CARBONELL, J., Institut des Sciences Nucléaires, Grenoble, France

Prof. CARL, H., Institut de Physique Théorique, Université de Lausanne, 1015 Dorigny, Switzerland

Prof. CASARTELLI, M., Sezione Teorica, Università di Parma, Parma, 43100, Italy

Miss CASASAYAS I MAS, J., Universidad de Barcelona, Facultad de Matemáticas, Barcelona, Spain

Prof. CASATI, G., Istituto di Scienze Fisiche, Università degli studi di Milano, 20133 Milano, Italy

Prof. CAWLEY, R., US Naval Surface Weapons Center, Silver Springs, MD 20910, USA

Prof. COSNARD, M., C.N.R.S., Laboratoire IMAG, Grenoble 38041, France

Mr. DANA, I., Technion, Dept. of Physics, Technion City, 32000 Haifa, Israel

Mr. DELSHAMS, A., Facultat de Matemátiques, Universitat de Barcelona, Barcelona-7, Spain

Mr. DE SOLÁ-MORALES, J., Facultat de Ciencies, Secció de Matemátiques, Universitat Autónoma de Barcelona, Bellaterra, Barcelona, Spain

Prof. DILÃO, R., Centro de Física de Materia Condensada, 1699 Lisboa, Portugal

Dr. EDERY, D., C.E.A., STGI-Fusion, Fontenay-Aux-Roses 92260, France

Prof. DIAS DE DEUS, J.V.C., Centro de Física de Materia Condensada, 1699 Lisboa, Portugal

Mr. ELGIN, J., Imperial College, Dept. Physics, London SW 7, Great Britain

Prof. EU, B. Ch., McGill University, Dept. of Chemistry, Montreal PQ Canada H3A 2K6

Prof. FALQUÉS SERRA, A., Universidad Politécnica Barcelona (E.T.S. Enginyers Camins), Barcelona-34, Spain

Prof. FARQUHAR, I.E., University of St. Andrews, School of Physical Sciences, North Haugh, St. Andrews, Fife, Scotland

Prof. FEIGENBAUM, M. J., Los Alamos Laboratory, Los Alamos NM 37545, USA

Mr. FONTICH JULIÁ, E., Universitat Politécnica de Barcelona, Barcelona-34, Spain

Mr. FONT, J., Facultat de Matemátiques, Universitat de Barcelona, Barcelona-7, Spain

Prof. FRISCH, U., CNRS, Observatoire de Nice, 06007 Nice, France

Prof. FROYLAND, J., Institute of Physics, University of Oslo, Blindern, Oslo 3, Norway

Dr. GAST, O.W., Theoretical Physics I, University of Münster, 44 Münster, Germany

Prof. GARRIDO, L., Universidad de Barcelona, Facultad de Física, Diagonal 647, Barcelona-28, Spain

Prof. GEISEL, T., Universität Regensburg, Institut für Physik I, 8400 Regensburg, Germany

Prof. GIACHETTI, R., I.N.F.N. Sezione di Firenze, 50125 Firenze, Italy

Prof. GIGLIO, M., CISE S.P.A. - 20100 Milano, Italy

Prof. GILBERT, A.D., Dept. of Mathematics, University of Edinburgh Edinburgh EH9 3JZ, Scotland

Dr. GILLOT, Ch., Institut National des Sciences Appliquées, 31077 Toulouse, France

Mr. GLENDINNING, P., University of Cambridge, DAMTP, Cambridge CB3 9EW, Great Britain.

Prof. GOMEZ MUNTANÉ, G., Universitat Autónoma de Barcelona, Facultat de Ciencies, Bellaterra, Barcelona, Spain

Prof. GONZALEZ-GASCON, F., Universidad de Salamanca, Facultad de Ciencias, Salamanca, Spain

Mr. GRAU, M., Facultad de Matemáticas, Universidad de Barcelona, Barcelona-7, Spain

Mr. GRUNBAUM, K., University of Roskjilde, IMFUFA, 4000 Roskjilde Denmark

Prof. GUCKENHEIMER, J., University of California, Dept. of Physics Santa Cruz 95064, USA

Prof. GUMOWSKI, I., Université Paul Sabatier, 31062 Toulouse, France

Mr. HAUBS, G., Institut für Theoretische Physik, I, Stuttgart-80, Germany

Dr. HENTSCHEL, H. G. E., Weizmann Institute of Science, Rehovot, Israel

Miss HERNANDEZ, A., Universidad de Barcelona, Facultad de Física, Diagonal 647, Barcelona-28, Spain

Prof. HUBERMAN, B.A., Xerox Corporation, Palo Alto Research Center, Palo Alto, CA 94304, USA

Dr. JAUSLIN, H.R., Université de Genève, Dept. de Physique Théorique, 32, Blvd. d'Ivoy, 1211 Genève, Switzerland

Prof. JOU MIRABENT, D., Universidad Autónoma, Facultad de Física, Bellaterra, Barcelona, Spain

Prof. KAPLAN, H., Syracuse University, Dept. of Physics, Syracuse, NY 13210, USA

Prof. KATOK, A., University of Maryland, Dept. of Mathematics, College Park, MD 20742, USA

Dr. KELLER, G., Institut für angewandte Mathematik, Universität Heidelberg, Im Neuenheimer Feld, 69 Heidelberg, Germany

Prof. KRUSKAL, M.D., Princeton University, Physics Dept. Princeton NJ 08544, USA

Dr. KUNICK, A., Rechenzentrum/ZOD 52, 8520 Erlangen, Germany

Dr. KURLAND, H.L., Boston University, Dept. of Mathematics, Boston Mass. 02215, USA

Prof. LIBCHABER, A., Groupe de Physqique des Solides, Ecole Normale Supérieure, 24 rue Lhomond, 75231 Paris, France

Prof. LLIBRE, J., Universitat Autónoma de Barcelona, Facultad de Ciencias, Bellaterra, Barcelona, Spain

Mr. LOPEZ DESA, J.M., Facultad de Matemáticas, Universidad Complutense de Madrid, Ciudad Universitaria, Madrid-3, Spain

Prof. LOVESEY, S.W., Rutherford Appleton Laboratory, Chilton, Oxfordshire OX11 0QX, Great Britain

Mr. LLEBOT, J.E., Dpto. de Termología, Universidad Autónoma de Barcelona, Bellaterra, Barcelona, Spain

Prof. LUENGO PASCUAL, L.M., Facultad de Informática, Jordi Girona Salgado 31, Barcelona-34, Spain

Miss LUIS, M.A., Facultad de Químicas, Universidad Complutense de Madrid, Ciudad Universitaria, Madrid-3, Spain

Dr. MAGNENAT, P., Observatoire de Genève, CH 1290 Sauverny, Switzerland

Mr. MALAGRIDA, I., Depto. de Paleontología, Facultad de Geología, Universidad de Barcelona, Spain

Mrs. MARTINEZ BARCHINO, R., Universitat Autónoma, Secció de Matemátiques, Bellaterra, Barcelona, Spain

Dr. MARTINEZ, J., Universidad de Valencia, Facultad de Matemáticas, Valencia, Spain

Prof. MASSAGUER, J.M., E.T.S.I., Universidad Politécnica de Barcelona, Jorge Girona Salgado 31, Barcelona-34, Spain

Mr. MAYER-KRESS, G., Institut für Theoretische Physik I, 7 Stuttgart-80, Germany

Prof. MAYNARD, R., Université de Grenoble, C.N.R.S. - CRTBT, Grenoble 38042, France

Prof. MIRA, Ch., Inst. National des Sciences Appliquées, Av. de Rangueil, 31077 Toulouse, France

Dr. MISGUICH, J., C.E.A., STGI - Fusion, 92260 Fontenay-Aux-Roses, France

Miss MOHEDANO, M.V., Facultad de Ciencias Matemáticas, Universidad Complutense, Ciudad Universitaria, Madrid-3, Spain

Mr. MORA, X., Secció de Matemátiques, Facultat de Ciencies, Universidad Autónoma de Barcelona, Bellaterra, Barcelona, Spain

Dr. MOROZ, I.M., School of Mathematics, University of Leeds, Leeds L52 9JT, Great Britain

Miss MÜLLENBACH, S., Inst. National des Sciences Appliquées, Av. de Rangueil, 31077 Toulouse, France

Dr. NAGASHIMA, T., Institute of Precission Mechanics, Faculty of Engineering, Hokkaido University, Japan

Prof. NEWHOUSE, S., University of North Carolina, Phillips Hall 039-A, Chapel Hill, NC 27514, USA

Dr. NIERWETBERG, J., Institut I, Theoretische Physik, Universität Regensburg, D 84 Regensburg, Germany

Prof. NOGUERA BATLLE, M., Facultat d'Informatica, U.P.B., Jordi Girona Salgado 31, Barcelona-34, Spain

Prof. NOHONHA DA COSTA, A., CMFC, Av. Prof. Gama Pinto, 2 1699 Lisboa, Portugal

Prof. OBERMAIR, G.M., Universität Regensburg, Fakultät für Physik, 84 Regensburg, Germany

Dr. PACKARD, N.H., IHES, 35 Route de Chartres, 91440 Bures-sur-Yvette, France

Prof. PERELLÓ i VALLS, C., Universitat Autónoma, Facultat de Ciencies Bellaterra, Barcelona, Spain

Dr. PERINI, U., CISE S.P.A., POB 12061, 20100 Milano, Italy

Dr. PESQUERA GONZALEZ, L., Universidad de Santander, Fac. de Ciencias, Santander, Spain

Miss PIGNATARO, T., Physics Dept., Princeton University, POB 708 Princeton, NJ 08540, USA

Prof. PISMEN, L.M., Institute of Applied Chemical Physics, CCNY, Convent Av. and 140th St. New York, NY 10031, USA

Prof. PROCACCIA, I., Dept. of Chemical Physics, Weizmann Institute of Science, Rehovot, Israel 76100

Dr. RAMASWAMY, R., Tata Institute of Fundamental Research, Homo Bhabha Rd., Bombay 400 005, India

Prof. RAÑADA, A., Universidad Complutense, Facultad de Ciencias Físicas, Ciudad Universitaria, Madrid-3, Spain

Mr. REICHERT, P., Institut für Physik, Universität Basel, Klingelbergstr. 82, 4056 Basel, Switzerland

Dr. RENZ, W., Technical University, RWTH Aachen, Templergraben 55, 5100 Aachen, Germany

Prof. RIELA, G., Universitá di Palermo, Istituto di Fisica, Via Archirafi, Palermo 90123, Italy

Prof. ROD, D.L., University of Calgary, Dept. Mathematics, Calgary Alberta T2N 1N4, Canada

Mr. RODRIGUEZ DIAZ, M.A., Universidad de Santander, Facultad de Ciencias, Santander, Spain

Dr. RODRIGUEZ MENDEZ, J.A., Facultad de Matemáticas, Universidad de Santiago de Compostela, Spain

Prof. ROJAS, E., Dpto. de Termología, Facultad de Ciencias, Universidad de Salamanca, Spain

Mr. ROGERS, J.B., University of Melbourne, Mathematics Dept. Parkville, Victoria, Australia 3052

Miss ROS, R.M., Universidad de Barcelona, Facultad de Física, Diagonal 647, Barcelona-28, Spain

Prof. RÖSSLER, O.E., Universität Tübingen, Auf der Morgenstelle 8, 74 Tübingen, Germany

Dr. RÖSSLER, (Mrs.), Universität Tübingen, Auf der Morgenstelle 8, 74 Tübingen, Germany

Mr. RUBÍ, J.M., Dpto. Termología, Universidad Autónoma, Bellaterra Barcelona, Spain

Prof. RUDNICK, J., University of California at Davis, Dept. of Physics, Davis, CA 95616, USA

Prof. SAENZ, A.W., Catholic University of America, Physics Dept., Washington, D.C. 20064, USA

Mr. SAGUÉS, F., Universidad de Barcelona, Facultad de Física, Diagonal 647, Barcelona-28, Spain

Prof. SANCHO, J.M., Universidad de Barcelona, Facultad de Física, Diagonal 647, Barcelona-28, Spain

Prof. SANDERS, J.A., Wiskundig Seminarium, Vrÿe Universiteit, PB 7161, 1007 MC Amsterdam, Holland

Prof. SAN MIGUEL, M., Universidad de Barcelona, Facultad de Física, Diagonal 647, Barcelona-28, Spain

Dr. SARAMITO, B., C.E.A., D.R.F.C. - S.T.G.I., Centre d'Etudes Nucléaires, BP 92260, Fontenay-aux-Roses, France

Dr. SCHILLING, R., Institute of Physics, University of Basel, Klingelbergstr. 82, CH 4056 Basel, Switzerland

Dr. SCHÖLL, E., Institut für Theoretische Physik, RWTH Aachen, 5100 Aachen, Germany

Prof. SCOTTI, A., Joint Research Center of the European Communities and GNSM-CNR, Istituto di Fisica, Università di Parma, Italy

Mr. SERRA, R., INB "Sant Cugat", Rambla Ribatallada 36, Sant Cugat del Vallés, Barcelona, Spain

Dr. SHINER, J.S., Biophysik. Chemie, Biozentrum, Universität Basel 4056 Basel, Switzerland

Dr. SIEGBERG, H.W., Universität Bremen, FB Mathematik, D 28 Bremen, Germany

Prof. SIGGIA, E., Institute for Theoretical Physics, University of California, Santa Barbara, CA 93106, USA

Prof. SIMÓ, C., Universidad de Barcelona, Facultad de Matemáticas Gran Via 585, Barcelona-7, Spain

Dr. SOUSA RAMOS, J., Centro de Física da Materia Condensada, Av. Prof. Gama Pinto 2, 1699 Lisboa, Portugal

Dr. STEEB, W.H., Universität Paderborn, Theoretische Physik, 4790 Paderborn, Germany

Dr. STEPHAN, W., Fakultät für Biologie, Universität Konstanz, PF 5560, 7750 Konstanz, Germany

Prof. TEBALDI, C., Istituto di Matematica Applicata, Facoltà di Ingegneria, Università di Bologna, Bologna 40136, Italy

Mr. THUAL, O., Observatoire de Nice, BP 525, 06007 Nice, France

Prof. TIRAPEGUI, E., Depto. de Física, Universidad de Chile, Avda. Blanco Encalada 2008, Santiago de Chile, Chile

Mr. TISHBY, N., Dept. of Theoretical Physics, The Hebrew University of Jerusalem, Jerusalem 91904, Israel

Mr. VALLÉS, J., Universidad de Barcelona, Facultad de Física, Diagonal 647, Barcelona-28, Spain

Mr. VAZQUEZ MARTINEZ, L., Universidad Complutense, Facultad de Física Ciudad Universitaria, Madrid-3, Spain

Prof. VELASCO MAILLO, S., Universidad de Salamanca, Facultad de Ciencias, Salamanca, Spain

Prof. WEISS, N., University of Cambridge, DAMPT, Silver St. Cambridge, CB3 9EW, Great Britain

Prof. WEST, B.J., La Jolla Institute, POB 1434, La Jolla, CA 92038, USA

Prof. WILCOX, C.H., University of Utah, Mathematics Dept. Salt Lake City, Utah 84112, USA

Prof. WILLIAMS, M., Virginia Polytechnic Institute, Dept. of Mathematics, Blacksburg, VA 24061, USA

Dr. WOLFF, W., Institut für Theoretische Physik, Universität zu Köln 5000 Köln-41, Germany

CONTENTS

SEMINARS AND COMMUNICATIONS

PROLOGUE

SOME IDEAS ABOUT STRANGE ATTRACTORS

L. Garrido

Facultad de Física
Universidad de Barcelona
Diagonal 657, Barcelona-28, Spain

C. Simó

Facultat de Matemàtiques
Universitat de Barcelona
Gran Via 585, Barcelona-7, Spain

1. INTRODUCTION

Strange attractors have appeared in the scientific literature quite recently [130,185,218,167,208,96]. Dissipative systems of differential equations in more than two dimensions can have bounded trajectories whose behavior does not converge to an equilibrium point nor to a periodic or quasiperiodic orbit [1]. They can be attracted by an object of complicated structure which attracts the neighbor points but has some inherent instability along it. The flow is essentially aperiodic.

The study of trajectories of differential equations can be made more simple thanks to the analysis of the sequence of points of intersection of the trajectory with a given hypersurface (a surface of section). Then we need to study the iterates of points under a given mapping (the Poincaré mapping associated to the flow with respect to the surface of section). The complicated structure attracting points of a neighborhood can appear much more clear when we analyze the related mapping. Therefore, it seems easier to learn about these attractors using the Poincaré map. In general we can study the attractors of diffeomorphisms without explicit reference to the flow which originates them. We note that the Poincaré map can be not defined on some points and not differentiable or even not continuous on others [28]. Without dimension reduction we can also use the map "flow acting along a fixed time" ϕ_τ .

Since the sixties there has been an increasing interest in strange attractors, S.A.,(even at an explosive rate), in their appearance in many applications and in some attempts to put order within chaos. However, there is not even an agreement in the definition of a S.A. It is not enough to say "strange attractor" = "attractor" + "strange" because there is a lack of definition of the word strange. Attractors different from fixed point, periodic or quasiperiodic orbit can still be mathematically (and physically) uninteresting. Sometimes we find in the literature sentences like: "it would be a S.A. provided it were an attractor" [117]. We shall call it a "potential strange attractor" and this plays an important role in the study of the formation, evolution and destruction of S.A.

In this preface we intend to give a working definition of S.A., some results about their origin, structure, and several examples. We will also provide the reader with a survey of types of S.A. (we do not claim our work to be exhaustive!). Some references are included. Further references can be found in these proceedings.

Before we define the S.A. we would like to add some comments. Despite the large number of papers on S.A. that appeared in the last years, what we know about them until now is essentially the result of numerical simulations (or direct physical experiments). However, little is known about the structure of a S.A., its origin or how it evolves. The simulations tell us that sometimes the S.A. coexists with punctual attractors [67,181], that several S.A. can coexist [181,77,78,23] or that several pieces of S.A. can collapse into one piece (Lorenz in [94,181,99,32]). The converse (splitting of a S.A.) can also happen [99,47]. Even doubts about the strange character or very high order periodicity of the attractor are in order [83].

In our opinion there is no systematic search for the relevant items in the numerical computations. By this we mean the critical or periodic points of the related Poincaré map, the spectra of the differential at these points, the invariant stable and unstable manifolds [98], etc. This is certainly difficult to obtain globally by numerical means in dimensions higher than three [23,70,71,72,59], for instance, when both manifolds have a dimension of at least two. In many cases one of the manifolds (mainly the unstable one) has dimension 1 or 2. Then numerical computation can give a good insight despite the fact that the stable manifold may have higher(possibly infinite) dimension. Anyway the computation of these manifolds seems a crucial step for the understanding of the phenomenon. Note that the invariant items can be of high dimension with complicated stable and unstable sets and that they can disappear under perturbation.

Several attempts to explain the structure of S.A. of dissipative three-dimensional flows proceed as follows: Take a 2-dimensional surface of section and define the Poincaré map. Because of dissipation this map is very close to a one dimensional map, the related Lorenz map [1]. Then one can apply the standard tools for 1-dimensional maps (scaling of bifurcations [64], itineraries [32], kneading theory [146,84], invariant measures [121,19,32,168], etc.) This gives information about the ω-limit of the flow. However, the Poincaré map is (hopefully!) a diffeomorphism, while the 1-dimensional map has no inverse. Therefore, some amount of information is lost in this step and this fact claims a careful analysis of two-dimensional (in general, higher dimensional) perturbations of 1-dimensional maps. (In examples related to the von der Pol equation this analysis has been effectively done[125]).

Some examples are known of two-dimensional diffeomorphisms not everywhere contracting having S.A. (see Curry -Yorke in [138,85,86]). The mapping T can have a region of expansion ($|DT|>1$) and a region of contraction ($|DT|<1$). Until now the investigation of the origin of this kind of S.A. has received little attention, but we believe that it is of the same type as those described below.

The search for S.A., its structure and evolution is an attempt to reduce the complexity of a higher dimensional flow. This is, to put some order in chaos!

As a last point of this introduction we return to the very first example of S.A. In a remarkable paper Lorenz [130] displayed the S.A. associated to a quadratic system of differential equations in $\mathbb{R}^3$. This system intends to be a very crude approximation of a system of partial differential equations. In fact other truncations are possible working in $\mathbb{R}^{14}$, $\mathbb{R}^{42}$, etc. instead of $\mathbb{R}^3$ [43,144]. The qualitative results so far obtained do not show much agreement. The same is true for several truncations of the Navier-Stokes equations for the two-torus [23,70,71,72]. Then the question is: even if we know for some families of equations that all the dynamics of an infinite dimensional system tends to a submanifold of finite dimension, how can we determine beforehand this dimension and how do we write down the "good finite-dimensional equations" in a faithful way? It seems to us that there is a lot to learn before we can trust in the meaning of the S.A. obtained by truncation of partial differential equations (or other functional differential equations). The exception are the cases when we can prove that really there are at most one or two unstable directions, the others being strongly compressive.

2. ON THE DEFINITION OF STRANGE ATTRACTOR

Let (X,f) be a discrete dynamical system, i.e. X is a compact metric space and f a homeomorphism. For many applications X is required to be a smooth manifold (of class $\mathscr{C}^r$ say) and f a smooth diffeomorphism. However, in some cases f has no inverse (we then call (X,f) a discrete semidynamical system) or f is even not continuous. For other applications X is not compact.

2.1 Definition

Let $A\subset X$ be a compact set positively invariant under f(i.e. $f(A)\subset A$). We say A is an attractor senso lato if there exists a $\mathcal{U}$, neighborhood of A, such that $A=\bigcap_{n\geq 0} f^n(\mathcal{U})$.

Note that the dynamics in A can be uninteresting (i.e., A can consist of fixed points) or that A can be split in several unrelated pieces. Therefore, we have to require something about the dynamics of f on A [83].

2.2 Definition

Let $A\subset X$ be an attractor senso lato. We call A an attractor if $f|_A$ is topologically transitive, i.e., given any two open sets V,W in A there exists $n\in\mathbb{N}$ such that $f^n W\cap V\neq\emptyset$.

We can ask for a dense orbit in A instead of topological transitivity. This is equivalent. In fact topological transitivity implies that the set of points with dense orbit is a G_δ dense set (a countable intersection of open sets).

An example of attractor can be given by taking X as a solid torus, i.e. $X=D^2\times S^1$ where D^2 is the two-dimensional closed disc. Let (r,α) be the polar coordinates on D^2 and z the angular one in S^1. We define f through $f(r,\alpha,z) = (r/2,\alpha,z+2\pi\omega \pmod{2\pi})$ where $\omega\notin\mathbb{Q}$. Then the attractor A is the line r=o and $f|_A$ is a rotation of angle incommensurable with 2π. We have a dense set and therefore topological transitivity. However, two points a,b at a distance δ in A have all the images satisfying $d(f^n a,f^n b)=\delta$. This is not strange and the example is again uninteresting.

In order to define a S.A. we require sensitivity with respect to initial conditions.

2.3 Definition

Let A be an attractor on (X,f). We say that A is sensible [82,32] or that it has sensitivity with respect to initial conditions if for all $x\in A$ the positive orbit $O_+(x)$ of x is Lyapunov unstable, i.e. there exists $\varepsilon>0$ such that for all neighborhood $\mathcal{U}\ni x$ there is some $y\in\mathcal{U}$ and $n\in\mathbb{N}$ such that $d(f^n x,f^n y)>\varepsilon$.

2.4 Proposition

If A is an attractor, $x\in A$ has dense orbit under f and $O_+(x)$

is Lyapunov unstable then $\forall y \in A$, $O_+(y)$ is Lyapunov unstable.

Proof. Let $\mathcal{U}$ be a neighborhood of y. There is some n such that $f^n x \in \mathcal{U}$. Let $V \subset \mathcal{U}$ be a neighborhood of $f^n x$. Choose $z \in f^{-n}V$ such that there exists $m > n$, $\varepsilon > 0$ with $d(f^m x, f^m z) > \varepsilon$. Then either $d(f^{m-n}y,\ f^{m-n}(f^n x)) > \varepsilon/2$ or $d(f^{m-n}y,\ f^{m-n}(f^n z)) > \varepsilon/2$. Therefore, the instability of $O_+(y)$ follows.

We do not know if the requirement "for all $x \in A$" in 2.3 can be weakened. A possible example with $O_+(x)$ Lyapunov unstable for some x with nondense orbit and Lyapunov stability for all the points with dense orbit taking in a subset of T^2 the map $\varnothing_{t=1}$ under the Cherry flow has been suggested to the authors by W.Melo.

2.5 Definition

We say that an attractor A is strange if it has sensitivity with respect to initial conditions.

2.6 Examples

a) As a first and simple example we consider the solenoide [185]. Let $X=D^2 x S^1$ with coordinates $(r,\alpha) \in D^2, z \in S^1$ as before. We define f_2 as $f_2(r,\alpha,z)=((\frac{1}{2})_z + (\frac{r}{4})_\alpha, 2z)$ where $(\frac{1}{2})_z$ or $(\frac{r}{4})_\alpha$ mean modulus and argument of points in D^2 seen as complex numbers. If we look for $f_2(D^2 x S^1)$ we find a smaller solid torus inside X which closes after two revolutions. Successive iterates under f_2 produce smaller and smaller solid torus which close after 2^k revolutions. The section through a meridian plane is shown in fig.1. Then the attractor is given by $S=\bigcap_{n\geq 0} f_2^n\ (D^2 x S^1)$ which has hyperbolic character, i.e., there is a splitting of the tangent space on the points of S which changes in a continuous way when we change the point in S, such that the vectors of one subspace grow exponentially and the ones in the other subspace decrease exponentially. Note that in the definition of S we can use f_a instead of f_2, with $2 < a \in \mathbb{N}$ and replacing 2z by az in the definition of f_2. We can even change a in each iteration. This gives rise to different solenoides that can be topologically no equivalent.

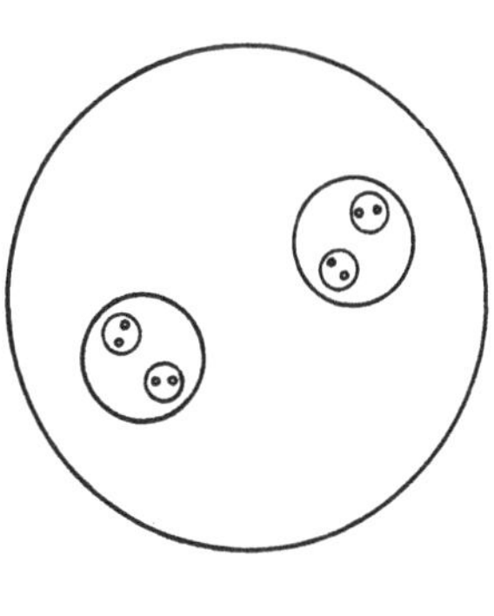

Fig. 1

b) The Lozi map is a piecewise linear map which has been introduced to simplify the Hénon map to be described later [131]. It is given by $f(x,y)=(1+y-a|x|,bx)$. It has been shown by Misiurewicz[94] that if $b\in(0,1)$, $a>0$, $2a+b<4$, $b<\frac{a^2-1}{2a+1}$ and $a\sqrt{2}>b+2$ then there is a hyperbolic fixed point H of saddle type given by $x=(a+1-b)^{-1}$, $y=bx$ such that the strange attractor is the closure of their unstable manifold $\overline{W_H^u}$ [194]. In the (a,b) space the allowed area is shaded in figure 2. For instance a=1.7, b=0.4 is in this area. Numerically the S.A. has been observed even for a=1.7, b=0.5, but we remark that the previous conditions are only sufficient. In fact, the attractor is hyperbolic. Here we allow for a piecewise continuous splitting.

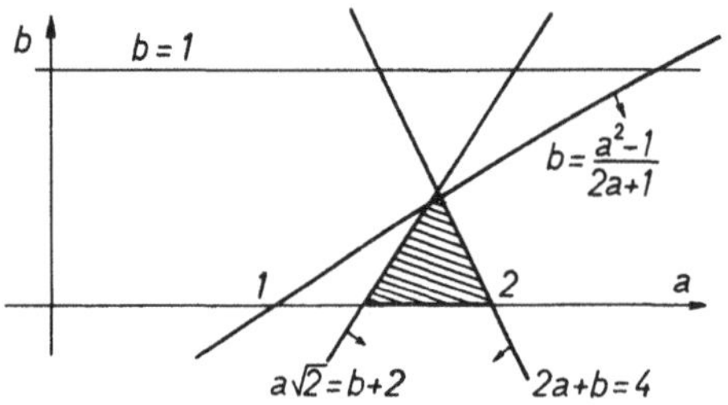

Fig. 2

c) The Hénon map $f(x,y)=(1+y-ax^2,bx)$, $|b|<1$ includes all the (non trivial) quadratic maps with constant jacobian (in fact, with non nul jacobian everywhere)[24,44,45,46,61,67,69,73,77,78,84,85,96,99,139,148,181,195,206]. It has been introduced in order to see under a moderate rate of compression the type of phenomena in the Lorenz attractor. Fig. 3 shows the way to produce the Hénon map as a composition of elementary maps. The folding for the Lorenz case (really for the Poincaré map when defined) is much stronger and the nice details are difficult to detect by simulation. For the values initially used by Hénon a=1.4, b=0.3 the direct simulation produces an attracting set which looks like the closure of the invariant manifold of a hyperbolic fixed point, just like in the case of Lozi's map (see figure 4). However, this has not been proved till now. See later for the difficulties in proving this.

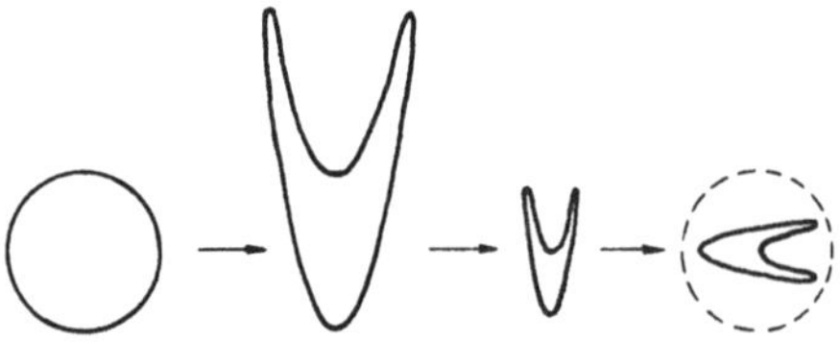

Fig. 3

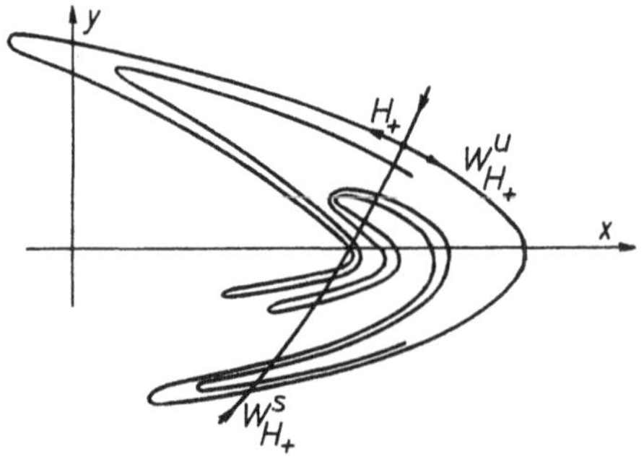

Fig. 4

d) Much information is available for S.A. satisfying the Axiom A condition [19,168,185,186,21], i.e., the set of nonwandering points $\Omega(F)$ which carries the asymptotic dynamics is hyperbolic and the periodic orbits are dense in it. However, it is difficult to check for a given diffeomorphism displaying what seems a S.A. whether it is Axiom A or not.

e) Note that some maps which have an attractor that seems complicated are really easy. For instance for the well known family of unimodal maps $f_\mu(x)=1-\mu x^2$ [30,31,32,64,80,81,82,85,86,87,111,113,115,142,145,146,147,149,150,169,211] a cascade of period doubling bifurcations accumulates on the value $\mu_\infty \approx 1.401155$. However, for μ_∞ we have an attractor which is not strange according to the given definition because of lack of sensitivity.

2.7 Definition

We say that a discrete dynamical system possesses chaos if it has some strange attractor. For flows we require that the Poincaré map related to some surface of section possesses chaos. (See [170,171] for other ideas).

One of the criteria often used to detect chaos consists in the study of the power spectra of some of the variables of the system (or of some function thereof)[32,38,57,56,137,175,197]. When we have an attracting fixed point, a periodic or quasiperiodic orbit the power spectra is merely a sum of peaks. The appearance of a broad band is quite often taken as synonymous of chaos [76,166,213,215]. That is the method used in numerical experiments or directly in physical experiments. Sometimes the output of the experiment is a time series which is analyzed afterwards. Takens (in [165]) has proposed methods for computing the (Haussdorff) dimension [195,51,59,60,61,73,220] of the limit set and the entropy from time series [156], but it seems that they are not reliable for higher dimensions.

As it was remarked in the introduction with respect to numerical simulations it seems to us that looking only for the power spectrum, a large amount of geometrical information is lost and no insight is obtained about the mechanisms leading to the creation/destruction of a particular S.A. It would be of interest to look into experiments for the geometrically important objects.

Let us return to the power spectrum and display an example of S.A. according to the given definition which does not show a continuous band.

Let $\varphi: A \to \mathbb{R}$ be a continuous map (f.i., a coordinate of the system). Then for $\omega \in [0,1)$ we define

$$F(\omega)=\lim_{n\to\infty}\frac{1}{n}\sum_{k=o}^{n-1} e^{-2\pi i\omega k}\,\varphi(f^k(x)) \qquad \text{for an } x\in A.$$

If $\bar{x}$ is a fix point then $F(\omega)=\delta_{\omega,o}\varphi(\bar{x})$. If $\{x_1,\ldots,x_m\}$ is a periodic orbit f $x_i=x_{i+1\,(\mathrm{mod.}\,m)}$ then

$$F(q/m)=\frac{1}{m}\sum_{k=1}^{m}\varphi(x_k)\,e^{-2\pi iqk/m}$$ and $F(\omega)=o$ if $\omega\notin\mathbb{N}/m$. For the map

$x\to e^{2\pi i\alpha}x$ in S^1, $F(\omega)=x_o\,\delta_{\omega,\alpha}$.

Now we consider the following case:

2.8 Example

Let $X=T^2\times[-1,1]$ with coordinates $x,y\in\mathbb{R}/\mathbb{Z}$, $z\in[-1,1]$. Then define $f(\binom{x}{y},z)=(\binom{x}{y}+k\binom{1}{\alpha}\psi(x,y),\ z/2)$. The attractor is obviously $T^2\times\{o\}$. We take $\alpha\notin\mathbb{Q}$, $\psi(x,y)=\sin^2\pi x+\sin^2\pi y$. Then, if k is small ($k<((1+|\alpha|)\pi)^{-1}$ is enough), f is an analytical diffeomorphism. The origin (o,o) is the only fixed point. The line $y=\alpha x$ is split by (o,o) in the unstable and stable manifolds of the origin W^u, W^s. They are dense and this guarantees the sensitivity. In fact, if $x\notin W^s$ there are points y in every neighborhood U of x such that $y\in W^s$. After a large number of iterations $f^n y$ is almost stopped near (o,o) where it enters and x not. If $x\in W^s$ just pick up an $y\notin W^s$. With respect to transitivity it is enough to show it for open sets like V,W as in figure 5 (other cases can reduce to this taking suitable preimages or images of V or W). Then the image of a segment ℓ in W under f^n is a curve with one end approaching (o,o) and the other near the displayed portion of W^u. Therefore, for all n greater than some n_o, $f^n W\cap V\neq\emptyset$. Hence $T^2\times\{o\}$ is a S.A. under f.

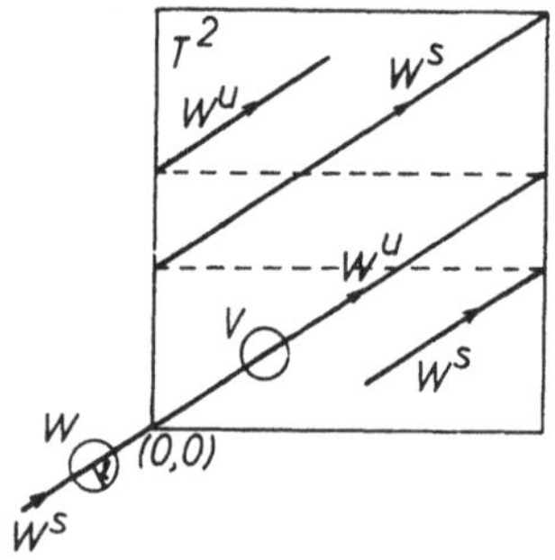

Fig. 5

However, the dynamics is quite slow near (o,o). It is easy to show the following result:

2.9 Proposition

Consider the system $f|_{T^2\times\{o\}}$ of example 2.8. Let $\varepsilon,\delta>o$ be arbitrarily small fixed numbers and B_ε the ball of radius ε centered at (o,o). Then there exists $N_0(\varepsilon,\delta)$ such that for all $N > N_0$ we have $\frac{1}{N}\#\{n \in [0,N-1] \cap \mathbb{Z} \mid f^n(q) \in B_\varepsilon\} > 1-\delta$ for all $q \in T^2 \times \{o\}$.

The proof is essentially related to the computation of $\int_0^\varepsilon \frac{1}{x} / \int_0^1 \frac{1}{x}$ using the method of equirepartition and stopping the computation in both terms after N steps. The value $N_0(\varepsilon,\delta)$ is clearly related to the continuous fraction expansion of α.

Note that if $\alpha = (\sqrt{5}-1)/2$ and $\varepsilon = 10^{-2}$ after 10^6 iterations only 66% of the points are in B_ε ; after 10^7 iterations we reach only 73%. To reach 95% some 10^{32} iterations are needed. This means that the process of convergence to an invariant measure concentrated in (o,o) is very slow.

Obviously, if we compute now the power spectrum we get $F(\omega) = \varphi(o,o)\, \delta_{\omega,o}$ according to 2.9. Using $\psi(x,y) = |\sin \pi x|^p + |\sin \pi y|^p$ with $p > 2$ we get the same result (with quicker convergence). For $1 < p < 2$ the invariant measure has the full torus as support and continuous bands in $F(\omega)$ are observed. Note that for p different from an even integer f is a piecewise analytical diffeomorphism but is not $\mathscr{C}^1$ for x=0 or y=0.

For the example 2.8 it is also clear that the Lyapunov exponents[155,158] for $f|_{T^2 \times \{o\}}$ are zero. This is due to the fact that the separation of nearby points is rather slow when they go near (o,o). Therefore, 2.5 allows for some "weak" strange attractors. However, for most cases encountered in the physical or numerical experiments it seems that the maximal Lyapunov exponent ℓ is positive when a S.A. is present, and that $F(\omega)$ shows continuous components. This would be true if f were Axiom A and we conjecture that it should be true when some absolutely continuous invariant measure exists in the S.A. which is left invariant under f. However, this has no sense if we cannot put some kind of Lebesgue measure on the S.A. If the S.A. is the closure of an unstable invariant manifold this can be done effectively.

3. MECHANISMS EXPLAINING THE STRANGE ATTRACTORS

Let us first take a basic example: the Hénon map. If $o < a < \frac{3}{4}(1-b)^2$ there are two fixed points $H_\pm$. H_+ is stable and H_- is unstable. The global picture is displayed in figure 6. The two branches of W_-^s bound a region R. Points inside R are attracted to H_+. Points outside R are carried under iterates of f near the outer branch of W_-^u and then escape to infinity. This dynamics is quite simple but it will be of interest for later discussion. Keeping b fixed and increasing <u>a</u> the point H_+ looses stability. Then a cascade of period doubling

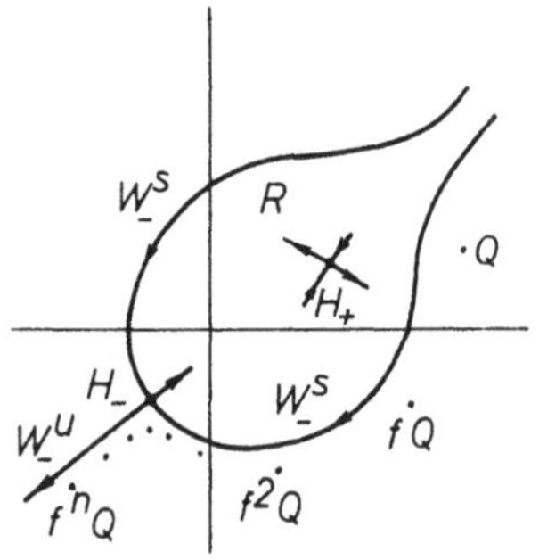

Fig. 6

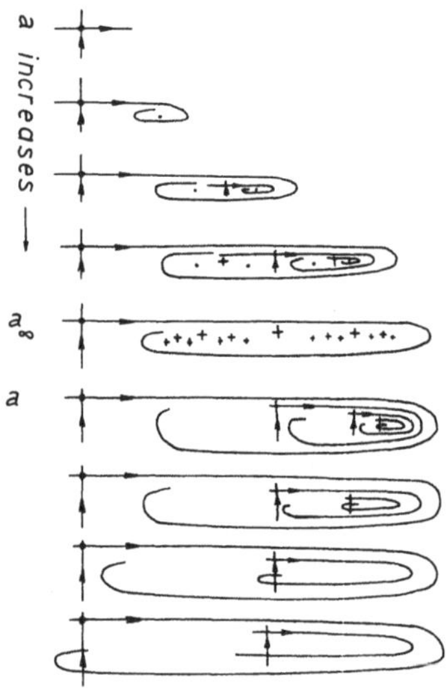

Fig. 7

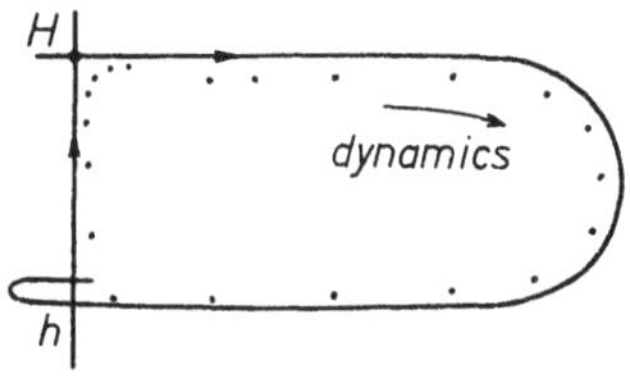

Fig. 8

bifurcations is found which accumulates on some value $a=a_\infty$. For a greater than a_∞ transversal homoclinic points of higher order appear. In fact they exist (for some 2^k-periodic point) for every $a > a_\infty$. Increasing values of a produce tangential and after transversal homoclinic points for periodic points of periods $2^{k-1}, 2^{k-2}, \ldots, 2^3, 2^2, 2, 1$, succesively. Let us call a_{c_1} the value for which H_+ has tangential homoclinic points.

Figure 7 shows the evolution of the invariant manifolds and the development of the homoclinics. When an unstable manifold of a fixed or periodic saddle point has transversal homoclinic points in a dissipative mapping (figure 8) a <u>potential strange attractor</u> $\overline{W^u}$ appears. The homoclinic is needed to have some feedback: Points spread by the unstable manifold passing near the homoclinic h are injected again near the saddle H. Of course this is a very simple model and the reader can ask about points <u>not</u> passing near h. However, $W^u_{H_+}$ is folded on itself and points near $W^u_{H_+}$ come again near H. See later for the discussion on the Newhouse phenomenon [151,83]. Once a S.A. is recognized as some $\overline{W^u_H}$ the computation of dimension, the Cantorian character of the set obtained by a transversal section of the attractor, and the obtension of invariant measures, Lyapunov exponents, etc. are possible.

Now we look for the creation of attractive regions for different periodic orbits [181]. Suppose that for a given a a periodic orbit appears with one of the eigenvalues equal to one. Decreasing a the orbit disappears. Increasing a a bifurcation is produced and a saddle B and a stable node A appear. Then the stable manifold W^s_B bounds a region R of points being attracted by A (figure 9) (with some plausible exceptions; higher order stable periodic orbits and small basins). We can

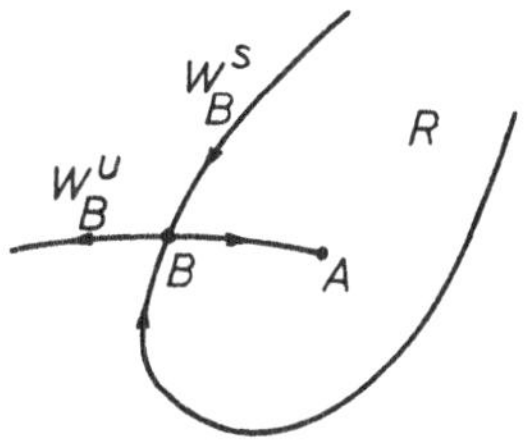

Fig. 9

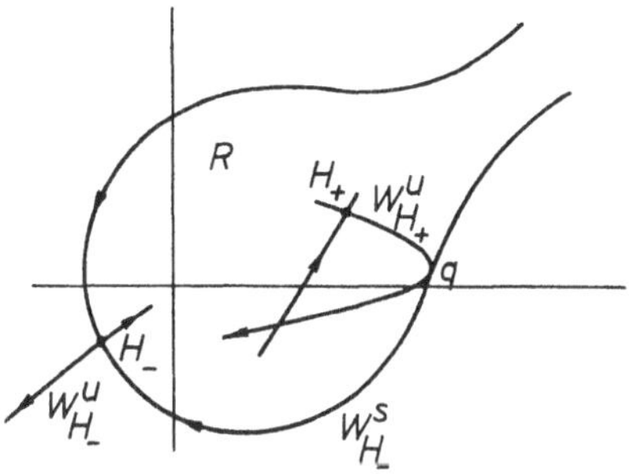

Fig. 10

compare figure 9 with figure 6.

Let us analyze the destruction of a S.A. [181,77,78,85,99,101]. The potential S.A. of figure 7 is in a region like the region R of figure 6. For some value $a=a_{c_2}$ a heteroclinic point q in $W^s_{H_-} \cap W^u_{H_+}$ appears (figure 10). "If" there is ergodicity on $W^u_{H_+}$ for values slightly larger than a_{c_1} as some part of $W^u_{H_+}$ is outside R, after a sufficiently large number of iterations a point goes to the outer part of $W^u_{H_-}$ and then is carried out along $W^u_{H_-}$ just like in figure 6. We note that the curve a=a(b) such that in q there is a tangential homoclinic point, can be obtained numerically. For values of a slightly larger a direct simulation can lead to the conclusion that there is no escape at all, but this is only due to the fact that the number of iterations is still small.

Let us suppose that going from a_{c_1} to a_{c_2} a situation like the one depicted in figure 11 is found. Then $W^u_{H_+}$, the potential S.A. is destroyed not by carrying the points to infinity but to a fixed or periodic point A, or even to another different S.A. In fact a countable set of windows in the a parameter appears for which the scheme of figure 11 is found. The point marked A can in turn give rise to a cascade of bifurcations and to some S.A. When W^u_A is cut by $W^s_{H_+}$ the two attractors are "mutually destroyed". This means that really $A \in W^u_{H_+}$ and we recover the initial S.A.

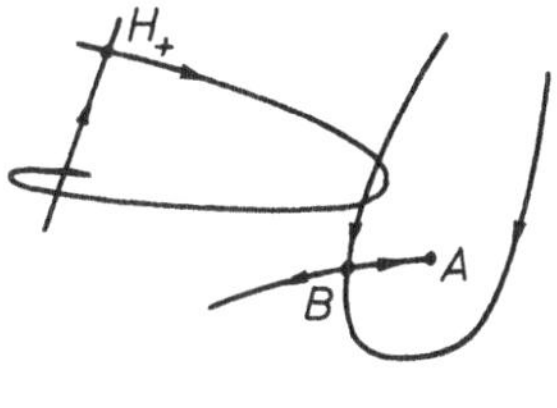

Fig. 11

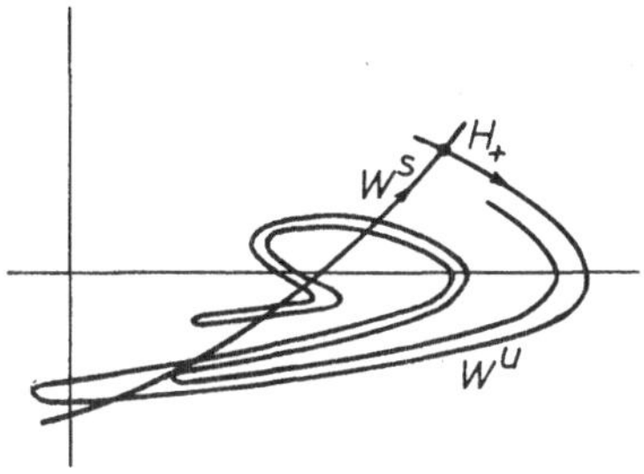

Fig. 12

Now we present the so called Newhouse phenomenon. Let f_μ be a family of $\mathcal{C}^3$ diffeomorphisms with a homoclinic tangency for $\mu=\mu_0$. Suppose that for $\mu>\mu_0$ the homoclinic point is transversal (if it is so for $\mu<\mu_0$, merely reverse sign of μ. A family with tangencies for all μ in some neighborhood of μ_0 is not generic). Then there exist values of $\mu>\mu_0$ arbitrarily near μ_0 for which f_μ has infinite sinks (attracting periodic orbits).

Two opposite comments are in order with respect to this result. The first one is that it may seem that this can produce the destruction of the S.A. only near the first tangency between $W^u_{H_+}$ and $W^s_{H_+}$, for the Hénon problem, f.i. This is not true because after the critical value a_{c_1} the homoclinic point certainly is transversal. However, the successive folds of $W^u_{H_+}$ can give rise to infinite values of a for which some fold produces tangential homoclinics (see figure 12). The second comment is that nothing is known about the basin of attraction [109] of these sinks. In fact we do not know of any simulation detecting them. Therefore, it is plausible that the basin be extremely small and does not produce destruction of the S.A., or alternatively, the period be so high that it is hard to distinguish between a true S.A. and such periodic orbit. In any case a careful analytical and numerical study seems interesting.

For the Lorenz model $\dot{x} = 10y-10x$ (see 3,4,13,17,68,84,90,117,130, 140, 188,195,219)

$$\dot{y} = rx-y-xz$$

$$\dot{z} = xy-8z/3$$

taking r as a parameter, a homoclinic orbit doubly asymptotic to the origin is found for $r_a \simeq 13.9265574075$. Increasing r a periodic orbit is found which emerges from the homoclinic one and dies at a subcritical Hopf bifurcation for $r_b = 470/19$. Let us look at the Poincaré section by the plane $z=r-1$. Between r_a and r_b there are four fixed points under the Poincaré map. Two of them, C_1 and C_2 are sinks related to critical points of the flow. The other two, P_1 and P_2 are the intersections of the two (symmetrical) unstable periodic orbits. For $r>r_a$, $\overline{W^u_{P_i}}$ are potential strange attractors which become destroyed by the action of C_i. But before reaching the value r_b (for some value $r\simeq 24.06$ according to simulations) the basin of C_i does not reach $\overline{W^u_{P_i}}$ and the S.A. is established.

When a potential S.A. is destroyed by some sink and for some value of the parameter, the sink disappears or turns out to become a saddle through bifurcation. This fact is a two-dimensional model of what in one-dimensional maps is called intermittency. It can be responsable for creation(or destruction, reversing sign of μ) of S.A.

Now we can ask how a differential equation can be produced such that a S.A. is likely displayed. Rössler (in 87), among others, has given easy methods. Consider a dynamical system in $\mathbb{R}^3$ with some slow variables and a fast variable. The motion is almost confined to a surface $f(x,y,z)=0$ if one of the equations reads $\varepsilon\dot{x} = f(x,y,z)$ with ε small. If the surface developes a fold (see figure 13) and the dynamics is such that points falling to the lower part of the fold are reinjected at a suitable place in the upper part, i.e. near an unstable focus, we have a good chance of finding S.A. When $\varepsilon \to 0$ the Poincaré map with respect to some surface of section turns out to be essentially one-dimensional. Then, under suitable hypothesis of expansiveness, some absolutely continuous invariant measure is found in the interval with a support that, when seen in two dimensions gives rise to an unstable manifold. This is the S.A. which we were looking for.

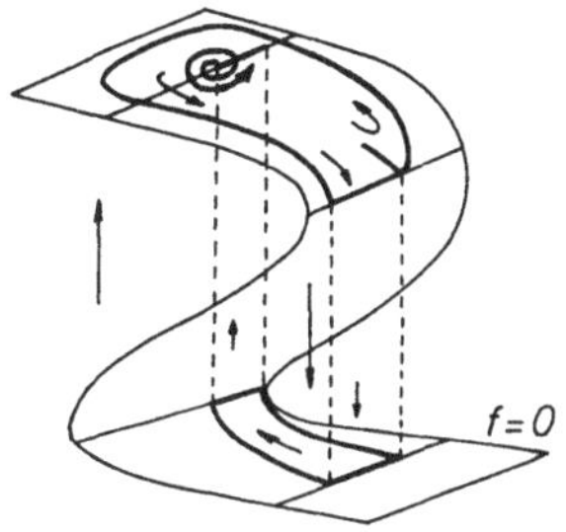

Fig. 13

A similar situation is found by perturbation of a system having a critical point with eigenvalues $\lambda_1 = a+bi$, $\lambda_2 = \bar{\lambda}_1$, $\lambda_3 = c$, $-a,b,c \in \mathbb{R}_+$ and with $2a+c < 0$, such that it has a homoclinic orbit to this point (figure 14). A theorem of Shilnikov [9] assures the existence of an infinity of periodic orbits for this system. The reader can check that the dynamics in figure 13 is a perturbation of the situation described in figure 14.

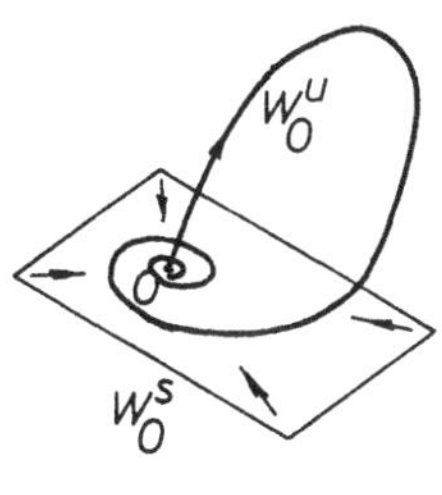

Fig. 14

4. EXAMPLES OF SYSTEMS WITH STRANGE ATTRACTORS

We present a short list of examples found in the literature. All of them give S.A. for an ordinary differential equation. S.A. for diffeomorphisms in $\mathbb{R}^2$ are the Hénon and Lozi maps already displayed. Another interesting example is the Plykin diffeomorphism [162,129,83] and their generalization. Other "artistic" S.A. for diffeomorphisms can be found in [85,94].

a) Lorenz model: Equations

$$\dot{x} = -\sigma x + \sigma y$$
$$\dot{y} = -xz + rx - y$$
$$\dot{z} = xy - bz \ .$$

with $\sigma=10$, $b=8/3$ and r taken as parameter.

For $r \in (24.06, 24.74)$ there is a S.A. and two attracting fixed points. For $r > 24.74$ (up to some higher value) the fixed points are saddles and we find only a S.A. Perturbations of the Lorenz model are topologically conjugated to some element of a biparametric family.

b) Duffing equation with forcing term [209,153,94,104]: Equation $\ddot{x}+K\dot{x}+x^3 = b_0+b_1 \cos t$.

Several values of parameter for which one has found a S.A. are $(K,b_0,b_1) = (0.05, 0, 12)$, $(0.05, 0.045, 0.16)$, $(0.1, 0, 10)$, $(0.2, 0.85, 1.2)$.

c) Several Rössler attractors [87].

Equations

$$\dot{x} = -(y+z)$$
$$\dot{y} = x+ay$$
$$\dot{z} = b+xz-cz,$$

for $a=0.343$, $b=1.82$, $c=9.75$ we have the so called funnel attractor; for $a=0.2$, $b=0.2$, $c=5.7$ the classical Rössler attractor.

Other families displaying S.A. when changing parameters are

$$\dot{x}=-y+ax-bz,\quad \dot{y}=x+1.1,\quad \dot{z}=(1-z^2)(x+z)-\varepsilon z,$$
$$\dot{x}=x-xy-z,\quad \dot{y}=x^2-ay,\quad \dot{z}=b(cx-z),$$
$$\dot{x}=xy-ax-z,\quad \dot{y}=-x+by+cz,\quad \dot{z}=d+exz+fx,$$
$$\dot{x}=-y-z,\quad \dot{y}=x,\quad \dot{z}=a(y-y^2)-bz,$$
$$\dot{x}=-y-z-w,\quad \dot{y}=x,\quad \dot{z}=a(y-y^2)-bz,\quad \dot{w}=c(z/c-z^2)-dw.$$

d) A dissipative Hénon-Heiles model [93].

Equations

$$\dot{x}=p\ ,\quad \dot{p}=-x^2-2xy+\lambda x-\varepsilon\nu p$$
$$\dot{y}=q\ ,\quad \dot{q}=-y^2-x^2+\lambda y-\varepsilon\nu q$$
$$\dot{\lambda}=-\varepsilon(\lambda+a(x^2+y^2-1))\ .$$

If $x=p\equiv 0$ we get the Lorenz model, essentially. For $a=6.5$, $\nu=4.125$, $\varepsilon=0.1$ a 5-dimensional S.A.

e) The Brussel model for a chemical reaction [116,201].

Equations

$$\dot{x}=A+x^2y-Bx-x+a\cos\omega t$$
$$\dot{y}=Bx-x^2y\ ,$$

that for A=0.4, B=1.2, a=0.12, ω=0.9 displays a S.A.

f) Shilnikov attractors [8,9]. Equations in general

$$\dot{x}=\rho x-\omega y+P(x,y,z)$$
$$\dot{y}=\omega x+\rho y+Q(x,y,z)$$
$$\dot{z}=\lambda z+R(x,y,z)$$

where P,Q,R are of second order in the neighborhood of the origin. Specific examples producing S.A. are

$\ddot{x}+\beta\dot{x}+x=y$, $\dot{y}=f_\mu(x)$ where $\beta>0$ and $f_\mu(x)=1+ax$ for $x\leq 0$, $f_\mu(x)=1-\mu x$ for $x>0$, and

$\dot{x}=y$, $\dot{y}=z$, $\dot{z}=-y-\beta z+f_\mu(x)$, where $f_\mu(x)=-\mu x-\mu-a$ for $x\leq -1$, $f_\mu(x)=ax$ for $|x|<1$, $f_\mu(x)=-\mu x+\mu+a$ for $x>1$.

g) Volterra attractors [6,10].

Equations $\dot{N}_i=N_i\sum_{j=1}^{3}\alpha_{ij}(1-N_j)$. Choose $(\alpha_{ij})=\begin{pmatrix} .5 & .5 & .1\\ -.5 & -.1 & .1\\ \mu & .1 & .1\end{pmatrix}$

and select $\mu=1.43$ to get a S.A.

h) Parametrically driven pendulum [12].

Equations $\ddot{x}+\nu\dot{x}+\omega_0^2(1+F\cos\Omega t)\sin x=0$. With the parameters $\nu=0.18$, $\omega_0=2$, $\Omega=2$, F=1.58 we obtain a S.A.

i) A forced negative-resistence oscillator [210].

Equations
$$\dot{x}=y$$
$$\dot{y}=\mu(1-x^2)y-x^3+B\cos\nu t.$$

For μ=0.2, and the following pairs of (B,ν) : (1,0.94), (1.2, 0.92), (2,0.6), (2.4, 0.7), (17, 4), a S.A. is found.

j) The Rikitake dynamo [35]: $\dot{x}=yz-\mu x$, $\dot{y}=(z-a)x-\mu y$, $\dot{z}=1-xy$.

k) The Robbins dynamo [27]: $\dot{x}=\alpha(y-x)$, $\dot{y}=zx-y$, $\dot{z}=R-xy-\nu z$ with $R>\alpha\nu(\alpha+\nu-3)/(\alpha-1-\nu)$.

l) A forced nonlinear oscillator [27]: $\ddot{x}+\gamma\dot{x}-x+x^3=f\cos\omega t$.

m) A particle under the action of a potencial plus electric field [106]: $\ddot{x}+\alpha\dot{x}+x-4x^3=\Gamma\cos(\frac{\omega}{\omega_0}t)$ (equivalent to the previous one) which for α=0.4, Γ=0.115, ω_0=1, ω=0.5529 displays a S.A.

n) A nonlinear oscillator with dissipation and forcing term [208]:

$$\ddot{x}-\mu(1-x^2)\dot{x}+x^3=B\cos\nu t.$$

For the values of the parameters μ=0.2, ν=4, B=17 a S.A. is found. This is the first example (known to the authors) where a discussion in terms of invariant manifolds and homo/heteroclinic points is sketched.

o) Josephson junctions [107]: $\ddot{x}+\beta\dot{x}+\gamma\sin x=\delta\cos\omega t$ display S.A. when varying ω and keeping the values β=0.5, γ=6.4, δ=3.8.

p) 5-mode truncated Navier-Stokes equations in T^2 [23].

General equations

$$\begin{aligned}
\dot{x}_1 &= -2x_1+4x_2x_3+4x_4x_5+r_1\\
\dot{x}_2 &= -9x_2+3x_1x_3\\
\dot{x}_3 &= -5x_3-7x_1x_2+r\\
\dot{x}_4 &= -5x_4-x_1x_5\\
\dot{x}_5 &= -x_5-3x_1x_4+r_5
\end{aligned}$$

If $r_1=r_5=0$ the system is symmetric. Then a S.A. is found for $r=29$, for instance. In the general asymmetric case $r_1, r_5 \neq 0$, for $r_1=r_5=1$ and several values of r, such as 23.45, 24,35,36,45, a S.A. is found again.

q) Glycolytic oscillations [200].

General equations:

$$\dot{x} = \sigma(t) - \sigma_M f(x,y)$$

$$\dot{y} = \sigma_M f(x,y) - k_s y$$

Sel'kov-Higgins model: $\sigma(t)=0.999+0.42 \cos \omega t$, $\sigma_M=k_s=1, f(x,y)=xy^2$. S.A. found for $\omega=1.69$.

Goldbeter-Lefever model: $\sigma(t)=12.5+12 \cos \omega t$, $\sigma_M=10^3, k_s=1$,

$$f(x,y)= \left[(1+y)^2 \frac{x}{\varepsilon+1} (1+\frac{x}{\varepsilon+1})\right] / \left[L(1+cx)^2+(1+y)^2 (1+\frac{x}{\varepsilon+1})^2\right] \text{ and}$$

$L=5.10^6$, $c=10^{-2}$, $\varepsilon=10^{-1}$.

The values $\omega=1.8$, 2.1 produce S.A.

r) A quadratic model for chemical kinetics [217]:

$$\begin{aligned}
\dot{x} &= x(a_1-k_1x-z-y)+k_2y^2+a_3\\
\dot{y} &= y(x-k_2y-a_5)+a_2\\
\dot{z} &= z(a_4-x-k_5z)+a_3 \ .
\end{aligned}$$

Pick up the values $k_1=0.25$, $k_2=10^{-3}$, $k_5=0.5$, $a_1=30$, $a_2=a_3=10^{-2}$, $a_4=16.5$, $a_5=10$ to detect a S.A.

s) A model for the interaction of perturbations [94]:

$$\begin{aligned}
\dot{x} &= 2y^2+\gamma x+(\delta-\alpha x)\, y\\
\dot{y} &= 2xy+\gamma y-(\delta-\alpha z)\, x\\
\dot{z} &= -2xz-2z
\end{aligned}$$

, offering a S.A. for $\gamma=0.2$, $\delta=0.5$, $\alpha=0.11$.

t) A model of the Belousov-Zhabotinskii reaction [203]:

$$\begin{aligned}
\dot{x} &= (1-\phi)x+y-xy-xz\\
\dot{y} &= -(1+\phi)y+z-xy+m\\
p\dot{z} &= -(1+p\phi)z+x-xy \ ,
\end{aligned}$$

where the parameters $m = m_o+\phi y_o$, $y_o=1$, $p = 90$ and $\phi = 0.048$

are taken. Varying m_0 a S.A. is found.

u) Glass-Mackey model for blood production [134,58,59].

Equations: $\dot{x}(t)= ax(t-\tau)(1+(x(t-\tau))^c)^{-1} -bx(t)$.

This is a differential delay equation displaying a S.A. for τ larger than 16.8 keeping a=0.2, b=0.1, c=10 fixed.

v) A model of baleen whales population [94].

Equations $\dot{N}(t)= -\mu N(t)+\nu N(t-\tau)(1-(N(t-\tau))^z)_+$, where $(f(t))_+ = f(t)$ if $f(t)>0$, $(f(t))_+ = 0$ otherwise. Fixing the values $\tau=2$, $\mu=1$, $\nu=2$ a S.A. is found for z=3.65, 3.69, 3.76 and 3.91 for example.

w) S.A. for the sunflower equation: $\dot{x}(t)+ \frac{a}{r} x(t) + \frac{b}{r} \sin(x(t-r)), r>0$ with suitable values of a,b,r have been found by one of the authors. (C.S.) (to be published elsewhere).

5. CONCLUSIONS

For two-dimensional diffeomorphisms or flows reducing essentially to them the evolution of S.A. can be described geometrically using bifurcations, homoclinic and heteroclinic points. However, many questions are left open:

1) Prediction of values of the parameters for which a S.A. appears or is suddenly destroyed.

2) Existence of invariant measures on the S.A. Ergodic or mixing properties of the diffeomorphism restricted to the S.A., with respect to this measure.

3) Examination of the geometry of the S.A. for higher dimensions. Mechanisms producing or destroying S.A. in this case: Study of homo/heteroclinic points of normally hyperbolic invariant or periodic objects.

We strongly recommend to look for the geometric structure in physical or numerical experiments. It seems to us that without this knowledge one cannot get a really deep insight in the problem of S.A.

REFERENCES

1. Abraham, R., Shaw, C.:"Dynamics: The Geometry of Behavior", vol.1 Aerial Press Inc. 1982 (following volumes to appear.)

2. Afraĭmovič, V.S., Šil'nikov, L.P.: "On some global bifurcations connected with the disappearance of a fixed point of saddle-node type", Soviet Math. Doklady 15, 1761-1765 (1974).

3. Afraĭmovič, V.S., Bykov, V.V., Šil'nikov, L.P.: "On the appearance and structure of the Lorenz-attractor", Dokl. Akad. Nauk. USSR 234, 336-339 (1977).

4. Aizawa, Y., Shimada, I.: "The Wandering motion on the Lorenz surface", Prog. Theor. Phys. 57, 2146-2147 (1977).

5. Allgower, E.L., Glashoff, K., Peitgen, H.O. (ed): "Numerical solution of nonlinear equations", Lect. Notes in Math. 878, Springer (1981).

6. Arneodo, A., Coullet, P., Tresser, C.: "Occurrence of strange attractors in 3-dimensional Volterra equations", Phys. Lett. 79A, 259-263 (1980).

7. Arneodo, A., Coullet, P., Tresser, C.: "A possible new mechanism for the onset of turbulence", Phys. Lett. 81A, 197-201 (1981).

8. Arneodo, A., Coullet, P., Tresser, C.: "Possible new strange attractors with spiral structure", Commun. Math. Phys. 79,573-579, (1981).

9. Arneodo, A., Coullet, P., Tresser, C.: "Oscillators with Chaotic Behavior: An illustration of a theorem of Šil'nikov", J. Stat. Phys. 27, 171-182 (1982).

10. Arneodo, A.,Coullet, P., Peyraud, J., Tresser, C.: "Strange attractors in Volterra equations for species in competition", J. Math. Biology 14, 153-157 (1982).

11. Arneodo, A.,Coullet, P., Tresser, C.: "On the relevance of period-doubling cascades at the onset of turbulence", to appear in Physica D.

12. Arneodo, A.,Coullet, P., Libchaber, A., Maurer, J., D'Humières, D., Tresser, C.: "About the observation of the uncompleted cascade in a Rayleigh-Bénard experiment", to appear in Physica D.

13. Arnold, V.I.: "Chapitres Supplémentaires de la Théorie des Equations Différentielles Ordinaires", Editions MIR, Moscow,1980.

14. Aronson, D.G., Chory, M.A., Hall, G.R., McGehee, R.P.: "Bi furcations from an invariant circle for two parameter families of maps of the Plane: A computer-assisted study", Commun. in Math. Phys. 83, 303-354 (1982).

15. Atten, P., Lacroix, J.C., Malraison, B.: "Chaotic motion in a Coulomb force driven instability: Large aspect ratio experiments". Phys. Lett. 77A, 255-258 (1980).

16. Auslander, J., Yorke, J.: "Interval maps, factors of maps and chaos", Tohoku Math. J. 32, 177-188 (1980).

17. Bernard, P., Ratiu, T. (ed): "Turbulence Seminar", Lect. Notes in Math. 615, Springer (1977).

18. Bountis, T.C.: "Period doubling and universality in conservative systems", Physica D (to appear).

19. Bowen, R., Ruelle, D.: "The ergodic theory of axiom A diffeomorphisms", Inventiones Math. 29, 181-202 (1975).

20. Bowen, R.: "A horseshoe with positive measure", Inventiones Math. 29, 203-204 (1975).

21. Bowen, R.: "On axiom A diffeomorphism", A.M.S., C.B.M.S. Regional Conference 35 (1978).

22. Bowen, R.: "Invariant measures for Markov maps of the interval", Comm. Math. Phys. 69, 1-17 (1979).

23. Baivé, D., Franceschini, V.: "Symmetry breaking on a model of five-mode truncated Navier-Stokes equations", J. Stat. Phys. 26, 471-484 (1981).

24. Bridges, R., Rowlands, G.: "On the analytic form of some strange attractors", Physical Lett. 63A, 189-190 (1977).

25. Campanino, M., Epstein, M.: "On the existence of Feigenbaum's fixed point", Comm. Math. Phys. 79, 261-302 (1981).

26. Chang , Shau-Jin, Wortis, M., Wright, J.A.: "Iterative properties of a one dimensional quartic map: Critical lines and tricritical behavior", Physical Review 24, 2669-2684 (1981).

27. Chillingworth, D.R.J., Holmes, R.J.: "Dynamical system and the models for reversals of the Earth's magnetic field", Math. Geology 12, 41-59 (1980).

28. Chirikov, B.V., Izraelèv, F.M.: "Some numerical experiments with a nonlinear mapping: Stochastic component", in "Transformations ponctuelles et leurs applications", 409-428, Colloques Internationaux du C.N.R.S. 229, and "Degeneration of turbulence in simple systems", Physica D 2, 30-37 (1981).

29. Clerc, R., Hartmann, Ch.: "Bifurcation mechanism of a second order recurrence leading to the appearance of a particular strange attractor", preprint, Univ. of Toulouse.

30. Collet, P., Eckmann, J.P.: "On the abundance of aperiodic behavior for maps of the interval", Comm. Math. Phys. 73, 115-160 (1980).

31. Collet, P., Eckmann, J.P., Lanford, O.: "Universal properties of maps of an interval", Comm. Math. Phys. 76, 211-254 (1980).

32. Collet, P., Eckmann, J.P.: "Iterated maps on the interval as dynamical systems", Birkhäuser, Boston, 1980.

33. Collet, P., Eckmann, J.P., Koch, H.: "Period doubling bifurcations for families of maps on $\mathbb{R}^n$", J. Stat. Phys. 25 (1981).

34. Collet, P., Crutchfield, J.P., Eckmann, J.P.: "Computing the topological entropy of maps", to appear in Physica D.

35. Cook, A., Roberts, D.: "The Rikitake two-disc dynamo system", Proc. Cambridge Philos. Soc. 68, 547-569 (1970).

36. Coullet, P., Tresser, C.: "Critical transition to "Stochasticity" for some dynamical systems", J. de Physique Lett. 41, L 255 (1980).

37. Crutchfield, J.P.: "Prediction and stability in classical mechanics, University of Santa Cruz Thesis, 1979.

38. Crutchfield, J.P., Farmer, J.D., Packard, N.H., Shaw, R.S., Jones, G., Donnelly, R.: "Power Spectral Analysis of a Dynamical System", Phys. Lett. 76A, 1 (1980).

39. Crutchfield, J.P., Huberman, B.A.: "Fluctuations and the onset of chaos", Phys. Lett. 77A, 407 (1980).

40. Crutchfield, J.P., Farmer, J.D., Huberman, B.A.: "Fluctuations and simple chaotic dynamics", to appear in Physics Reports.

41. Crutchfield, J.P., Nauenberg, M., Rudnick, J.: "Scaling for external noise at the onset of chaos", Phys. Rev. Lett. 46, 933 (1981).

42. Crutchfield, J.P., Packard, N.H.: "Symbolic dynamics of one-dimensional maps: entropies, finite precision and noise", to appear in Intl. J. Theor. Phys.

43. Curry, J.: "A generalized Lorenz system", Comm. Math. Phys. 60, 193-204 (1978).

44. Curry, J.: "On the Hénon transformation", Comm. Math. Phys. 68, 129-140 (1979).

45. Curry, J.: "On computing the entropy of the Hénon attractor", J. Stat. Phys. 26, 683-695 (1981).

46. Daido, H.: "Analytical conditions for the appearance of homoclinic and heteroclinic points of a 2-dimensional mapping", Prog. Theor. Phys. 63, 1190-1201 (1980).

47. Daido, H.: "Universal relation of a band-splitting sequence to a preceding period doubling one", Phys. Lett. 86A, 259-262 (1981).

48. Dell'Antonio, G., Doplicher, S., Jona-Lasinio, G. (ed.): "Mathematical problems in theoretical physics", Lect. Notes in Physics 80, Springer (1978).

49. Derrida, B., Pomeau, Y.: "Feigenbaum's ratios of 2-dimensional area preserving maps", Phys. Lett. 80A, 217-219 (1980).

50. Donnelly, R.J., Park, K., Shaw, R.S., Walden, R.W.: "Early nonperiodic transition in Couette flow", Phys. Rev. Lett. 44, 987 (1980).

51. Douady, A., Oesterlé, J.: "Dimension de Hausdorff des attracteurs", C.R. Acad. Sci. Paris, 290, 1135-1138 (1980).

52. Easton, R.: "A topological conjugacy invariant involving homoclinic points for diffeomorphisms of two-manifolds", preprint, Univ. of Boulder, Colorado.

53. Eckmann, J.P.: "Roads to turbulence in dissipative dynamical systems", Rev. Mod. Phys. 53, 643-654 (1981).

54. Eckmann, J.P.: "Renormalization group analysis of some highly bifurcated families", preprint, Univ. of Genève.

55. Eckmann, J.P.: "Intermittency in the presence of noise", J. Phys. A 14, 3153-3168 (1981).

56. Eckmann, J.P.: "A note on the power spectrum of the iterates of Feigenbaum's function", Commun. Math. Phys. 81, 261-265 (1981).

57. Farmer, J.D.: "Spectral broadening of period-doubling bifurcation sequences", Phys. Rev. Lett. 47, 179 (1981).

58. Farmer, J.D.: "Order within chaos", Univ. of Santa Cruz, Ph.D. Thesis (1981).

59. Farmer, J.D.: "Chaotic attractors of an infinite-dimensional dynamical system", Physica D 4, 366-393 (1982).

60. Farmer, J.D. "Information dimension and the probabilistic structure of chaos", to appear in Z. Naturforschung.

61. Farmer, J.D., Ott, E., Yorke, J.A.: "The dimension of chaotic attractors", to appear in Physica D.

62. Farrell, F.T., Jones, L.E.: "New attractors in hyperbolic dynamics, J. Diff. Geometry 15, 107-133 (1980).

63. Fatou, P.: "Sur les équations fonctionelles", Bull. Soc. Math. de France, 47, 161-270 (1919), 48, 33-95 & 208-314 (1920).

64. Feigenbaum, M.: "Quantitative universality for a class of nonlinear transformations", J. Stat. Phys. 19, 25-52 (1978).

65. Feigenbaum, M.: "The onset spectrum of turbulence", Phys. Lett. 74A, 375 (1979).

66. Feigenbaum, M.: "The transition to aperiodic behavior in turbulent systems", Commun. Math. Phys. 77, 65-86 (1980).

67. Feit, S.: "Characteristic exponents and strange attractors", Commun. Math. Phys. 61, 249-260 (1978).

68. Franceschini, V.: "Feigenbaum sequence of bifurcations in the Lorenz model", J. Stat. Phys. 22, 397-407 (1980).

69. Franceschini, V., Russo, L.: "Stable and unstable manifolds of the Hénon mapping", preprint, Univ. of Modena.

70. Franceschini, V.: "Two models of truncated Navier-Stokes equations on a two-dimensional torus", Phys. of Fluids (to appear).

71. Franceschini, V.: "Truncated Navier-Stokes equations on a two-dimensional torus", preprint, Los Alamos.

72. Franceschini, V.: "Bifurcation of tori and phase-locking in a dissipative system of differential equations", preprint, Los Alamos.

73. Frederickson, P., Kaplan, J.L., Yorke, E.D., Yorke, J.: "The Lyapunov dimension of strange attractors", J. Diff. Equations (to appear).

74. Fujisaka, H., Yamada, T.: "Limit cycles and chaos in realistic models of the Belousov-Zhabotinskii reaction system, Z. Physik B, 37, 265-275 (1980).

75. Garrido, L. (ed.): "Systems far from equilibrium", Lect. Notes in Physics 132 (1980).

76. Gollub, J.P. Swinney, H.L.: "Onset of turbulence in a rotating fluid", Phys. Rev. Lett. 35, 921 (1975).

77. Grebogi, C., Ott, E., Yorke, J.A.: "Chaotic attractors in crisis", Phys. Rev. Lett. 48, 1507-1510 (1982).

78. Grebogi, C., Ott, E., Yorke, J.A.: "Crisis, sudden changes in chaotic attractors and transient chaos", preprint, Univ. of Maryland.

79. Grmela, M., Marsden, J.E. (ed.): "Global analysis", Lect. Notes in Math. 755, Springer(1979).

80. Grossmann, S., Thomae, S.: "Invariant distributions and stationary correlation functions", Z. Naturforschung 32A, 1353-1365 (1977).

81. Guckenheimer, J.: "Bifurcations of maps of the interval", Inventiones math. 39, 165-178 (1977).

82. Guckenheimer, J.: "Sensitive dependence on initial conditions for one-dimensional maps", Commun. Math. Phys. 70, 133-160 (1979).

83. Guckenheimer, J., Moser, J., Newhouse, S.: "Dynamical Systems", Birkhäuser, Boston, 1980.

84. Guckenheimer, J., Williams, R.: "Structural stability of Lorenz attractors", Pub. Math. I.H.E.S. 50, 60-72 (1980).

85. Gumowski, I., Mira, C.: "Dynamique Chaotique", Editions Cepadues, Toulouse, 1980.

86. Gumowski, I., Mira, C.: "Recurrence and discrete dynamical systems", Lect. Notes in Math. 809, Springer, 1980.

87. Gurel, O., Rössler, O.E. (ed.): "Bifurcation theory and applications in scientific disciplines", Annals New York Acad. Sci. 316, (1979).

88. Haken, H.: "Synergetics, An Introduction", 2nd ed. Springer series in Synergetics, 1, (1978).

89. Haken, H. (ed.): "Synergetics, A Workshop", Springer series in Synergetics 2, (1977).

90. Haken, H., Wunderlin, A.: "New interpretation and size of strange attractors of the Lorenz model of turbulence", Phys. Lett. 62A, 133-134 (1977).

91. Haken, H. (ed.): "Chaos and order in Nature", Springer (1981).

92. Haken, H. (ed.): "Evolution of ordered and chaotic patterns in systems treated by the natural sciences and mathematics", Springer (1982) (to appear).

93. Helleman, R.H.G.: "Self-generated chaotic behavior in nonlinear mechanics", in "Fundamental Problems in Statistical Mechanics V", Ed.: E.G.D. Cohen, 165-233, North Holland (1980).

94. Helleman, R.H.G. (ed.): "Nonlinear dynamics", Annals New York Acad. Sci. 357 (1980).

95. Helleman, R.H.G., Iooss, G. (ed.): "Chaotic behavior in deterministic systems", North-Holland, (1982) (to appear).

96. Hénon, M.: "A two-dimensional mapping with a strange attractor", Commun. Math. Phys. 50, 69-77 (1976).

97. Herring, C., Huberman, B.A.: "Dislocation motion and solid-state turbulence", Appl. Phys. Lett. 36, 975-977 (1980).

98. Hirsch, M.W., Pugh, C.C., Shub, M.: "Invariant manifolds", Lect. Notes in Math. 583, Springer (1977).

99. Hitzl, D.L.: "Numerical determination of the capture/escape boundary for the Hénon attractor", preprint Lockheed, Palo Alto(1981).

100. Holmes, P.J.: "Strange phenomena in dynamical systems and their physical implications", Appl. Math. Modelling, 1, 362-366(1977).

101. Holmes, P.J.: "A nonlinear oscillator with a strange attractor", Phil. Trans. Roy. Soc. London, Ser. A 292, 419-448 (1979).

102. Holmes, P.J.: "Averaging and chaotic motions in forced oscillations", SIAM J. Appl. Math. 38, 65-80 (1980).

103. Holmes, P.J., Moon, F.C.: "A magneto-elastic strange attractor", J. Sound vibration 65, 275-296 (1979).

104. Holmes, P.J. (ed.): "New approaches to nonlinear problems in dynamics",SIAM, Philadelphia (1980).

105. Hoppenstedt, F.C. (ed.): "Nonlinear oscillations in Biology", A.M.S. Lectures in Appl. Math. 17, Providence (R.I.) (1979).

106. Huberman, B.A., Crutchfield, J.P.: "Chaotic states of anharmonic systems in periodic fields", Phys. Rev. Lett. 43, 1743-1747 (1979).

107. Huberman, B.A., Crutchfield, J.P., Packard, N.H.: "Noise phenomena in Josephson junctions", Appl. Phys. Lett. 37, 750-752 (1980).

108. Huberman, B.A., Rudnick, J.: "Scaling behavior of chaotic flows" Phys. Rev. Lett. 45, 154-156 (1980).

109. Hurley, M.: "Attractors: Persistence and density of their basins", Trans. Amer. Math. Soc. 269, 247-271 (1982).

110. Izraelev, F.M., Rabinovich, M.I., Ugodnikov, A.D.: "Approximate description of three dimensional dissipative systems with stochastic behavior", Phys. Lett. 86A, 321-325 (1981).

111. Jakobson, M.V.: "Absolutely continuous invariant measures for one-parameter families of one-dimensional maps". Commun. Math. Phys. 81, 39-88 (1981).

112. Jeffreis, C., Pérez, J.: "Direct observation of crisis of the chaotic attractor in a nonlinear oscillation", preprint, Univ. of Berkeley (1982).

113. Jonker, L., Rand, D.: "Bifurcations in one-dimension, I, II," Inventiones math. 62, 347-365 and 63, 1-15 (1981).

114. Jorna, S.(ed.): "Topics in Nonlinear Dynamics", Amer. Inst. Phys. Conf. Proc. 46, (1978).

115. Julia, G.: "Mémoire sur l'itération des fonctions rationelles", J. de Math. sér. 7, 4, 47-245 (1918).

116. Kai, T., Tomita, K.: "Stroboscopic phase-portrait of a forced nonlinear oscillator", Prog. Theor. Phys. (to appear).

117. Kaplan, J.L., Yorke, J.A.: "Preturbulence: A regime observed in a fluid flow model of Lorenz", Commun. Math. Phys. 67, 93-108, (1979).

118. Keener, J.P.: "Chaotic cardiac dynamics", in Lect. in Appl. Math. 19, 299-325, "Mathematical Aspects of Physiology", ed. F.C. Hoppensteadt, AMS (1981).

119. Kidachi, H.: "On a chaos as a mode interaction phase", Prog. Theor. Phys. 65, 1584-1594 (1981).

120. Knobloch, E.: "Chaos in the segmented disc dynamo", Phys. Lett. 82A, 439-440 (1981).

121. Lasota, A., Yorke, J.: "On the existence of invariant measures for piecewise monotonic transformations", Trans. Amer. Math. Soc. 184, 481-488 (1973).

122. Laval, G., Gresillon, D. (ed.): "Intrinsic stochasticity in plasmas", International Workshop, Cargèse, Les Editions de Physique, Orsay (1979).

123. Leipnick, R.B., Newton, T.A.: "Double strange attractors in rigid body motion with linear feedback control", Phys. Lett. 86A, 63-67 (1981).

124. Leven, R.W., Koch, B.D.: "Chaotic behavior of a parametrically excited damped pendulum", Phys. Lett. 86A, 71-74 (1981).

125. Levi, M.: "Qualitative analysis of the periodically forced relaxation oscillators", Mem. A.M.S. 244, (1981).

126. Li, T.Y., Yorke, J.A.: "Period three implies chaos", Americ. Math. Monthly, 82, 985-992 (1975).

127. Li, T.Y., Misiurewicz, M., Pianigiani, G., Yorke, J.: "Odd chaos", Phys. Lett. 87A, 271-273 (1982).

128. Libchaber, A., Maurer, J.: "Une expérience de Bénard-Rayleigh de géometrie réduite; multiplication, accrochage et démultiplication de fréquences", J. de Physique 41,(Coll. C3),51-56 (1980).

129. Lopes, A.D.: "An example of interpolation of an attractor", preprint, Inst. Mat. Porto Alegre.

130. Lorenz, E.N.: "Deterministic nonperiodic flow", J. Atmosph. Sci. 20, 130-141 (1963).

131. Lozi, R.: "Un attracteur étrange (?) du type attracteur de Hénon", J. de Physique 39 (Coll. C5), 9-10 (1978).

132, Lozi, R.: "Sur un modèle mathématique de suite de bifurcations de motifs dans la réaction de Belousov-Zhobotinsky", C.R. Acad. Sci. Paris 294, 21-26 (1982).

133. L'vov, V.S., Predtechensky, A.A.: "On Landau and stochastic pictures in the problem of transition to turbulence, Physica D, 2 38-51 (1981).

134. Mackey, M.C., Glass, L.: "Oscillators and chaos in physiological control systems", Science 197, 287-289 (1977).

135. Manneville, P.: "Intermittency in dissipative dynamical systems", Phys. Lett. 79A, 33-35 (1980).

136. Manneville, P., Pomeau, Y.: "Different ways to turbulence in dissipative dynamical systems", Physica D, 1, 219-226 (1980).

137. Manneville, P.: "Intermittency, self-similarity and 1/f spectrum in dissipative dynamical systems", J. Physique 41, 1235-1243 (1980).

137. Markley, N.G., Martin, J.C., Perrizo, W. (ed.): "The structure of attractors in dynamical systems", Lect. Notes in Math. 668, Springer, (1978).

139. Marotto, F.R.: "Chaotic behavior in the Hénon mapping", Commun. Math. Phys. 68, 187-194 (1979).

140. Marsden, J.E., McCracken, M.: "The Hopf bifurcation and its applications", Appl. Math. Sci. 19, Springer (1976).

141. Marzec, C.J., Spiegel, E.A.: "Ordinary differential equations with strange attractors", SIAM J. Appl. Math. 38, 387-421 (1980).

142. May, R., Oster, G.: "Bifurcations and dynamic complexity in simple ecological models", The Amer. Natur. 110, 573-599 (1976).

143. Mayer-Kress, G., Haken, H.: "Intermittent behavior of the logistic system", Phys. Lett. 82A, 151-155 (1981).

144. McLaughlin, J., Martin, P.: "Transition to turbulence in a statically stressed fluid system", Phys. Rev. A 12, 186-203 (1975).

145. Metropolis, M., Stein, M.L., Stein, P.R.: "On finite limit sets for transformations of the unit interval", J. Combinatorial Theory 15, 25-44 (1973).

146. Milnor, J., Thurston, W.: "On iterated maps of the interval", preprint, Univ. of Princeton.

147. Mira, C.: "Accumulation de bifurcations et "structures boîtes emboîtées" dans les recurrences et transformations ponctuelles", VII ICNO, Berlin (1975).

148. Misiurewicz, M., Swecz, B.: "Existence of a homoclinic point for the Hénon map", Commun. Math. Phys. 75, 285-291 (1980).

149. Myrberg, P.J.: "Iteration der reellen Polynome zweiten Grades, I, II, II", Ann. Acad. Sci. Fenn. 256A, 1-10 (1958); 268A, 1-10 (1959), 336A, 1-10 (1963).

150. Myrberg, P.J.: "Iteration der Polynome mit reellen Koeffizienten" Ann. Acad. Sci. Fenn. 374A, 1-18 (1965).

151. Newhouse, S.: "Diffeomorphisms with infinitely many sinks", Topology 12, 9-18 (1974).

152. Nitecki, Z., Robinson, C. (ed.): "Global theory of dynamical systems", Lect. Notes in Math. 819, Springer (1980).

153. Ogura, H., Ueda, Y., Yoshida, Y.: "Periodic stationarity of a chaotic motion in the system governed by Duffing's equation", Prog. Theor. Phys. 66, 2280-2283 (1981).

154. Oono, Y., Osikawa, M.: "Chaos in nonlinear differential equations, I", Prog. Theor. Phys. 64, 54-67 (1980).

155. Oseledec, V.I.: "A multiplicative ergodic theorem. Lyapunov characteristic numbers for dynamical systems", Trans. Moscow Math. Soc. 19, 197-231 (1968).

156. Packard, N.H., Crutchfield, J.P., Farmer, J.D., Shaw, R.S.: "Geometry from a time series", Phys. Rev. Lett. 45, 712 (1980).

157. Peitgen, H.O., Walther, H.O. (ed.): "Functional differential equations and approximation of fixed points", Lect. Notes in Math. 730, Springer (1979).

158. Pesin, Ya. B.: "Characteristic Lyapunov exponents and smooth ergodic theory", Russ. Math. Surveys, 32, 55-115 (1977).

159. Peters, H.: "Chaotic behavior of nonlinear differential-delay equations", preprint, Univ. of Bremen.

160. Pikowsky, A.S., Rabinovich, M.I.: "Stochastic oscillations in dissipative systems". Physica D 2, 8-24 (1981).

161. Pixton, D.: "Planar homoclinic points", J. of Diff. Eq. 44, 365-382 (1982).

162. Plykin, R.: "Sources and sinks for A-diffeomorphisms", Math. Sb. 23, 233-253 (1974).

163. Pomeau, Y., Manneville, P.: "Intermittent transition to turbulence in dissipative dynamical systems", Commun. Math. Phys. 74, 189-197 (1980).

164. Pounder, J.R., Rogers, T.D.: "The geometry of chaos :dynamics of a nonlinear second order difference equations", Bull. Math. Biol. 42, 551-597 (1980).

165. Rand, D.A., Young, L.S. (ed.): "Dynamical systems and turbulence, Warwick 1980", Lect. Notes in Math. 898, Springer (1981).

166. Roux, J.C., Rossi, A., Bachelart, S., Vidal, C.: "Representation of a strange attractor from an experimental study of chemical turbulence", Phys. Lett. 77A, 391-393 (1980).

167. Ruelle, D., Takens, F.: "On the nature of turbulence", Commun. Math. Phys. 20, 167-192 (1971).

168. Ruelle, D.: "A measure associated to axiom A attractors", Amer. J. of Math. 98, 619-654 (1976).

169. Ruelle, D.: "Applications conservant une mésure absolument continue par rapport à dx sur [o,1]",Comm. Math. Phys. 55,47-51 (1977).

170. Ruelle, D.: "Strange attractors", The Mathematical Intelligencer 2, 126-137 (1980).

171. Ruelle, D.: "Small random perturbations of dynamical systems and the definition of attractors", Commun. Math. Phys. 82, 137-151 (1981).

172. Ruelle, D.: "Do there exist turbulent crystals?", preprint, I.H.E.S.

173. Šarkovskii, A.N.: "Coexistence of cycles of a continuous map of a line into itself", Ukr. Mat. Z. 16, 61-71 (1964).

174. Scholz, H.J., Yamada, T., Brand, H., Graham, R.: "Intermittency and chaos in a laser system with modulated inversion", Phys. Lett. 82A, 321-323 (1981).

175. Shaw, R.S.: "Strange attractors, chaotic behavior, and information flow", Z. Naturforschung 36A, 80 (1981).

176. Shaw, R.S.: "On the predictability of mechanical systems", Univ. of Santa Cruz, Ph.D. Thesis (1980).

177. Shaw, R.S., Anderek, C.D., Reith, L.A., Swinney, M.L.: "Nonlinear superposition of traveling waves in circular Couette flow", Phys. Rev. Lett. (to appear).

178. Shimada, I., Nagashima, T.: "The iterative transition phenomenon between periodic and turbulent states in a dissipative dynamical system", Prog. Theor. Phys. 59, 1033-1035 (1978).

179. Shimada, I.: "Gibbsian distribution on the Lorenz attractor", Prog. Theor. Phys. 62, 61-69 (1979).

180. Shimizu, T."Asymptotic form of a strange attractor", Phys. Lett. 84A, 85 (1981).

181. Simó, C.: "On the Hénon-Pomeau attractor", J. Stat. Phys. 21, 465-493 (1979).

182. Simonov, A.A.: "An investigation of bifurcations in some dynamical systems by the methods of symbolic dynamics", Soviet Math. Dok. 19, 759-763 (1978).

183. Sinai, J.G. Vul, E.B.: "Hyperbolicity conditions for the Lorenz model", Physica D, 2, 3-7 (1981).

184. Smale, S.: "Diffeomorphisms with many periodical points", in Differential and Combinatorial Topology, 63-80, Princeton Univ. Press (1965).

185. Smale, S.: "Differentiable dynamical systems", Bull. Amer. Math. Soc. 73, 747-817 (1967).

186. Smale, S. (ed.): "Global analysis", Proc. Sympos. Pure Math. 14, Amer. Math. Soc. (1970).

187. Steeb, W.H., Kunick, A.: "Lagrange functions of a class of dynamical systems with limit cycle and chaotic behavior", Phys. Rev. A 25, 2889-2892 (1982).

188. Steeb, W.H.: "Continuous symmetries of the Lorenz and the Rikitake two-disc dynamo system", J. Phys. A 15, L 389-390 (1982).

189. Steeb, W.H., Erig, W., Kunick, A.: "Chaotic behavior and limit cycle behavior of anharmonic systems with periodic external perturbations", preprint, Univ. of Paderborn.

190. Steeb, W.H., Kunick, A.: "On the Painlevé property of anharmonic systems with an external period field", preprint, Univ. of Paderborn.

191. Stefan, P.: "A theorem of Šarkovskii on the existence of periodic orbits of continuous endomorphisms of the real line", Commun. Math. Phys. 54, 237-248 (1977).

192. Swinney, H.L., Gollub, J.P.: "The transition to turbulence", Physics Today 31, 41-49 (1978).

193. Takeyama, K.: "Dynamics of the Lorenz model of convective instabilities II", Prog. Theor. Phys. 63, 91-105 (1980).

194. Tél, T.: "On the construction of stable and unstable manifolds of two-dimensional invertible maps", preprint, Eötvös Univ.

195. Temam, R. (ed.): "Turbulence and Navier-Stokes Equation", Lect. Notes in Math. 565, Springer (1976).

196. Testa, J., Held, G.A.: "Period doubling, bifurcations, chaos and periodic windows of the cubic map", preprint, Univ. of Berkeley.

197. Thomae, S., Grossmann, S.: "Correlations and spectra of periodic chaos generated by the logistic parabola", J. Stat. Phys. 26, 485-504 (1981).

198. Thurston, W.: "On the geometry and dynamics of diffeomorphisms of surfaces", preprint, Univ. of Princeton.

199. Tomita, K., Kai, T.: "Stroboscopic phase-portrait and strange attractors", Phys. Lett. 66A, 91-93 (1978).

200. Tomita, K., Daido, H.: "Possibility of chaotic behavior and multi-basins in forced glycolytic oscillations", Phys. Lett. 79A, 133-137 (1980).

201. Tomita, K., Kai, T.: "Chaotic response of a nonlinear oscillator" J. Stat. Phys. (to appear).

202. Tomita, K., Tsuda, I.: "Towards an interpretation of Hudson's experiment on the Belousov-Zhabotinskii Reaction: chaos due to delocalization", Prog. Theor. Phys. 64, 1138-1160 (1980).

203. Tomita, K., Tsuda, I.: "Chaos in the Belousov-Zhabotinskii reaction in a flow system", Phys. Lett. 71A, 489-492 (1979).

204. Tomita, K.: "Chaotic behavior of deterministic orbits. The problem of turbulent phase", preprint, Univ. of Kyoto.

205. Tresser, C., Coullet, P., Arneodo, A.: "On the existence of hysteresis in a transition to chaos after a single bifurcation", J. de Physique Lettres 41, L 243-246 (1980).

206. Tresser, C., Coullet, P., Arneodo, A.: "Topological horseshoe and numerically observed chaotic behavior in the Hénon mapping", Lett. to the Editor, J. Phys. A13, L 123-127 (1980).

207. Tsuda,I. "Self-similarity in the Belusov-Zhabotinsky reaction" Phys. Lett. 85A, 4 (1981).

208. Ueda, Y., Hayashi, C., Akamatsu, N.: "Computer simulation of nonlinear ordinary differential equations and nonperiodic oscillations", Electronics and Communications in Japan 56A, 27-34 (1973).

209. Ueda, Y.: "Randomly transitional phenomena in the system governed by Duffing's equation", J. Stat. Phys. 20, 181-196 (1979).

210. Ueda, Y., Akamatsu, N.: "Chaoticaly transitional phenomena in the forced negative-resistance oscillator", IEEE, Trans. on Circuits and Systems 28, 217-224 (1981).

211. Ulam, S., Von Neumann, J.: "On combinations of stochastic and deterministic processes", Bull. Amer. Math. Soc. 53,1120 (1947).

212. Velsen, R.V., Oberman, C.R.: "Statistical properties of chaotic dynamical systems which exhibit strange attractors", Physica D, 4, 183-196 (1982).

213. Vidal, Ch. et al.: "Étude de la transition vers la turbulence chimique dans la réaction de Belousov-Zhabotinskii", C.R. Acad. Sci. Paris 289 C, 73-77 (1979).

214. Walther, H.O.: "Homoclinic solutions and chaos in $\dot{x}(t)=f(x(t-1))$", Nonlinear Analysis: Theory, Methods and Appl. 5, 775-788 (1981).

215. Wegmann, K., Rössler, D.: "Different kinds of chaotic oscillations in the Belousov-Zhabotinskii reaction", Z. Naturforschung 33A, 1179-1183 (1978).

216. Wersinger, J.M., Finn, J.H. Ott, E.: "Bifurcation and "strange" behavior in instability saturation by nonlinear three-wave mode coupling", Phys. Fluids 23, 1142-1154 (1980).

217. Willamowski, K.D., Rössler, O.E.: "Irregular oscillations in a realistic abstract quadratic mass action system", Z. Naturforschung 35A, 317-318 (1980).

218. Williams, R.: "One dimensional nonwandering sets", Topology 6, 473-487 (1967).

219. Williams, R.: "The structure of Lorenz attractors", Pub. Math. I.H.E.S. 50, 73-99 (1980).

220. Young, L.S.: "Capacity of attractors", Ergod. Th. and Dynam. Syst. 1, 381-388 (1981).

221. Zaslavsky, G.M.: "The simplest case of a strange attractor", Phys. Letters 69A, 145-147 (1978).

CHAOTIC DYNAMICS IN HAMILTONIAN SYSTEMS WITH DIVIDED PHASE SPACE

Boris V. Chirikov

Institute of Nuclear Physics,
630090 Novosibirsk, USSR

1. INTRODUCTION

The subject of this lecture is related to a peculiar dynamical phenomenon in classical mechanics commonly termed among physicists as the chaotic, or stochastic, motion. Until recently the mathematicians used to speak just about ergodic properties of a dynamical system. However, nowadays the term "random motion" becomes also popular. I would like to emphasize from the beginning that the problem we are going to discuss is purely dynamical, without any random element either in the equations of motion or in the initial conditions. Hence, the term - dynamical, or intrinsic chaos. Below we restrict ourselves to only Hamiltonian dynamics for which the invariant measure (phase space volume) is known beforehand, unlike dissipative systems.

The interest in the dynamical chaos is twofold. First, it is a fundamental phenomenon in physics which, in particular, gives, at last, a long-awaited model for the true random process. Second, no matter how strange the random dynamics may appear it turns out to be fairly wide-spread in many fields of science and technology as, in particular, the present Conference demonstrates.

In a rare occasion, when chaos comprises all the phase space of a dynamical system or, at least, a whole invariant surface of the motion, a fairly simple statistical description is possible as contrasted to most complicated dynamical pictures of motion. In many cases, however, the situation is not that simple. A typical example is the so-called divided phase space,divided into regions of both chaotic and regular motions separated by highly intricate borders. It is the structure of that chaos border which considerably complicates statistical description of the motion. Even though the mathematical theory of dynamical systems admits divided phase space and, moreover, terms it by a special notion - ergodic component - not much is actually known thus far on the dynamical behavior therein. Below we are going to consider a number of selected questions related to this topic. I choose an old Poincaré problem, which is still not solved completely, to discuss some recent developments in this field. A general review of the modern mathematical theory can be found in [1-3], while related physical theory is surveyed, e.g., in [4,5].

2. POINCARE'S PROBLEM

We begin with a "simple" example considered by Poincaré [6] in his attempt to understand profound difficulties arising in the study of nonlinear dynamics, in general, and of the famous three body problem, in particular. Much later, this example has proven to typify a fairly general situation in Hamiltonian dynamics (see [4] and Section 3b below).

Consider the motion of the ordinary pendulum under a high frequency parametric perturbation as described by the Hamiltonian

$$H(p,\varphi,\theta)=\frac{p^2}{2}+\omega_0^2\cdot\cos\varphi\cdot(1+\varepsilon\cdot\cos\theta) \tag{2.1}$$

Here φ is the angular position of pendulum ($\varphi = 0$ corresponds to the unstable equilibrium); $p=\dot{\varphi}$ is the angular momentum and ω_0 is the frequency of small oscillation. The perturbation is characterized by two small parameters: that of strength ε, and of adiabaticity $1/\lambda=\omega_0/\Omega$, $\Omega=\dot{\theta}$. The motion of the unperturbed ($\varepsilon = 0$) pendulum, as is well known, is periodic for any initial conditions with one important exception corresponding to the value of $H=\omega_0^2$. The latter trajectory is called separatrix since it separates the pendulum oscillation ($H<\omega_0^2$) from its rotation ($H>\omega_0^2$). In what follows the separatrix is going to play a leading part in dynamical chaos. The motion period T is increasing indefinitely when approaching separatrix. In immediate vicinity of the latter

$$T\approx\frac{1}{\omega_0}\ln\frac{32}{|w|}\,;\quad w=\frac{H}{\omega_0^2}-1 \tag{2.2}$$

The separatrix motion is, thus, aperiodic, and it has continuous Fourier spectrum which may be characterized by the integral [4]:

$$A_m(\lambda)=\omega_0\int_{-\infty}^{\infty}e^{i\left(\frac{m}{2}\varphi_s(t)-\Omega t\right)}dt\approx\frac{4\pi}{\Gamma(m)}(2\lambda)^{m-1}\cdot e^{-\pi\lambda/2} \tag{2.3}$$

The last expression holds for [6] $\lambda\gg 1$; $\Gamma(m)$ is the gamma function with any positive real m, and

$$\varphi_s(t)=4\cdot\arctan\left(e^{\omega_0 t}\right)-\pi \tag{2.4}$$

is the separatrix motion (in case of $m<0$ $A_m=A_{|m|}\cdot e^{-\pi\lambda}$).

What is the impact of perturbation on the pendulum motion? The first move would be to consider the perturbation as completely nonresonant because of the condition $\lambda \gg 1$. Then, in the first approximation of the asymptotic theory [7] the perturbation can be neglected, or averaged out. Yet, in the second approximation it changes the effective potential:

$$U(\varphi) = \omega_0^2 \cdot \cos\varphi \rightarrow \omega_0^2 \cdot \left(\cos\varphi - \frac{\varepsilon^2}{8\lambda^2} \cos 2\varphi\right) \tag{2.5}$$

and shifts the frequencies at both stable and unstable equilibria.

Now, let us inspect the perturbation more carefully. Is it really completely nonresonant? And is the change (2.5) its only effect? Certainly, it is not on the separatrix, as is obvious from the boundless spectrum (2.3). Hence, in some vicinity around separatrix we also cannot neglect the perturbation even in the first approximation. That the motion here is very sensitive to perturbation, which makes it highly intricate, has been found out and well recognized already by Poincaré. He was very close to the discovery of chaotic dynamics although he did never use this sort of language, instead speaking just about homoclinic solutions, or trajectories. One of the problems he has left to future researchers was to find out the dimension, structure, and measure of the homoclinic region near separatrix.

3. SOLUTION OF THE POINCARÉ PROBLEM

a. Separatrix mapping

First, we construct a mapping describing the motion near separatrix in finite time steps. It is natural to choose the motion period T as the time step. Then, the change in energy w over this step is given by the integral of the type (2.3) while the change in perturbation phase θ is determined by the dependence (2.2). Thus, we arrive at the separatrix mapping $(w, \theta) \rightarrow (\bar{w}, \bar{\theta})$[4]:

$$\bar{w} = w + \xi \cdot \sin\theta; \qquad \bar{\theta} = \theta + \lambda \cdot \ln\frac{32}{|\bar{w}|} \tag{3.1}$$

The new perturbation parameter ξ is given by the expression

$$\xi = -4\pi\varepsilon\lambda^2 e^{-\pi\lambda/2} \tag{3.2}$$

While ξ is proportional to small parameter ε, it cannot be expand in powers of adiabaticity parameter $1/\lambda$. Hence, as is commonly believed, the expression (3.2) as well as the map (3.1) go beyond the asymptotic perturbation theory. However, one can argue in a different way: it is not so much a fault of asymptotic theory but, rather, our own failure to choose the proper, adequate perturbation parameter. In other words, the true small parameter of the adiabatic perturbation is not the usual one $1/\lambda$, which enters the original Hamiltonian, but another one which explicitly takes account of weak resonances present in spite of adiabatic conditions. An important point of this philosophy relates to the fact that there is no principal difficulty in evaluating this ξ . The evaluation actually follows the usual asymptotic procedure of successive approximations since the unperturbed separatrix motion (2.4) is used. The really crucial difference from earlier unsuccessful approaches to Poincaré's and similar problems lies in seeking out the resonances even if they do appear to be absent.

Parameter ξ immediately gives the so-called splitting of separatrix, i.e. a gap between the two branches of separatrix going up and down in time (the first corresponds to $w=0$, and the second does so to $\bar{w}=0$, the maximal gap being $2|\xi|$). This effect has also been discovered by Poincaré [6] (Section 401). In our time it was further studied by Melnikov [8], Shilnikov [9] and others.

Separatrix splitting is a very important dynamical phenomenon. Yet, it does not tell us anything about a long-term evolution of the system. Are variations of w restricted or unbounded?

Before we proceed further we transform (3.1) introducing a new variable $y=w/\xi$, that is we take half of separatrix splitting as the unit for w . Ignoring a constant phase shift in the second equation (3.1) we arrive at the reduced map

$$\bar{y} = y + \sin\theta; \qquad \bar{\theta} = \theta - \lambda\cdot \ln|\bar{y}| \tag{3.3}$$

b. The standard map

For treating the separatrix mapping (3.3) analytically we introduce another approximate model [4] by linearizing the second equation (3.3) in y around one of resonant values of $y=y_r$ where $\lambda\cdot \ln y_r = 2\pi r$, and r is any integer. We get the map

$$\bar{I} = I + K\cdot \sin\theta; \qquad \bar{\theta} = \theta + \bar{I} \tag{3.4}$$

which is called the standard map since it is the final reducing step for a number of particular problems in nonlinear dynamics [4]. The new momentum $I=(y_r-y)\lambda/y_r$, and the perturbation parameter:

$$K = -\frac{\lambda}{y_r} \tag{3.5}$$

The standard map provides a local (in y) description for the previous model (3.3) under the condition: $|y_r - y_{r-1}| \ll y_r$, or $\lambda \gg 2\pi$. Note an additional symmetry of this map: $I \to I + 2\pi r$, which is not present in (3.3). It is just this symmetry that considerably simplifies the motion analysis since it makes the motion structure periodic in momentum I.

We replace, further, a discrete system (3.4) by the completely equivalent continuous one with Hamiltonian [4]

$$H(I,\theta,t) = \frac{I^2}{2} + K \cdot \sum_{r=-\infty}^{\infty} \cos(\theta - 2\pi r t) \tag{3.6}$$

which has an infinite series of (integer) resonances $I = I_r = 2\pi r$. If we single out one of them, say, $r = 0$, and ignore (average out) all the others, we just come back to the pendulum whose motion we intended to study in this way. It is easy to see that leaving two more terms in series (3.6) ($r = \pm 1$) we completely recover the original problem (2.1) with the parameters:

$$\omega_0^2 = K; \quad \Omega = 2\pi; \quad \lambda = \frac{2\pi}{\sqrt{K}}; \quad \varepsilon = 2 \tag{3.7}$$

Yet, it is not a vicious circle but a spiral of cognition! In a more formal language it is called renormalization.

Now, let us mention, first of all, that the dynamics of a single nonlinear resonance can be described as a pendulum motion, or in the "pendulum approximation". As is shown in [4], this approximation is applicable under fairly broad conditions. Moreover, the original problem (2.1) relates to the dynamics of several (three) resonances and, hence, does include also the resonance interaction. Here, precisely, lies the importance of the Poincaré example and of the Poincaré problem.

Renormalized system (3.6) is not completely equivalent to the original one (2.1) in that the former has infinitely many resonances instead of three only for the latter. At the first glance, the problem becomes, thus, much more complicated, yet this is not the case. Just due to periodicity in I, the standard map, unlike the perturbed pendulum, has a sharp critical value of its parameter $|K| = K_{cr}$ which separates the bounded and unbounded variation of I. What is this critical value? First, we may just refer to the numerical simulation [4] which gives $K_{cr} = 0.989 \approx 1$ to the accuracy within a few percent. Using a completely different approach, based on a combination of analytical as well as numerical procedures, Greene [10] has found $K_{cr} = 0.971635$. This latter result has been confirmed also in [11]. The accuracy of this value is open to criticism [12], yet, at any event, it is fairly close to the above numerical result.

The critical K value can be also estimated, in order of magnitude, from a simple resonance overlap criterion [4]

$$s = \frac{(\Delta\omega)_r}{\delta\omega_r} \approx \frac{4\sqrt{K}}{2\pi} \approx \frac{4\Omega_\phi}{2\pi} \sim 1 \qquad (3.8)$$

where Ω_ϕ is the frequency of small phase oscillation on a resonance, $(\Delta\omega)_r$ is the resonance width, and $\delta\omega_r$ is the spacing of resonances under consideration. Even though Eq.(3.8) gives the correct order it considerably overestimates $K_{cr} \sim 2.5$ because only integer resonances ($\omega_r = I_r = 2\pi r$) are taken into account. Meanwhile, in higher approximations of perturbation theory the full set of resonances ($\omega_{rq} = 2\pi r/q$) does appear which obviously lowers K_{cr}. A partial consideration of those higher order resonances results in a much better estimate [4]: $K_{cr} \approx 1.1$.

Below the threshold, that is for $|K| < 1$ (we neglect the above discrepancies in K_{cr}), the I variation is strictly bounded by the resonance width: $|\Delta I| \lesssim 4\sqrt{K}$. Above the threshold the motion is generally (depending on initial conditions) unlimited in I and chaotic [4]. From Eq.(3.5) we immediately see that the motion near separatrix is chaotic within the layer $|y| \lesssim \lambda$, or:

$$|w| \lesssim w_s = \lambda|\xi| = 4\pi\varepsilon\lambda^3 e^{-\pi\lambda/2} \qquad (3.9)$$

This relation resolves [4] the Poincaré problem as to the dimension of a homoclinic region. Thus, the whole homoclinic structure generated by the two branches of split separatrix is chaotic and occupies a layer whose width is about λ times the separatrix splitting. That layer is commonly termed as the stochastic layer.

c. Numerical evidence

The first numerical verification of estimate (3.9) was undertaken [4] using the standard map as a model. Indeed, we have seen above that the latter is essentially equivalent to the original system (2.1) with parameters (3.7). As to the other resonances in (3.6), their contribution is exponentially small according to (3.9). There is an additional complication with the standard map related to the fact that parameter $\varepsilon = 2$ is no longer small. On the other hand, numerical simulation is much simpler, of course, for a map than for a continuous system like (2.1). The first numerical experiments showed, however, that Eq.(3.2) is not exact for the map, and an additional factor has to be introduced:

$$\xi \to \xi R_e\,; \qquad R_e \approx 2.15 \tag{3.10}$$

Even though this factor can be calculated analytically as an effect of higher approximations [4], its actual evaluation seems to be formidable and constitutes an unsolved problem. This shows also that the above assumed condition $\varepsilon \ll 1$ is generally essential for the validity of Eqs. (3.2) and (3.9). Taking into account the factor (3.10), we arrive at the expression

$$w_s = 64\pi^4 R_e \frac{e^{-\pi^2/\sqrt{K}}}{K^{3/2}} \tag{3.11}$$

to be compared with numerical data. In a completely different approach this estimate has been confirmed in [13] except for the correction (3.10). New and more accurate data, obtained by Vecheslavov, are presented in Fig.1 as the dependence of w_s on motion time (the number of map iterations) for both the outer (curve 1) and the inner (curve 2) parts of stochastic layer (K = 0.5). Note unusually big fluctuations which we are going to discuss below (Section 4c). The values of w_s were calculated from the mean motion period T_m for a single trajectory in the layer, using the relation [4]

$$w_s = 32 \cdot \exp\left(1 - \sqrt{K} \cdot T_m\right) \tag{3.12}$$

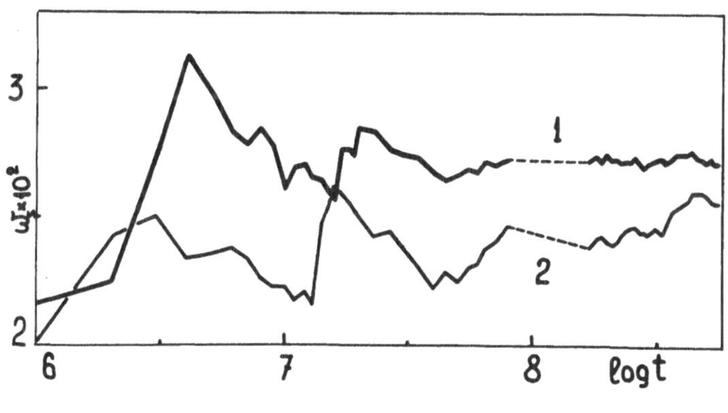

Fig. 1 Stochastic layer half-width vs. motion time

It is obtained via simple averaging of Eq.(2.2) assuming the uniform distribution of trajectory over the layer. Eq.(3.11) gives in this case w_s = .0329. Thus, the accuracy of simple estimate (3.11) is about 20 percent for this not a very big λ = 8.89.

The agreement can be improved by taking account of: i) frequency shift (2.5); ii) nonuniform equilibrium distribution in the layer (see Fig.2 as an example [14]); iii) the effect of marginal resonance inside the layer [4], the final accuracy achieved being about 4 percent. Note a slight asymmetry of the layer in Fig.1 which indicates the accuracy of asymptotic relation (2.2).

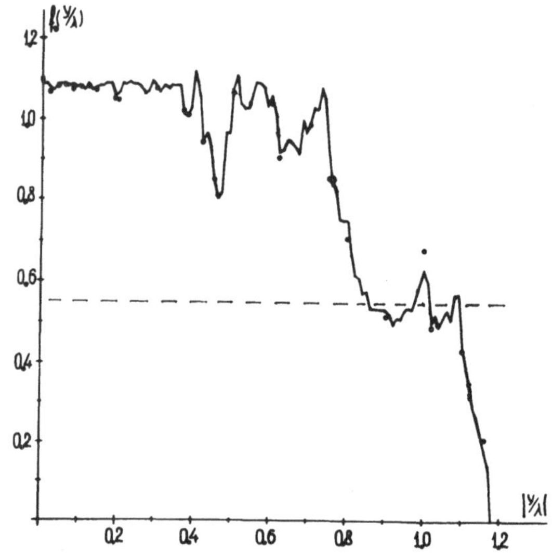

Fig. 2 Equilibrium distribution of a single trajectory in stochastic layer: $t = 10^7$ (broken line); $t = 4\text{x}10^6$ (circles).

4. ON STRUCTURE OF THE CHAOS BORDER

Chaotic motion, particularly that in a stochastic layer, is, in principle, undistinguishable from a true random process according to the algorithmic theory of dynamical systems [3]. The random means here unpredictable, or uncomputable, which appears to be in conformity with our intuitive ideas of what the random is like. However, the randomness does not yet determine the statistical properties of motion. As is well known, the most fundamental of them is correlation.

a. Diffusion near the border

Consider, first, the standard map (3.4). The "force" correlation is defined by

$$C_F(\tau) = \langle \sin\theta(t+\tau) \cdot \sin\theta(t) \rangle \tag{4.1}$$

where averaging is performed either along a trajectory (in motion time t) or over an ergodic component of the motion. For sufficiently large K , when regular component of the motion is negligible, this correlation is known to decay fairly fast [15], so that the sum

$$S_F = \sum_{\tau=1}^{\infty} C_F(\tau) \tag{4.2}$$

does certainly converge. Hence, a simple statistical description of the chaotic motion is possible by means of the diffusion equation:

$$\frac{\partial f(I,t)}{\partial t} = D \cdot \frac{\partial^2 f(I,t)}{\partial I^2} \tag{4.3}$$

where the diffusion rate

$$D(|K|) = \lim_{t\to\infty} \frac{\langle(\Delta I)^2\rangle}{2t} = \frac{K^2}{4}\left[1 + 4S_F(|K|)\right] \tag{4.4}$$

For large $|K|$ the correlation correction $S_F \sim |K|^{-1/2}$ (4.2) vanishes, and the diffusion rate approaches its limiting, uncorrelated value $D^{\infty} = K^2/4$. In the opposite case $|K| \to 1$ [4] the correlation dominates, and diffusion rate rapidly decreases

$$D(|K|) \approx a \cdot (|K| - 1)^{\alpha} \tag{4.5}$$

where $a \approx 1/5$, and $\alpha \approx 2.55$ according to numerical simulation.

For separatrix mapping (3.3) the diffusion becomes inhomogeneous since the dependence $D(|K|)$ turns into $D(|y|)$ according to Eq.(3.5). Generally, the diffusion equation includes an additional (drift) term. Indeed, the Fokker-Plank-Kolmogorov (FPK) equation can be written in the form (see, e.g. [16]):

$$\frac{\partial f}{\partial t} = -\frac{\partial Q}{\partial y}; \qquad Q = -D(y)\frac{\partial f}{\partial y} + U(y) f \tag{4.6}$$

Here $Q(y,t)$ is the flux, and $U(y)$ is the drift velocity related to equilibrium distribution $f_0(y)$ by the expression

$$U(y) = \frac{d}{dy} D(y) + D(y) \frac{d}{dy} \ln f_0(y) \tag{4.7}$$

Inspection of Fig. 2 shows that there are two regions within a stochastic layer where the drift can be neglected:

i) near the layer center where $f_0(y)$ = const exactly (variations of f_0 seen in Fig. 2 are due to fluctuations) and where $D(y) \approx D^{\infty}$ = 1/4;

ii) near the layer border where $f_0(y) \not\approx$ const approximately only (see below), and where

$$D(y) \approx a \cdot \left(1 - \frac{|y|}{\lambda}\right)^{\alpha} \tag{4.8}$$

(see Eqs. (4.5) and (3.5)). Since the border line is of a complicated shape, the y variable above (as well as θ) is assumed to have been transformed in such a way as to "straighten out" this line ($|y_B| = \lambda$).

In a model like (2.1) the diffusion spreads across the layer, and is obviously restricted by a finite layer width. Neglecting so far the slow diffusion (4.8) at the layer edges, it takes $t_r \sim \lambda^2$ iterations for a trajectory to get across the layer, or for a distribution function to relax. Since, however, Eq.(4.4) still holds, a long time correlation does arise due to the boundary conditions. How simple the nature of that correlation may appear, it led to a paradox (or, rather, misunderstanding) [17-19] that the mixing precludes the diffusion instead of implying it. A formal reason for such a surprise conclusion is in that the mixing does provide existence of the limit in (4.4), while the paradox is a result of too literal understanding of this limit. It reminds us of an additional (besides the mixing) condition for the diffusion description of relaxation in a chaotic system to be applicable. Namely, there must exist two different time scales of the motion

$$t_c \ll t_r \tag{4.9}$$

that of correlation decay (t_c) on which the limit (4.4) is asymptotical, and the other one of relaxation (t_r) on which the same limit is local. For example, the motion in a stochastic layer has $t_c \sim 1$, and $t_r \sim \lambda^2$, so that the condition (4.9) requires $\lambda \gg 1$.

The long time correlation within stochastic layer is of a primary importance in many-dimensional systems where the diffusion along the layer (the so-called Arnold diffusion [4]) does generally occur. For the latter diffusion to be long-range, it has to be independent of the diffusion across the layer (due to different perturbation terms involved, for example) to get rid of that correlation.

Now, what would be the impact of the slow diffusion (4.8) on the motion in stochastic layer? It turns out to be crucial if the exponent $\alpha > 2$. Assume the following diffusion equation near the layer border

$$\frac{\partial f}{\partial t} = \frac{\partial}{\partial x} x^{\alpha} \frac{\partial f}{\partial x} \tag{4.10}$$

where we have introduced a new variable $x = 1 - y/\lambda$ $(y > 0)$, and rescaled t appropriately. First, let us try to find the eigenfunctions, that is to solve the equation

$$\frac{d}{dx} x^{\alpha} \frac{df_{\varkappa}}{dx} + \varkappa^2 f_{\varkappa} = 0 \tag{4.11}$$

It admits a solution via the cylindrical functions $Z_p(z)\ (\alpha \neq 2)$:

$$f_{\varkappa}(x) = x^{\frac{1-\alpha}{2}} Z_p\left(\frac{2\varkappa x^{1-\frac{\alpha}{2}}}{\alpha-2}\right); \quad p = \left(\frac{\alpha-1}{\alpha-2}\right)^2 \tag{4.12}$$

If $\alpha < 1$ the solution is regular at $x = 0$, and the relaxation is exponential. However, for $\alpha \geqslant 1$ the solution is generally singular, and one would expect a nonexponential relaxation.

The general solution of this diffusion problem is not known. However, we may analyze a particular self-similar solution to Eq.(4.10) which, as is easily verified, reads:

$$\varphi_s' \equiv \frac{d\varphi(s)}{ds} = C \cdot s^{-\alpha} \cdot \exp\left(-(\alpha-2)^{-2} \cdot s^{2-\alpha}\right) \tag{4.13}$$

Here C is an arbitrary constant, and $s = x \cdot t^{1/(\alpha-2)}$. At $x = 0$ the flux

$$Q(x,t) = -x^{\alpha}\frac{\partial\varphi}{\partial x} = -x^{\alpha-1} s\varphi_s' = -\frac{C}{t^{\frac{\alpha-1}{\alpha-2}}} \cdot \exp\left(-(\alpha-2)^{-2} \cdot s^{2-\alpha}\right) \tag{4.14}$$

is always zero, while density $\varphi(s)$ may be non-zero (for $C < 0$). Due to the self-similar nature of this solution the second boundary condition cannot be imposed at any fixed x (e.g., at the layer center, $x = 1$). However, asymptotically as $t \to \infty$ it doesn't matter since the diffusion mainly proceeds in an ever narrowing region at the layer edge. The size x_D of this region ($s \sim 1$) scales with t as $x_D \propto t^{-1/(\alpha-2)}$, while for $s \to \infty$ the flux (4.14) becomes independent of x.

If the initial density at the layer edge is less than that at equilibrium, the relaxation corresponds to a negative (i.e. toward the edge) flux ($C > 0$), and to the boundary conditions:

$$\varphi(0,t) = 0; \quad \varphi(\infty,t) = \beta C \approx f_0 = 1; \quad \beta = \frac{\Gamma\left(\frac{\alpha-1}{\alpha-2}\right)}{(\alpha-2)^{\frac{\alpha}{\alpha-2}}} \tag{4.15}$$

where equilibrium distribution f_0 is assumed to be constant, and $\Gamma(z)$ is the gamma function. Asymptotically as $t \to \infty$, and except the diffusion region $\sim x_D$, the relaxation proceeds as follows

$$|\varphi(x,t) - \varphi(x,\infty)| \to \int_t^{\infty} |Q|\, dt = \frac{\alpha-2}{\beta \cdot t^{\frac{1}{\alpha-2}}} \tag{4.16}$$

In the opposite case a similar positive flux sets in ($C < 0$), and Eq.(4.16) remains unchanged. Thus, the slow diffusion ($\alpha > 2$) near the chaos border results in a power-type relaxation.

Since the time correlation of a pair of functions depends on the relaxation for one of them, we would expect, generally, the same power law (4.16) for the correlation as well. The latter may be faster though, if the relaxing function is close to equilibrium one near the border already from the beginning.

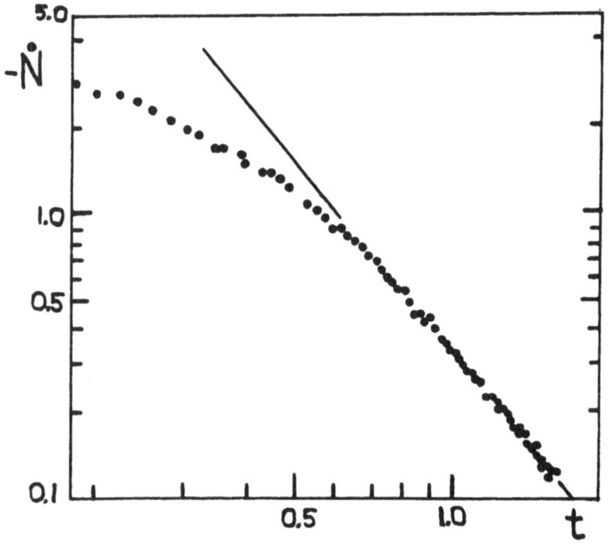

Fig. 3 Electron current out of magnetic trap (arbitrary units) vs. time (in msec); straight line: $\dot{N} \propto t^{-2.2}$

There is an interesting experiment on the behavior of electrons in a magnetic trap [20] which appears to confirm a power-type relaxation. The authors [20] observed a nonexponential dependence on time for the electron current $J(t) = -e\dot{N}$ out of the trap, due to a chaotic motion of electrons in inhomogeneous magnetic field, and did fit it by a doubly exponential function. On the other hand, the chaotic region of that electron motion is known to always have the border [4]. If one rescales the data [20] in the log-log plot, as shown in Fig. 3, they perfectly fit, for a sufficiently large time, the power dependence $\dot{N} \propto t^{-q}$ with exponent $q \approx 2.2$. This is to be compared to the flux (4.14): $Q \propto t^{-\frac{\alpha-1}{\alpha-2}}$, whence $\alpha \approx 2.83$. Remarkably, this value is not far away from that for the standard map ($\alpha \approx 2.55$, see (4.5)). It indicates some universal behavior near the chaos border.

For further studies of this behavior the Poincaré recurrencies proved to be very useful [14].

b. Poincaré recurrencies

Consider separatrix map (3.3), and follow a single trajectory while it crosses successively the symmetry line $y = 0$. The motion time interval between two successive crossings we shall call the recurrence time τ. As motion proceeds the distribution of τ values tends to a limiting function $F(\tau)$ defined as the probability for a recurrence to occur later than τ. Obviously, $F(1) = 1$ (for the map), and generally $F(\tau) \to 0$ as $\tau \to \infty$. An exception from the latter is, for example, the asymptotic motion (2.4) along the unperturbed pendulum separatrix. Note that in case of the motion with discrete spectrum (quasiperiodic or almost periodic motions) $F(\tau) \equiv 0$ at any τ greater than some τ_1, while in chaotic motion $F(\tau) \neq 0$ for all τ. Poincaré recurrencies do not imply, thus, quasiperiodicity as is stated sometimes.

In stochastic layer motion the asymptotic behavior of $F(\tau)$ as $\tau \to \infty$ relates to the structure of the layer border. Such an approach was actually used in [21] where the power dependence

$$F(\tau) \sim \tau^{-p}; \quad \tau \geqslant 1 \tag{4.17}$$

has been found with $p = 1/2$. As was pointed out in [14], it corresponds to the free homogeneous diffusion until the layer border is reached, that is for $\tau \lesssim \lambda^2$. At larger $\tau >> \lambda^2$ the dependence $F(\tau)$ approximately remains of a power-type but the exponent p changes; according to numerical data [14], the mean p for various λ is $\langle p \rangle \approx 3/2$. Besides, apparently irregular variations of $p(\tau)$ are present which do not depend on trajectory and, hence, relate to the border structure rather than to fluctuations in motion.

c. Scaling

As was mentioned above, there are numerical indications suggesting some universal behavior near the chaos border in the phase space. Now we are going to consider a theoretical model for this alleged universality. That the resonance structure determining transition to chaos is hierarchic has been known already since quite long ago (see, e.g. [22,4]). Yet, only in the pioneering work due to Greene [10] that structure has been exploited to evaluate a critical perturbation for the standard map. Hierarchic and scaling behavior at the transition to chaos was further studied extensively in many papers (see, e.g. [11,13] and references therein). A distinctive feature of our problem (see also [23]) is in that the perturbation strength here is not a parameter, as for standard map, but rather a function of dynamical variables (mainly, momentum y for separatrix map (3.3)). This leads just to a chaos border in the phase space rather than to a critical perturbation strength.

Assume the following scaling hypothesis: near the chaos border any two of dynamical variables (v, u) are interrelated by a power dependence:

$$v \propto u^{P_{uv}} \tag{4.18}$$

where P_{uv} is scaling parameter, and $P_{uv} \cdot P_{vu} = 1$. Choosing one variable (u) as the fundamental scaling unit we have

$$v \propto u^{P_v} \tag{4.19}$$

Such a scaling hypothesis is essentially identical to that in the fluctuation theory of phase transitions [24] which leads to some similarity of these two problems. However, important distinctions should not be missed. The scaling in phase transitions is continuous and essentially statistical (fluctuation scaling), while in our problem scaling is discrete (see below), and does relate to both chaotic as well as purely regular components of motion on both sides of the chaos border. What makes the two problems similar is a crucial impact of an infinite sequence of scales (continuous or discrete) upon the behavior at transition.

Transform (x, θ) variables in such a way as to provide: $x \propto |\omega(x) - \omega_b|$ near the border, $2\pi\omega(x)$ being the motion frequency of system (3.3) under consideration, and $\omega_b = \omega(0)$ the frequency at the border $x = 0$. Hence: $P_x = P_\omega$, or, choosing $(\omega - \omega_b)$ as the fundamental scaling unit ($P_\omega \equiv 1$), $P_x = 1$. Note that in original variables the exponent P_x would depend on θ (see [11]). The measure of chaotic component $\mu \propto x$ since at the border the resonances are just about to overlap in all scales (comp. Fig. 2), whence $P_\mu = 1$.

To proceed further we need to relate these scales to that of time. It can be done via the overlap parameter s (3.8). The width $(\Delta\omega)_q$ of a high order resonance $\omega_q = r/q$ depends on its phase oscillation frequency Ω_q as [4,5]: $q(\Delta\omega)_q \sim \Omega_q$, while the resonance spacing $\delta\omega_q \sim q^{-2}$. The latter follows from the total number of resonances, within a given interval of ω, which is proportional to q^2. In a more formal way it is also implied from the best approximation of a given irrational number (ω_b in our case) by the convergents of the continued fraction representation [25]:

$$\left|\omega_b - \frac{r}{q}\right| \sim \frac{C(\omega_b)}{q^2} \tag{4.20}$$

Hence, at the chaos border

$$s_q = \frac{(\Delta\omega)_q}{\delta\omega_q} \sim q^2 (\Delta\omega)_q \sim q\,\Omega_q \sim 1 \tag{4.21}$$

The overlap parameter s_q is related to the Greene residue [10]: $R_q \sim s_q^2$. For standard map with $|K| = K_{cr}$, which corresponds to the chaos border in map (3.3), $R_q \to 1/4$ as $q \to \infty$ [10] in accordance with estimate (4.21).

Suppose that a given scale is essentially determined by some resonance ω_q. Then, the associated time scale would be $T_q \sim \Omega_q^{-1}$, and $(\Delta\omega)_q \propto |\omega_q - \omega_b|$. Whence, $p_T = p_q = -1/2$. The scaling for diffusion rate near the border is, hence, $D \propto (\Delta\omega)_q^2 / T_q \propto x^2/T \propto x^{2.5}$, and the diffusion parameter $\alpha = 5/2$ which is close to the numerical values given above.

Since resonance width $(\Delta\omega)_q \propto V_q^{1/2}$, where V_q is the corresponding Fourier amplitude of the limiting perturbation in the Hamiltonian (see below), the scaling (4.21) implies $V_q \propto q^{-4}$, i.e. the perturbation has two continuous derivatives only. This is precisely the critical smoothness of perturbation for the map [26,4]. It means the following: If the initial perturbation $V^0(\theta)$ is an analytic function, its Fourier amplitudes, as is well known, fall off exponentially, like $V_q^0 \propto \exp(-\sigma^0 q)$, for example. However, as we proceed to higher approximations the amplitudes grow, or parameter σ decreases [27]: $\sigma^0 \to \sigma(K)$. At critical perturbation the dependence V_q on q becomes, as everything else, of power-type, that is [10] $\sigma(K_{cr}) = 0$. On the other hand, as is also known [26], the initial perturbation needs not to be analytic for a chaos border to exist, instead it suffices for $V^0(\theta)$ to be only smooth, that is $V_q^0 \propto q^{-p_0}$ provided $p_0 > p_{cr}$. Otherwise, the motion is chaotic for any non-zero perturbation strength.

As was mentioned above, the scaling near the chaos border is discrete. It means that there exists a denumerable sequence of principal scales which is determined by a sequence of resonances $\omega_{q_n} = r_n/q_n$ converging to the border: $r_n/q_n \to \omega_b$ as $q_n \to \infty$. The resonance sequence depends on arithmetical properties of irrational ω_b, for example, on its representation as a continued fraction: $\{\omega_b\} = [b_1, b_2, \ldots, b_n, \ldots]$ where $b_i \geq 1$ are integers, and brackets denote the fractional part. According to Greene's conjecture [10], ω_b is the "golden mean", i.e. $\{\omega_b\} = g_1 = [1, 1, \ldots, 1, \ldots] = (\sqrt{5}-1)/2 \approx 0.618$.

It is not known whether this is true for the standard map but generally it does not hold [12,23]. A much weaker hypothesis that ω_b has a "golden tail", i.e. $\{\omega_b\} = [b_1, \ldots, b_n, 1, \ldots, 1, \ldots]$ seems plausible. The main problem is to match the arithmetic of ω_b to the critical value of K which depends on x (comp. [23]). Apparently, the discrete scaling accounts for p(τ) variations mentioned above.

Finally, let us estimate the contribution to Poincaré recurrences from internal chaos borders of resonance stochastic layers. That there are many such layers within the main layer is immediately seen in Fig. 2 from a low equilibrium density near the border. It also follows from the limiting value of Greene residue R = 1/4 which means that the resonance centers near the border are not destroyed.

Let the time scale of a given resonance be T_q. Then the mean sojourn time in its region of measure $\mu_q \propto x_q$ is, due to ergodicity, $N_q T_q / t \propto x_q$, where N_q is the number of entries into this region, and t is the total motion time. Assume the universal distribution of Poincaré recurrences $F(\tau) \propto \tau^{-p}$ with some, unknown so far, p. Particularly, this implies the probability $F_q \propto (T_q/\tau)^p \propto (q/\tau)^p$ $(\tau \gtrsim q)$ for any internal chaos border of a resonance stochastic layer. Then, the contribution to Poincaré recurrences in the main layer from a particular resonance would be

$$F^{(q)} \propto N_q F_q / N \propto \frac{x_q q^{p-1}}{\tau^p} \cdot \frac{t}{N} \propto \frac{q^{p-3}}{\tau^p} \qquad (4.22)$$

where $N \propto t$ is the total number of recurrences. Now we need to sum up the contributions of all undestroyed resonances which do retain their stochastic layers. The number of those resonances can be estimated as follows. Define the border zone $x_z(q)$ as $\sigma(x_z) \cdot q \sim 1$ where $\sigma(x)$ is the exponential factor of the perturbation Fourier amplitudes introduced above. Assuming a linear dependence $\sigma(x) \propto x$ near the border we arrive at the scaling $x_z \propto q^{-1}$ for the border zone size. The latter implies that for a given q just one resonance gets into this zone, so we are to merely sum up contributions (4.22) over q :

$$F' \sim \sum_q^{\tau} F^{(q)} \propto \frac{1}{(p-2)\tau^2}, \; p \neq 2; \qquad F' \propto \frac{\ln \tau}{\tau^2}, \; p = 2$$

From universality $F'(\tau) \sim F(\tau)$, and p = 2. First of all, this would imply that the main contribution to Poincaré recurrences were not due to the diffusion near the main layer border but from a laby-

rinth of infinite hierarchies of internal chaos borders where the trajectory spends most of its recurrence time. If confirmed, it would
tory spends most of its recurrence time. If confirmed, it would
also mean that near the chaos border the above scaling hypothesis holds only approximately, to logarithmic accuracy. This also would change the behavior of both relaxation as well as correlation near the chaos border as compared to estimates in Section 4a based upon the diffusion equation (4.10). In any event, a power-type relaxation inevitably leads to big fluctuations in motion which are clearly seen, for example, in Fig. 2.

Certainly, the problem of the chaos border structure needs and deserves further studies.

Acknowledgements

I would like to express my sincere gratitude to D.L.Shepelyansky, Ya.G.Sinai, V.V.Vecheslavov and F.Vivaldi for many interesting and helpful discussions on the problems touched upon in this lecture.

REFERENCES

1. V.I.Arnold and A.Avez, Ergodic Problems of Classical Mechanics, Benjamin (1968).
2. I.P.Kornfeld, Ya.G.Sinai, S.V.Fomin, Ergodic Theory, Nauka, 1980 (in Russian).
3. V.M. Alekseev and M.V. Yakobson, Physics Reports, 75, 287 (1981).
4. B.V.Chirikov, Physics Reports, 52, 263 (1979).
5. A.J. Lichtenberg and M.A. Lieberman, Regular and Stochastic Motion, Springer-Verlag (1982).
6. H. Poincaré, Les méthods nouvelles de la méchanique céleste, Vol. II (1893), Sections 225-232; Vol. III (1899), Section 401.
7. N.N. Bogoliubov and Yu.A. Mitropolsky, Asymptotic Methods in the Theory of Nonlinear Oscillations, Hindustan Publ. Corp., Dehli, 1961.
8. V.K. Melnikov, Dokl. Akad. Nauk SSSR, 144, 747 (1962) (in Russian).
9. L.P. Shilnikov, Mat. sbornik, 77, 461 (1968) (in Russian).
10. J.M. Greene, J. Math. Phys. 20, 1183 (1979).
11. L.P.Kadanoff, Phys. Rev. Lett. 47, 1641 (1981); S.J. Shenker and L.P. Kadanoff, J. Stat. Phys., 27, 631 (1982).
12. B.V. Chirikov, F.M. Izrailev, D.L. Shepelyansky, in Soviet Scientific Reviews, Section C, Vol. 2 (1981), p. 209.
13. D.F. Escande, Large-Scale Stochasticity in Hamiltonian Systems, Intern. Conf. on Plasma Physics, Göteborg (1982).
14. B.V. Chirikov, D.L. Shepelyansky, Statistics of the Poincaré Recurrences and the Structure of Stochastic Layer of a Nonlinear Resonance, Preprint 81-69, Institute of Nuclear Physics, Novosibirsk (1981) (in Russian).
15. C. Grebogi and A.N. Kaufman, Phys. Rev., A24, 2829 (1981).
16. E.M. Lifshits, L.P. Pitaevsky, Physical Kinetics, Nauka (1979) (in Russian).
17. J.L. Lebowitz, in Statistical Mechanics, New Concepts, New Problems, New Applications, Univ. of Chicago Press (1972).
18. R. Balescu, Equilibrium and Nonequilibrium Statistical Mechanics, Wiley, New York (1975), Appendix.

19. G.E. Norman, L.S. Polak, Dokl. Akad. Nauk SSSR, 263, 337 (1982) (in Russian).
20. D. Bora, P.I. John, Y.C. Saxena and R.K. Varma, Plasma Physics, 22, 653 (1980).
21. S.R. Channon and J.L. Lebowitz, Ann. N.Y. Acad. Sci., 357, 108 (1980).
22. J.M. Greene, J. Math. Phys., 9, 760 (1968).
23. J.M. Greene, in Nonlinear Dynamics and the Beam-Beam Interaction, A.I.P. Conf. Proc., N°57 (1979), p. 257.
24. A.Z. Patashinskii and V.L. Pokrovskii, Fluctuation Theory of Phase Transitions, Pergamon (1979).
25. A.Ya. Khinchin, Continued Fractions, Fizmatgiz, Moscow (1961) (in Russian).
26. J. Moser, Stable and Random Motions in Dynamical Systems, Princeton Univ. Press (1973).
27. V.I. Arnold, Usp. mat. nauk, 18, N°6, 91 (1963)(in Russian).

PERIODIC AND QUASI-PERIODIC ORBITS FOR TWIST MAPS

A. Katok*

Department of Mathematics
University of Maryland
College Park, MD 20742
U.S.A.

1. INTRODUCTION

In these notes we present a motivation for and a survey of recent results which in a certain sense fill the gap between the classical work of G. D. Birkhoff on periodic orbits for mechanical systems with two degrees of freedom and celebrated KAM theory which deals with the existence of invariant tori for small perturbations of integrable Hamiltonian systems. S. Aubry [Au1] and J. Mather [M1] discovered that invariant tori never disappear completely. Certain "shades" always remain in the form of invariant solenoids carrying motions which can be described as quasi-periodic although not in the completely classical sense. Those motions are exactly the same as those in so-called Denjoy minimal sets on the two-dimensional torus [D]. However, Denjoy examples can appear only for dynamical systems of class at most C^1, but not C^2, so that they were usually considered as a mere curiosity, at least from the point of view of applications. In contrast, the Aubry-Mather solenoidal motions inevitably appear for large classes of non-integrable dynamical systems, including small perturbation of integrable Hamiltonian systems with two degrees of freedom (see Section 3). In [K1] the author showed that those motions appear as limits of certain periodic orbits which were already known to Birkhoff at least in particular cases [B1]. Both periodic and quasi-periodic orbits can be derived from a certain variation principle which generalizes the one used by Birkhoff for the billiard ball problem [B1].

Some of the periodic and solenoidal orbits are always accompanied by other orbits asymptotic to them [M2], [K2], [K3], [Au2]. Those orbits are derived from a minimax variational principle combined with general continuity arguments.

Mather's work was in part inspired by an earlier work by Percival

---

*Partially supported by NSF Grant MCS 8204024.

[P1], [P2] who used variational arguments in his numerical computations. On the other hand, as one of the outcomes of the development described in these notes, Mather [M2] obtains a necessary and sufficient condition for the existence of an invariant torus with a given rotation number in the form of vanishing of certain quantity (see Theorem 2 in Section 7). This condition can be verified with arbitrary degree of precision by a computer so that it makes possible to prove rigorously the non-existence of the invariant torus.

We shall describe from a unified point of view the principal results of [M1], [K1], [M2], [K2], [K3]; the last three preprints represent different preliminary versions of a joint work by J. Mather and the author. We refer to the introduction to [M2] for a more detailed discussion of relationships between Aubry's approach and results and ours.

2. TWIST MAPS

In this section we set up the stage for the subsequent discussion leaving the motivation for the next two sections.

Let $A = S^1 \times [0,1] = \{(\varphi,r),\ \varphi \in \mathbb{R}/\mathbb{Z},\ 0 \le r \le 1\}$ be the standard annulus (or a closed cylinder).

A twist map is a homeomorphism $f\colon A \to A$ such that the image $f(I_\varphi)$ of any "radius" $I_\varphi = \{\varphi\} \times [0,1]$ bends in the same direction. More systematically, let $S = \mathbb{R} \times [0,1] = \{(x,y),\ x \in \mathbb{R},\ 0 \le y \le 1\}$ be the strip which is the universal covering of the annulus A and $T\colon S \to S$ be the unit translation $T(x,y) = (x+1,y)$.

For any homeomorphism $f\colon A \to A$ its lift to S is defined up to a power of T; conversely, if $F\colon S \to S$ and F commutes with T then F is a lift of a homeomorphism of A. Let us write such an F in the coordinate form: $F(x,y) = (F_1(x,y), F_2(x,y))$.

Definition 2.1. A homeomorphism $f\colon A \to A$ is called a twist homeomorphism (or a twist map) if it preserves orientation, preserves boundary components of A and if for a lift F of f and for any $x \in \mathbb{R}$ the function $F_1(x,y)$ is a strictly monotone function of y.

Obviously for a twist map all functions $F_1(x,y)$ for different x must be either increasing or decreasing. According to that, we may speak of right of left twist maps. If f is a right twist map, then f^{-1} is a left twist map and vice versa. For definiteness we shall always assume that f is a right twist map.

Let us consider the restrictions f_0 and f_1 of a twist map f to the boundary components $A_0 = S^1 \times \{0\}$ and $A_1 = S^1 \times \{1\}$. Since f_0 and f_1 are homeomorphisms of a circle we can define for them rotation numbers. Each of those numbers is defined up to an integer that must be the same for both components. Thus, we define the twist interval $[\rho_o(f), \rho_1(f)]$ where

$$\rho_o(f) = \lim_{n\to\infty} \frac{F_1(F^{n-1}(x,0))-x}{n}$$

and

$$\rho_1(f) = \lim_{n\to\infty} \frac{F_1(F^{n-1}(x,1))-x}{n} \quad \text{for every } x \in \mathbb{R}.$$

It is also useful to introduce a certain measure for the "quality" or uniformity of the twist. In general, this can be accomplished through the notion of twist module [K1]. To avoid unnecessary technicalities we shall restrict ourselves to the most characteristic special case.

Definition 2.2. A map $f: A \to A$ is called a Lipschitz twist map if both f and f^{-1} satisfy the Lipschitz condition and if, in addition, there exists a constant $c \geq 0$ such that for every $x \in \mathbb{R}$ and $0 \leq y_1 < y_2 \leq 1$

$$F_1(x,y_2) - F_1(x,y_1) \geq c(y_2 - y_1) \tag{1}$$

For example, any diffeomorphism f of the annulus such that $\frac{\partial F_1}{\partial y} > 0$ is a Lipschitz twist map.

The results about the existence of various kinds of special orbits (Theorems 1 and 2 and the discussion in §7, but not in §6) require an additional assumption the f preserves the area or, more generally, has a sufficiently good invariant measure. It is enough to assume that this measure has the form $\rho(r,\varphi)drd\varphi$ where ρ is continuous in A and positive inside A. We shall refer to such twist maps simply as measure-preserving (Lipschitz) twist maps.

Situations fairly similar to the twist maps of the annulus often appear in various concrete problems (see Section 3). Let us describe these variations.

First, we can assume that f is defined, invertible and continuous only on the open annulus $S^1 \times]0,1[$. If (1) is still satisfied together with Lipschitz condition in every closed annulus $S^1 \times [\varepsilon,1-\varepsilon]$ the results described in Sections 5-7 remain true.

Next, instead of the annulus one can consider the standard disc D

(closed or open) and call a twist map of D any homeomorphism $f: D \to D$ which fixes the center of D and moves every radius in a manner similar to the one we have described for the annulus.

This situation can be reduced to the previous one by "blowing-up" the center of the disc and thus making the rest of it into an annulus without the "bottom" boundary component. Moreover, if a twist map f of the disc is differentiable then the map of the annulus obtained that way can be naturally extended to the bottom boundary component.

Yet another situation appears for a map f of the annulus if the functions $F_1(x,y)$ are monotone in y but the boundary components are not invariant. In this case f can always be extended to a twist map of a larger annulus. In order to make this extension useful we need to keep the preservation of measure. Suppose that f preserves the area $d\varphi dr$ and we want the extension also to be area-preserving. f maps A into the cylinder $C = S^1 \times \mathbb{R}$ so that $f(A)$ may overlap with the upper half-cylinder $A^+ = S^1 \times]1,\infty)$ or with the lower one $A^- = S^1 \; (-\infty,0[$. Obviously, if f allows an extension to an area-preserving twist homeomorphism $\overline{f}: S^1 \times [a,b] \to S^1 \times [a,b]$ then the area above A_1 but below $f(A_1)$ must be equal to the area above $f(A_1)$ but below A_1. It turns out that the extension is always possible under that condition but we shall not go into details.

We can apply the results about the twist maps to the extended map $\overline{f}$ and then consider orbits of f staying within the original annulus A, which coincide with the orbits of f.

3. MOTIVATING EXAMPLES

3.1. A neighborhood of an elliptic periodic orbit

Let us consider an autonomous Hamiltonian system with two degrees of freedom. Suppose that the system has a periodic orbit γ lying on an energy hypersurface E. Let us denote by D a small disc in E transversal to γ which crosses γ at a point x_0. Since the hypersurface E is invariant we can define the Poincaré map (the first return map) $P: D_0 \to D$ on a smaller disc $D_0 \subset D$. This map fixes the point x_0 and by the Liouville theorem it preserves the area induced by the invariant 3 dimensional volume in E. The orbit γ is called <u>elliptic</u> if the differential DP at x_0 has a pair of complex conjugate eigenvalues, say $e^{\pm 2\pi i \alpha}$. This means that the linear approximation to P near x_0 is a rotation. If α is irrational then better approximations are described in terms of so-called Birkhoff invariants $\rho_2, \rho_4 \ldots$. (See e.g. [A3], Appendix 7). Namely, for any integer n one can introduce polar coordinates φ, r on D near x_0 such that

P preserves the area $rd\varphi dr$ and has a form

$$P(\varphi,r) = (\varphi + 2\pi\alpha + \rho_2 r^2 + \rho_4 r^4 + \cdots + \rho_{2n} r^{2n} + g(\varphi,r), r + h(\varphi,r))$$

where the functions g and h are of order higher than $2n$. If at least one of ρ_{2n}'s is different from zero then P satisfies the twist condition near x_0. The procedure mentioned at the end of the previous section allows to extend P to an area-preserving twist map of a disc.

3.2. Small perturbations of an integrable Hamiltonian system

Let us consider a Hamiltonian system with two degrees of freedom which has a first integral indenpendent of the enrgy H. Then any non-critical energy hypersurface splits into invariant two-dimensional surfaces and if such a surface is compact it is a torus carrying quasi-periodic motions ([A2], Chapter VI). In a neighborhood of any such torus one can introduce so-called "action-angle" coordinates $(I_1, I_2, \varphi_1, \varphi_2)$ so that the Hamiltonian H depends only on I_1, I_2. This means that in the cross-section $\varphi_1 = 0$, $H = \text{const}$ the Poincaré map P has the following form in coordinates (φ_2, I_2)

$$P(\varphi_2, I_2) = (\varphi_2 + \omega(I_2), I_2) \qquad (2)$$

so that if the frequency function ω is monotone then P is a twist map.

A small perturbation of our Hamiltonian system is in general non-integrable but the construction of the Poincaré map on a new energy hypersurface in a neighborhood of an invariant torus of the unperturbed system can be carried through. If in (2) $\omega' \neq 0$ then the twist condition persists for any small C^1 perturbation. Naturally, we usually encounter the situation described at the end of Section 2 with the annulus not invariant but nearly invariant.

3.3. Billiard ball problem

This famous classical problem was praised by Birkhoff as an ideal test-ground for the qualitative methods in mechanics due to the complete absence of formal difficulties so formidable in many other mechanical problems. We consider the motion of a point billiard ball with the unit speed inside a smooth convex domain D in the plain. The ball moves along a straight line and then bounces off the boundary $\Gamma = \partial D$ according to the rule "the angle of incidence is equal to the angle of reflection."

A natural Poincaré map appears as a first-return map to the boundary.

The set S_Γ of linear elements of Γ is parametrized by the coordinates (ℓ,θ) where ℓ is the normalized length measured along Γ and θ is the angle between the positively oriented tangent line to Γ and the given linear element. Thus, ℓ is a cyclic coordinate and θ changes between 0 and π so that S_Γ is an annulus. The Poincaré map $P: S_\Gamma \to S_\Gamma$ $P(\ell,\theta) = (L(\ell,\theta),\Phi(\ell,\theta))$ is a twist map with the twist interval $[0,1]$ because for any ℓ the function $L(\ell,\theta)$ changes monotonically from 0 to 1. P preserves the measure $\sin\theta\, d\ell\, d\theta$.

3.4. One-dimensional conservative systems with periodic external force

Let us consider a (generalized) pendulum

$$\ddot{x} = f(x), \tag{3}$$

where x is an angular coordinate. This system is equivalent to the vector-field $(\dot{x},f(x))$ on the $(x,\dot{x})$ cylinder. Thus, for any sufficiently small time t say for $t \le t_o$ the t-shift on the cylinder is a twist map. If we add a periodic external force $g(t)$

$$\ddot{x} = f(x) + g(t) \tag{4}$$

with a period τ then the dynamics for the period is described by an area-preserving map of the cylinder onto itself. If the force is small enough in C^1 topology then this map is C^1 close to the τ-shift for (3) so that if $\tau < t_o$ the dynamics for (4) is described by a twist map.

4. INTEGRABLE TWIST MAPS AND INVARIANT CIRCLES

As we already noticed, there is a simple class of "integrable" twist maps defined by

$$f(\varphi,r) = (\varphi + \alpha(r),r) \tag{5}$$

where $\varphi(r)$ is a monotone function. Actually, the maps which appear in 3.1 and 3.2 are small perturbations of maps of that type. For integrable maps, the annulus splits into a family of invariant circles and every orbit is either periodic (if $\alpha(r)$ is rational) or dense on the corresponding circle. The twist interval here is equal to $[\alpha(0), \alpha(1)]$.

In a sense, integrable twist maps have the simplest orbit structure among all twist maps because the dynamics for such a map essentially reduces to one-dimensional dynamics on the invariant circles.

By an "invariant circle" for a general twist map we shall always mean an invariant set homeomorphic to a circle which separates the boundary components. Since invariant circles divide the annulus into smaller invariant annuli their existence provides important information not only about the orbits lying on the circles but about other orbits as well.

In fact, those invariant annuli must have rather regular structure as follows from Birkhoff's Invariant Circle Theorem ([B2] §44, [B3] §3, see also [H], Chapter 2). Any invariant circle for a Lipschitz twist map f has a form graph ψ where the function $\psi: S^1 \to [0,1]$ satisfies the Lipschitz condition with the constant which depends only on those for f and f^{-1} and on the constant c in (1).

For any invariant circle Ω the rotation number $\rho(\Omega)$ is defined. This number obviously belongs to the twist interval.

Proposition 4.1. For any measure-preserving twist map and for any irrational rotation number there exist at most one invariant circle with this rotation number.

We shall prove this proposition in Section 8.

A. N. Kolmogorov [Ko], [A1], discovered in the early fifties that any small area-preserving analytic perturbation of an integrable twist maps always have many invariant circles, namely the circles with irrational rotation numbers which are not too well approximated by rational numbers. Naturally, those circles are close to corresponding circles for the unperturbed map. Subsequent development came about in the works of V. I. Arnold [A2], J. Moser [Mo1], [Mo2], H. Rüssman [R1] and recently M. Herman [H] and again Rüssman [R2]. Both the minimal necessary smoothness of the perturbation and its size were studied. The question of smoothness was more or less settled by Herman's proof that many circles still persist for all perturbations sufficiently small in $C^{3+\varepsilon}$ topology ([H], Chapter 1 and 5) but every circle may disappear for a perturbation arbitrarily small in $C^{3-\varepsilon}$ topology ([H], Chapter 3 and 4.) The situation is much less clear with the size of the perturbation. There is a huge gap between theoretical estimates in KAM theory and numerical evidence (see e.g. [G]).

5. SPECIAL ORBITS

We now proceed to the main goal of this paper: to find orbits and, more generally, sets of orbits for general twist maps which behave as if they belong to invariant circles. We begin with the definitions

Definition 5.1. An orbit $\Gamma = \{f^n(\varphi_o, r_o) = (\varphi_n, r_n),\ n \in \mathbb{Z}\}$ for a twist map f is called a special orbit if there exists a homeomorphism $g: S^1 \to S^1$ such that $g^n(\varphi_o) = \varphi_n$ for all integer n.

Definition 5.2. A positive (negative) semiorbit $\Gamma^+(\Gamma^-) = \{f^n(\varphi_o, r_o) = (\varphi_n, r_n),\ n = 0,1,\ldots\ (n=0,-1,\ldots)\}$ is called a positive (negative) special semi-orbit if there exists a homemorphism $g: S^1 \to S^1$ such that $g^n(\varphi_o) = \varphi_n$ for all $n \geq 0$ $(n \leq 0)$.

Definition 5.3. A closed f-invariant set $E \subset A$ is called a special set if

(i) E intersects every interval I_φ at most at one point so that $E = \text{graph}\ \psi$ where $\psi: K \to [0,1]$ and K is a closed subset of S^1

(ii) There exists a homeomorphism $g: S^1 \to S^1$ such that $g|_K = \pi \circ f \circ (\text{id} \times \psi)$ where $\pi(\varphi, r) = \varphi$.

Special semi-invariant sets are defined in a similar way. Obviously, every orbit belonging to a special set is a special orbit. Every special orbit and special set carries a cyclic order given by the angular coordinate. This order is preserved by f. This property characterizes special orbits and sets.

Proposition 5.1. If an orbit (closed invariant set) for a twist map satisfies (i) and f preserves the order then this orbit (invariant set) is special.

As always in this paper we present the arguments only for Lipschitz twist maps. The following simple lemma plays a fundamental role not only in the proof of the proposition but in many other places.

Lemma 5.1. Let $F: S \to S$ be a lift of a Lipschitz twist map f, $(x_i, y_i) = F^i(x_o, y_o)$, $(x_i', y_i') = F^i(x_o', y_o')$ and $x_i' > x_i$ for $i = -1,0,1$. Then there exists a constant $M > 0$ such that

$$|y_o' - y_o| < M(x_o' - x_o).$$

Proof. Let us assume that $y_o' < y_o$ and denote $F(x_o', y_o) = (\tilde{x}, \tilde{y})$. We have from (1)

$$\tilde{x} > x_1' + c(y_o - y_o'). \tag{6}$$

On the other hand, since f is a Lipschitz map

$$x_1' > x_1 > \tilde{x} - L(x_0' - x_0) \tag{7}$$

where L is the Lipschitz constant for f. From (6) and (7) we obtain

$$y_0 - y_0' < Lc^{-1}(x_0' - x_0).$$

If $y_0' > y_0$ the argument goes the same way with f^{-1} instead of f. □

Proof of Proposition 5.1. It follows immediately from the lemma that the function ψ from (i) satisfies the Lipschitz condition. In the case of a single orbit ψ simply map φ_n into r_n. Thus, the map $g = \pi \circ f \circ (\mathrm{id} \times \psi)$ preserves order in its domain K and satisfies the Lipschitz condition and the same is true for $\pi \circ f^{-1} \circ (\mathrm{id} \times \psi)$ which obviously coincides with g^{-1}. Consequently, g can be extended to an invertible map of the closure $\bar{K}$ which also preserves order. Extension to the complementary intervals to K is then obvious. □

There is another immediate corollary of Lemma 5.1.

Corollary 5.1. The closure of every special orbit is a special set.

Definition 5.4. Two special sets are called compatible if their union is a special set. Two (or more) special orbits are called compatible if the closure of their union is the special set.

The rotation number $\rho(\Gamma)$ of a special orbit $\Gamma = \{(\varphi_n, r_n)\}$ is defined as the rotation number of the homeomorphism g from Definition 5.1 or directly as

$$\rho(\Gamma) = \lim_{n\to\infty} \frac{x_n - x_0}{n} \quad \text{where} \quad (x_0, y_0) \quad \text{is a lift of} \quad (\varphi_0, r_0) \quad \text{and}$$

$(x_n, y_n) = F^n(x_0, y_0)$.

The rotation number $\rho(E)$ of a special set E can be defined on the rotation number of any orbit belonging to E. Possible behaviour of orbit of circle homeomorphisms was described by Poincaré and Denjoy (see e.g. [N]). Since the projection to the circle provides a natural order-preserving one-to-one correspondence between a special orbit and an orbit of a circle homeomorphism Poincaré-Denjoy classification can be translated to special orbtis. We begin with a short survey of that classification.

Every homeomorphism of the circle with a rational rotation number equal, say, $\frac{p}{q}$ has a certain number of orbits of period q. Elements of every such orbit are permuted in the same way as on any orbit of the rotation by $\frac{2\pi p}{q}$. Any other orbit Γ is asymptotic in positive and negative directions to periodic orbits which we denote Γ^{ω} and Γ^{α}. If the homeomorphism has only one periodic orbit then Γ^{ω} coincides with Γ^{α}, otherwise those two orbits are always different.

Thus, there are three possible types of special orbits with the rotation number $\frac{p}{q}$; which we shall call types $1_{p/q}$, $2_{p/q}$ and $3_{p/q}$ accordingly: periodic, double asymptotic to a periodic orbit and asymptotic in positive and negative directions to different periodic orbits. Orbits of type $1_{p/q}$ are called in [K1] Birkhoff periodic orbits of type (p,q); sometimes we shall use that name.

Every circle homeomorphism with an irrational rotation number α is either conjugate to the rotation by $2\pi\alpha$ so that every orbit is dense and is in an order preserving correspondence with an orbit of that rotation or such a homeomorphism has an invariant nowhere dense minimal set (Denjoy set) and any orbit outside of this set is double asymptotic to that set. Let us note that in the latter case the orbits are also in one-to-one order preserving correspondence to the orbits of the rotation, but the inverse correspondence is not continuous.

Thus, there are 3 possible types of special orbits with irrational rotation number α:

1_{α} An orbit with dense projection to the circle; the closure of such an orbit is an invariant circle with dense orbits;

2_{α} A recurrent orbit whose closure E is a Cantor set which projects to the circle into a Denjoy type set. Every orbit from E is dense in E;

3_{α} A wandering orbit whose α- and ω- limit set E is a closure os an orbit of type 2_{α}. The projection of such an orbit belongs to complimentary intervals to a Denjoy set.

We leave a similar classification of special semi-orbits to the reader.

Now we are able to summarize the principal results of [K1] [K2] [K3]:

Theorem 1. Let $f: A \to A$ be a measure-preserving Lipschitz twist map. Then for every rational number $\frac{p}{q}$ from the twist interval there exist at least two different compatible special orbits of type $1_{p/q}$. Moreover, there is always either an orbit of type $2_{p/q}$ or an orbit of type $3_{p/q}$ or the whole circle consisting of the orbits of type $1_{p/q}$.

For every irrational α from the twist interval there is either an invariant circle with the rotation number α (this circle is filled either by dense orbits of type 1_α or by orbits of types 2_α and 3_α) or an invariant minimal Cantor set E filled by orbits of type 2_α, plus at leat one orbit of type 3_α compatible with E, and consequently, asymptotic to E in both directions.

Furthermore, for every $\varphi \in S^1\setminus\pi(E)$ there is at least one special positive semi-orbit and one negative special semi-orbit which begin at I_φ compatible with E.

In the next two sections we shall give a brief sketch of main ideas involved into the proof of Theorem 1.

6. CONTINUITY OF SPECIAL ORBITS

Proposition 6.1. Let $\Gamma_m = \{f^n(z_m): n \in \mathbb{Z}\}$, $m = 1,2,\ldots$ be special orbits and $z_m \to z$. Then $\Gamma = \{f^n(z)\}$ is also a special orbit and the rotation number $\rho(\Gamma) = \lim \rho(\Gamma_m)$.

Proof. Let $f^n(Z_m) = (\varphi_m^n, r_m^n)$ and let ψ_m be the function defined on $\pi\Gamma_m$ which assigns r_m^n to φ_m^n. By Lemma 5.1 all the function ψ_m satisfy the Lipschitz condition with constant M and consequently can be extended to function $\tilde{\psi}_m: S^1 \to [0,1]$ with the same property. The sequence of functions ψ_m is compact in uniform topology, so, passing to a subsequence, if necessary, we can assume that $\tilde{\psi}_n$ converges to a Lipschitz function $\tilde{\psi}$. Since graph $\tilde{\psi}_m \supset \Gamma_m$, graph $\psi \supset \Gamma$. The preservation of order on Γ follows immediately from the convergence. Thus, Γ is a special orbit. Finally, let $\rho = \lim \rho(\Gamma_m)$ (again, if necessary, we pass to a subsequence). For every k the order of first k points for the rotation by $2\pi\rho$ is the same as for the rotation by $2\pi\rho(\Gamma_m)$ for any sufficiently large m. This implies that $\rho = \rho(\Gamma)$. As a by-product of this argument we see that if $z_m \to z$ and Γ_m are special orbits then $\rho(\Gamma_m)$ converge. □

Counterparts of Proposition 6.1 for special sets and special semi-orbits can be proved in exactly the same way.

Corollary 6.1. If a Lipschitz twist map f has special orbits with any rotation number from a dense subset of the twist interval, it has special orbits with all rotation numbers from that interval.

In particular, as soon as Birkhoff periodic orbits are constructed for all admissible rational rotation numbers, the existence of orbits of type 1_α for irrational α is guaranteed. Now we can play the game in the opposite direction and consider the approximation of a rational number $\frac{p}{q}$ by irrational numbers α_n. The corresponding special sets are either circles and then in the limit we have an invariant circle with rotation number $\frac{p}{q}$ or they are Cantor sets with "holes". Taking a limit of a sequence of "holes" we see that both ends of the limit "hole" cannot be periodic. This allows us to obtain special orbits of type $2_{p/q}$ or $3_{p/q}$.

The hardest part of the argument is the construction of orbits of type 3_α in the absence of invariant circles. This requires a careful combination of continuity arguments and variational arguments which are used already for the proof of the existence of Birkhoff periodic orbits.

7. THE LAGRANGIAN AND VARIATIONAL ARGUMENTS

For notational convenience we shall work in the universal covering instead of the annulus. We begin with a rational number $\frac{p}{q}$ belonging to the twist interval. The idea for the construction of a Birkhoff periodic orbit of type (p,q) and then of the second compatible orbit is to consider the space of possible positions for angular coordinates for the desired orbit and then to introduce a certain functional whose critical points correspond to real orbits.

Let us denote $F_1(x,0) = g_o(x)$ and $F_1(x,1) = g_1(x)$ and consider the space of all non-decreasing sequences $\varphi: \mathbb{Z} \to \mathbb{R}$ such that

$$\varphi(n+q) = \varphi(n) + 1$$

and

$$g_o(\varphi(n)) \leq \varphi(n+p) \leq \varphi(n)$$

and identify every sequence $\varphi(n)$ with $\varphi(n) + k$ for every integer k. We denote the factor-space by $\Phi_{p,q}$.

Let $x, x' \in \mathbb{R}$ and $g_o(x) \leq x' \leq g_1(x)$. By the twist property the image of the interval $I = \{x\} \times [0,1]$ intersects the interval $I' = \{x'\} \times [0,1]$ at exactly one point, say $(x', h(x,x'))$. Let us denote by $T(x,x')$ the "triangle" bounded by the bottom boundary component S_o of S, the interval I' and the curve $F(I)$.

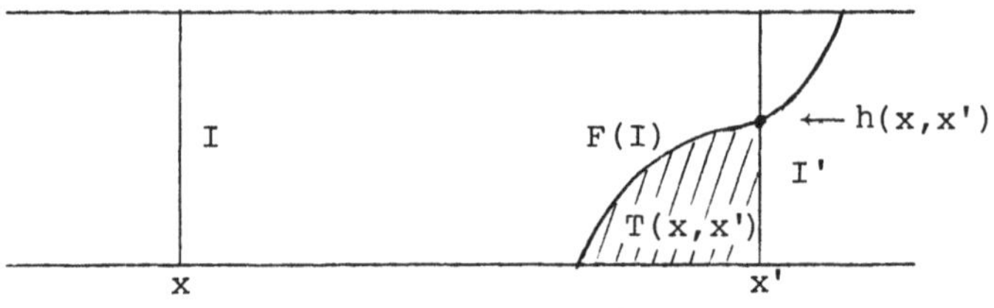

Now we can define the Lagrangian $L_{p,q}$, namely, for $\varphi \in \Phi_{p,q}$.

$$L_{p,q}(\varphi) = \sum_{n=o}^{q-1} \mu(T(\varphi(n),\varphi(n+p)))$$

where μ is the lift to S of the given f-invariant measure on the annulus.

We shall need a few more nottions, namely

$$h_1^{\varphi}(n) = h(\varphi(n-p),\varphi(n))$$

$$\psi_1(n) = (\varphi(n),h_1^{\varphi}(n))$$

$$\psi_2(n) = F^{-1}\psi_1(n+p) \overset{\text{def}}{=\!=} (\varphi(n),h_2^{\varphi}(n)).$$

If

$$h_1^{\varphi}(n) = h_2^{\varphi}(n) \tag{8}$$

then

$$F\psi_1(n) = \psi_1(n+p)$$

so that if (8) is satisfied for every n then the projection to the annulus A of the graph of the map $\psi_1: \mathbb{Z} \to S$ is a periodic orbit of period q which is in fact a Birkhoff orbit of type (p,q).

Lemma 7.1 ([K2], Section 4). If the functional $L_{p,q}$ reaches its local minimum at $\varphi \in \Phi_{p,q}$ then

$$h_1^{\varphi}(n) = h_2^{\varphi}(n)$$

for all integer n.

The idea of the proof of this key lemma is essentially local. Namely, one shows that if $h_1^{\varphi}(n) \neq h_2^{\varphi}(n)$ then by changing the value of φ only at n in such a way that perturbed sequence still belongs to $\Phi_{p,q}$ one can decrease the value of $L_{p,q}$. Several variations of that argument are used in the proof of other parts of Theorem 1.

Since the space $\Phi_{p,q}$ is compact, Lemma 7.1 immediately implies the existence of at least one Birkhoff periodic orbit. We shall call any orbit determined by a sequence ψ minimazing $L_{p,q}$ on the space

$\Phi_{p,q}$ a minimal Birkhoff orbit of type (p,q) and denote the minimal value of the functional $L_{p,q}$ by $L^{min}_{p,q}$.

Let us assume that the invariant measure has the form $\rho(r,\varphi)drd\varphi$. Then the differential of the functional $L_{p,q}$ is determined by the partial derivatives (see [K2])

$$\frac{\partial L_{p,q}}{\partial \varphi(h)} = \int_{h_2^{\varphi}(n)}^{h_1^{\varphi}(n)} \rho(\varphi(n),t)dt.$$

This formula shows, in particular, that any critical point of $L_{p,q}$ determines a Birkhoff periodic orbit of type (p,q).

To obtain a second Birkhoff orbit compatible with a given minimal orbit determined by $\psi \in \Phi_{p,q}$ we consider the following subspace $\Phi^{\psi}_{p,q}$ of $\Phi_{p,q}$

$$\Phi^{\psi}_{p,q} = \{\varphi \in \Phi_{p,q} : \psi(n) \le \varphi(n) \le \psi(n+1),\ n \in \mathbb{Z}\}.$$

$L_{p,q}$ has two minima on this space, namely, ψ and $\tilde{\psi}$ given by $\tilde{\psi}(n) = \psi(n+1)$. Both of them obviously lie on the boundary of $\Phi^{\psi}_{p,q}$ and $L_{p,q}(\psi) = L_{p,q}(\tilde{\psi}) = L^{min}_{p,q}$.

Proposition 7.1 ([K2], Section 4). The functional $L_{p,q}$ has at least one critical point φ inside $\Phi^{\psi}_{p,q}$.

This point can be constructed as a "mountain pass." Namely, let us consider various smooth paths $\alpha: [0,1] \to \Phi^{\psi}_{p,q}$ such that

$$\alpha(0) = \psi \quad \text{and} \quad \alpha(1) = \tilde{\psi}$$

and let

$$L^{min\,max}_{p,q} = \inf_{\alpha} \max_{0\le t\le 1} L_{p,q}(\alpha(t)).$$

The infinum in the last formula is always achieved for some α and some value t such that $\alpha(t)$ is a critical point of $L_{p,q}$ which we shall denote by ψ'. This critical point determines another Birkhoff orbit Γ' of type (p,q) compatible with the minimal orbit Γ, determined by ψ. We shall call the orbit Γ' a minimax orbit associated with Γ and denote

$$\Delta L_{p,q} \overset{\text{def}}{=} L_{p,q}^{\text{minmax}} - L_{p,q}^{\text{min}}.$$

Let α be an irrational number from the twist interval and $\frac{p_n}{q_n}$ be a sequence of rational numbers, converging to α. Now we can explain how the orbits of type 3_α arise. Let Γ_n and Γ'_n be correspondingly minimal and minimax orbits of type (p_n, q_n). Applying Proposition 6.1 to a sequence Γ_n or its subsequence and taking a closure of the limit orbit Γ we obtain a special set E with rotation number α. E is either a circle or it contains both orbits of type 2_α and 3_α or, finally it is a minimal Cantor set. Only the last case needs an extra consideration. We then consider the limit of Γ'_n in the same manner as before and obtain another special set E' compatible with E. If $E' \nsubseteq E$ then $\tilde{E} = E \cup E'$ is nonminimal, i.e., it contains orbits of type 3_α. Thus, to finish the proof of the existence of the special orbits mentioned in Theorem 1 it remains to consider the case when $\tilde{E} = E$ is a minimal Cantor set. We shall describe two ways to treat this case. One of them is based on the following result [M2] which also has an independent interest because it provides the criterion for the non-existence of invariant tori mentioned in the introduction.

Theorem 2

As $\frac{p_n}{q_n} \to \alpha$ $\lim\limits_{n\to\infty} \Delta L_{p_n,q_n}$ always exists and an invariant circle with rotation number α exists if and only if this limit is equal to 0.

The proof of this theorem given in [M2] is rather involved and is considerably longer than the complete proof of Theorem 1 in [K1], [K2], [K3] which uses another method of treating the remaining case.

According to this method we first construct special semi-orbits described in Theorem 1. The construction is based on conditional minimization of the functionals $L_{p,q}$ ([K2], Section 2). The space $\Phi_{p,q}^{\psi}$ is naturally fibered by smaller spaces $\Phi_{p,q}^{x,\psi}$ where $\psi(0) \le x \le \psi(1)$.

$$\Phi_{p,q}^{x,\psi} = \{\varphi \in \Phi_{p,q}^{\psi}, \varphi(0) = x\}.$$

Lemma 7.2 ([K2], Section 2). If the functional $L_{p,q}$ restricted to the space $\Phi_{p,q}^{x,\psi}$ reaches its local minimum at φ then for that φ and for $n \not\equiv 0 \pmod q$

$$h_1^\varphi(n) = h_2^\varphi(n).$$

The proof of this lemma is completely parallel to those of Lemma 7.1. Lemma 7.2 shows that the conditional minimum of $L_{p,q}$ on $\Phi^{x,\psi}_{p,q}$ determines a piece of orbit of length q which begins and ends at the same given radius, say I_φ and is compatible in a natural sense with the given minimal Birkhoff orbit. Let us denote the beginning of that piece by (φ,r) and its end $f^q(\varphi,r)$ by (φ,r').

Considering again a sequence $\frac{p_n}{q_n} \to \alpha$ and taking limits of (φ,r_n) and $f^{q_n}(\varphi,r_n) = (\varphi,r'_n)$ we obtain correspondingly positive and negative semi-orbits beginning at I_φ and compatible with the special orbit which appears as the limit of minimal Birkhoff orbits. If these two semi-orbits coincide they form together a special orbit of type 3_α. The difference $|r_n - r'_n|$ can be estimated through $\Delta L_{p_n,q_n}$. This reduces the construction of the orbit of type 3_α to the following proposition [K3].

Proposition 7.2. If $\tilde{E}$ is a minimal set then

$$\lim \Delta L_{p_n,q_n} = 0.$$

8. CONCLUDING REMARKS

8.1. Our variational arguments are restricted to finite-dimensional situation which appears for rational rotation numbers, while the case of irrational rotation numbers is dealt with by approximation. Alternatively, one can consider variational problems directly for irrational rotation numbers. In this case one has to consider infinite-dimensional spaces and analytical parts of the arguments become more complicated. This is the original approach of Mather [M1]; in a later paper [M2] he combines it with approximation arguments in order to prove Theorem 2.

8.2. Now we can prove Proposition 4.1 concerning the uniqueness of invariant circles.

First we show that for any measure preserving twist map any invariant circle Ω with an irrational rotation number is disjoint from any other invariant circle Ω'. For, the intersection $\Omega \cap \Omega'$ is f-invariant so that if there is a dense orbit on Ω then $\Omega' = \Omega$. Otherwise $\Omega \cap \Omega'$ is projected to the circle into a closed set with infinitely many complementary intervals. Since Ω = graph ψ and Ω' = graph ψ' then over every such interval I $\psi' \neq \psi$ so that one of the functions is strictly greater than the other. But then the domain between the two graphs over I is disjoint from all its images; but this contradicts the preservation of measure. Thus $\Omega \cap \Omega' = \emptyset$. Passing to the universal covering we consider the lifts of both circles which are the graphs of periodic functions which we shall still denote ψ and ψ'. Then $|\psi - \psi'| > 0$ and by compactness $|\psi - \psi'| > \delta > 0$. But then by the twist condition the shift by F on one of the graphs is uniformly bigger than on the other. That immediately implies that the rotation numbers are different.

The uniqueness of invariant circles naturally leads to the question about the uniqueness of minimal special sets. Mather [M5] shows that those set may be non-unique. His examples involve large perturbation of integrable twist maps. It is possible that for small C^1 perturbations of integrable twist maps a minimal special set with given rotation number is unique.

8.3. Let us describe a simple method to determine that special sets do not pass through certain areas. We assume that f is differentiable. Suppose the point (φ_o, r_o) is a non-isolated point in a special set E. Let us take the preimage and the image of the point (φ_o, r_o) and denote them (φ_{-1}, r_{-1}) and (φ_1, r_1). If (φ_o', r_o') is another point from the same special set with $f^{\pm 1}(\varphi_o', r_o') = (\varphi_{\pm 1}', r_{\pm 1}')$ then all 3 differences $\varphi_i' - \varphi_i$ $i = -1,0,1$ have the same sign and that should be possible for φ_o' arbitrary close to φ_o. This condition can be easily expressed in terms of the universal covering. Let (x,y) be a lift of (φ_o, r_o). Then the above condition implies that

$$\frac{\partial F_1}{\partial y}(F^{-1}(x,y)) \geq -\frac{\partial \hat{F}_1}{\partial y}(F(x,y)) \tag{9}$$

where $\hat{F}_1(x,y)$ is the first coordinate of F^{-1}.

There are numerous applications of this simple observation. For example, if this condition is violated for some x and every $0 \leq y \leq 1$ then no invariant circle exists and, moreover, no minimal special set passes through this vertical interval. This implies the non-existence of invariant circles in a billiard ball problem if the curvature of Γ vanishes at some point [M3], and for so-called standard examples [M4] for the values of the parameter $k > 2$. Condition (9) can be refined by considering several positive and negative iterates of the given point instead of one and applying a similar argument.

REFERENCES

[A1] V. I. Arnold, Proof of A. N. Kolmogorov's Theoren on the preservation of Quasi-Periodic Motions under small perturbations of the Hamiltonian, Russian Math. Surveys, 18, N. 5, 13-40 (1963).

[A2] V. I. Arnold, Small Divisor Problem in Classical and Celestial Mechanics, Russian Math. Surveys, 18, N.6, 85-191 (1963).

[A3] V. I. Arnold, Mathematical Methods of Classical Mechanics, Springer Verlag, Graduate Text in Math. 1978.

[Au1] S. Aubry. Theorey of the Devil's Staircase, Seminar on the Riemann Problem and complete integrability, 1978-79. Ed. D. G. Chudnovsky, Springer, Lecture Notes in Math., in press.

[Au2] S. Aubry, P. Y. Le Daeron, G. Andre, Classical Ground-States of a one-dimensional Model for Incommensurate Structure, submitted to Comm. Math. Phys.

[B1] G. D. Birkhoff, On the Periodic Motions of Dynamical Systems, Acta Mathematica, 50, 359-379. Reprinted in Collected Mathematical Papers, vol. II, Amer, Math. Soc. N.Y, 333-353 (1950).

[B2] G. D. Birkhoff, Surface transformations and their dynamical applications, Acta Math., 43, 1-119 (1922) Reprinted in Collected Mathematical Papers, vol. II, Amer. Math. Soc. N.Y, 111-229 (1950).

[B3] G. D. Birkhoff, Sur Quelques, Courbes Fermeés Remarquable, Bull. Soc. Math. de France, 60, 1-26 (1932); Reprinted in Collected Mathematical Papers, Amer. Math. Soc., N.Y., 418-443 (1950).

[D] A. Denjoy, Sur les courbes définies par les équations différentielles à la surface de tore, J. Math. Pure et Appliq, 11, 333-375 (1932).

[G] J. M. Greene, The calculation of KAM surgaces, Ann. New York Acad. Sci., 357, 80-89 (1980).

[H] M. R. Herman, Handwritten manuscript 1981

[K1] A. Katok, Some remarks on Birkhoff and Mather twist map theorems, Ergodic Theory and Dynamical Systems, 2, 2 (1982).

[K2] A. Katok, More about Birkhoff periodic orbits and Mather sets for twist maps, Preprint IHES, 1982.

[K3] A. Katok, Continuation of the preprint "More about Birkhoff periodic orbits and Mather sets for twist maps, 1982.

[Ko] A. N. Kolmogorov, On Quasi Periodic Motions under small perturbations of the Hamiltonian, Doklady AN SSSR, 98, 4, 527-530 (1954) (in Russian).

[M1] J. N. Mather, Existence of quasi-periodic orbits for twist homeomorphism of the annulus, Topology, 1982.

[M2] J. N. Mather, A Criterion for non-existence of invariant circles. Preprint, 1982.

[M3] J. N. Mather. Glancing Billiards, Ergodic Theory and Dynamical Systems, 2, 3-4 (1982).

[M4] J. N. Mather, Non-existence of Invariant circles. Preprint, 1982.

[M5] J. N. Mather, Non-uniqueness of Solutions of Percival's Euler-Lagrange Equation. Preprint, 1982.

[M6] J. N. Mather, Concavity of the Lagrangian for Quasi-Periodic orbits, to appear in Comm. Math. Helv.

[Mo1] J. Moser, On Invariant Curves of Area-Preserving Mappings of an Annulus. Nach. Akad. Wiss. Göttingen, Math.-Phys. Kl, IIa, N 1, 1-20 (1962).

[Mo2] J. Moser. Stable and Random Motions in Dynamical Systems, Princeton Univ. Press, 1973.

[N] Z. Nitecki, Differentiable Dynamics, MIT Press, 1971.

[P1] I. C. Percival, Variational Principles for Invariant Tori and Cantori, in Symposium on Nonlinear Dynamics and Beam-Beam Interactions. Amer. Inst. of Phys. Conf. Proc. N. 57, ed. M. Month, J. C. Herrara, 302-310, 1980.

[P2] I. C. Percival, J. Phys. A: Math. Nucl. Gen 12, L57, 1979.

[R1] M. Rüssmann, Über invariante Kurven differenzierbarer Abbildungen eines Kreisringer, Nachr. Akad. Wiss. Göttingen, Math.-Phys. K 1, II N 5, 67-105 (1970).
[R2] H. Rüssmann, On the existence of invariant curves of twist mappings of an annulus. Preprint, 1982.

MACROSCOPIC BEHAVIOR IN A SIMPLE CHAOTIC HAMILTONIAN SYSTEM

Otto E. Rössler

Institute for Physical and Theoretical Chemistry
University of Tübingen
7400 Tübingen, W. Germany

ABSTRACT

Nonlinear Hamiltonian systems of only two degrees of freedom readily produce complicated Poincaré-type (or, synonymously, chaotic) behavior. Nonlinear Hamiltonian systems of many degrees of freedom, on the other hand, produce not only chaos, but also modulate systematic motions (so-called dissipative structures). A two-degrees-of-freedom system is described which shows two types of 'macroscopic' behavior: On one hand, there is a 'main regime' which is characterized by a random distribution of both positional variables. From this 'equilibrium' the system only very rarely departs once it has come close to it. On the other hand, when the system is started sufficiently far away, it spends a certain time in a characteristic 'transient regime' which too (like the main regime) is the same for both time directions. One of the two particles hereby shows a statistical directional preference. This preference is the same as when the system is run as an open system ('temporarily open regime'). The simple nature of the system encourages further quantitative and qualitative investigations.

1. INTRODUCTION

Poincaré (1890) first described complicated trajectorial behavior (with homoclinic points in a cross-section) in a Hamiltonian system with two degrees of freedom. This kind of behavior was later analyzed further by Hadamard (1898) and Birkhoff (1927). For reviews see Arnold Avez (1968), Moser (1973), and Abraham and Marsden (1978). More recently, it became increasingly apparent that the so-called 'random' motions in Hamiltonian systems are not fundamentally different from those found in non-Hamiltonian (dissipative) dynamical systems (for which the name 'chaos' has become a successful unifying label; cf. Rössler, 1979, for a review). Aside from the similarities, there are also differences, however. An example is the recent finding of differing scaling laws for period-doubling bifurcations in the two classes of systems (Green et al., 1981). On the other hand, a property thought to be characteristic for the non-dissipative class (the absence of sinks -

that is, attractors which irreversibly attract an open-dense neighborhood) turned out to be shared by a class of dissipative systems (those with 'maximal chaos'; Rössler, 1980).

Recently, Prigogine and Stengers (1981) suggested that the qualitative behavior of simple, volume of flow in state space preserving, chaotic systems may help understand the qualitative behavior of much larger statistical mechanical systems (which according to a theorem of Sinai, 1970, are strongly mixing - and hence chaotic too). A certain observable (the 'degree of choppedness' as it could be called) was found to increase in both directions of time (as entropy does in larger systems). More complicated types of macroscopic behavior (comprising what Prigogine called dissipative structures) have so far no analogue in the simpler class, however (cf. Farquhar, 1972).

A well-known example of a Hamiltonian system of two degrees of freedom with nontrivial behavior that one might try to start out with is Hénon and Heiles' (1964) system. Its Hamiltonian is

$$H = \frac{p_1^2}{2} + \frac{p_2^2}{2} + \frac{s_1^2}{2} + \frac{s_2^2}{2} + s_1^2 s_2 - \frac{s_2^3}{3} . \qquad (1)$$

This system has one disadvantage, however: not all physically meaningful solutions of the same energy content H stay bounded for all times. Another example which avoids this difficulty is the 'two beads in a one-dimensional box' system, with associated Hamiltonian

$$H = \frac{p_1^2}{2} + \frac{p_2^2}{2m_2} + \frac{\alpha}{s_1} + \frac{\beta}{s_2 - s_1} + \frac{\gamma}{1 - s_2} . \qquad (2)$$

Here p_1, p_2 are the momenta of two beads with unequal masses ($m_1 = 1$ and $m_2 < 1$), while s_1, s_2 are their positions (between zero and one, with $s_2 > s_1$). When the three constants (α, β, γ) are all chosen smaller and smaller, the system approaches an elastic collision problem for rigid balls which can be solved algebraically.

One physical question that may be worth studying in the context of such simple systems is: What is the average distribution of kinetic energy over the two balls as a function of not only time but also space?

In the following, a special 2-beads case will be considered for which this spatial question can be answered.

2. COLLISION WITH A FRICTIONLESS TRAP DOOR

Figure 1 shows the physical setup. There are two tubes arranged in the form of a letter T. One contains a frictionless bullet. The other contains another such bullet, but with the property that part of it (that portion which is capable of protruding into the horizontal tube during part of its spontaneous oscillation) has the form of an asymmetric wedge. The front ends of the bullets are assumed perfectly elastic, and the ends of the two tubes (including the bottom of the horizontal tube) perfectly reflecting. An approximate physical implementation is in principle possible.

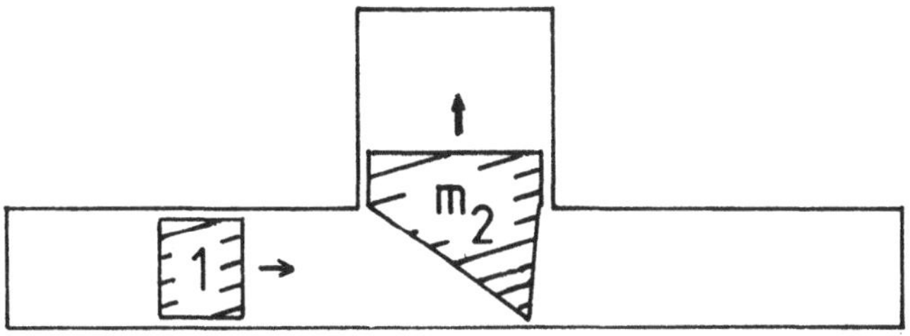

Fig.1 A special collision-type problem

This system can be considered an extremal simplification to a three-dimensional statistical mechanical system first indicated by Feynman et al. (1965): two gas-containing compartments separated by a membrane whose pores are on one side occluded by little (hanging)trap doors. (A similar system was proposed independently by U. Wais; personal communication 1973.) Whereas the mode of action of Feynman's diffusive rectifier is too complicated to permit a quantitative analysis, the present system can be investigated in detail.

Intuitively, what one would expect is that the horizontal bullet should be 'trapped' on the right-hand side of the T maze. This is necessarily the case when the vertical 'bullet' is resting close to the lower extreme position initially and, moreover, possesses a completely vertical right-hand edge throughout. In this case, the system clearly is degenerate (decoupled), however. This non-generic situation can be avoided if a slight slanting is assumed to apply to the right-hand portion of the wedge.

In this case, depicted in Figure 1, the horizontal bullet necessarily escapes from the right-hand compartment after a finite time. One nonetheless still has the feeling that the amount of time spent by the horizontal bullet in the right-hand compartment should on the average exceed that spent in the left one. Pictorially speaking, the 'pressure' (defined as the average number of 'hits' applied to the outer wall) should be larger on the right.

This impression is unwarranted, of course. The reason is that by its very Hamiltonian nature, the system is bound to function in both directions of time in the same fashion. Any tendency toward establishing a higher pressure in one compartment is tantamount to the opposite tendency. The vertical bullet therefore does not function as a Maxwellian demon. Also, linking the two ends of the horizontal tube into a circle will not generate a perpetual motion machine (turning undirected motion into one with a directional preference).

Nonetheless, something reminiscent of such behavior can be expected, as a transient. Starting out with a 'cool' gate, one may well find that for a while the desired behavior is indeed observable. Only later, when the trap door has picked up enough energy, can it be expected to as readily dispense energy (that is, pick up energy in negative time) as to comply with its further application (and pick up even more in positive time).

This is a testable question: Does the system of Figure 1 - which functions as a statistical 'bullet trap' under macroscopic conditions (that is, as an open system, with friction acting as an energy sink) - still function in the same manner as a transient, provided the initial conditions are close to the open regime? While the property of 'transitory openness' (or pseudo-openness, respectively) is frequently found in many-degrees-of-freedom systems, it has apparently not yet been described for a 2-degrees-of-freedom system.

3. A COLLISION MODEL

By making the assumptions of collision theory, a first quantitative description of the behavior of the system of Figure 1 can be achieved. As a first constraint, kinetic energy has to be conserved (that is, $\frac{1}{2} v_1^2 + \frac{1}{2} m v_2^2 = \text{const.}$). Secondly, a modified conservation of momentum law applies (modified, because the rigid walls of the assumed 'infinite mass' box absorb some of the momentum in the collisions). Using symmetry arguments ("At what ratio of momenta is there no exchange of energy between the two bullets?"), the following conservation of momentum law can be found: $p_1 + f' p_2 = \text{const.}$, where f' is the magnitude of the slope of the side of the wedge involved in the interaction.

Using these two axioms, one obtains for the momenta and velocities and kinetic energies 'after collision,' as functions of the same variables 'before collision,' the following simple laws:

$$\begin{aligned}
p_1' &= p_1 \frac{1-mf^2}{1+mf^2} + p_2 \frac{2f}{1+mf^2} \\
p_2' &= p_1 \frac{2mf}{1+mf^2} - p_2 \frac{1-mf^2}{1+mf^2} \\
v_1' &= v_1 \frac{1-mf^2}{1+mf^2} + v_2 \frac{2mf}{1+mf^2} \\
v_2' &= v_1 \frac{2f}{1+mf^2} - v_2 \frac{1-mf^2}{1+mf^2} \\
e_1' &= e_1 \frac{(1-mf^2)^2}{(1+mf^2)^2} + e_2 \frac{4mf^2}{(1+mf^2)^2} + p_1p_2 \frac{2f(1-mf^2)}{(1+mf^2)^2} \\
e_2' &= e_1 \frac{4mf^2}{(1+mf^2)^2} + e_2 \frac{(1-mf^2)^2}{(1+mf^2)^2} - p_1p_2 \frac{2f(1-mf^2)}{(1+mf^2)^2} \; .
\end{aligned} \tag{3}$$

Here p = momentum, v = velocity, e = kinetic energy, unprimed = before, primed = after collision, index 1 = horizontal bullet, 2 = vertical bullet, $m = m_2$, $f = f'$.

It is possible to feed these equations into a computer and solve for trajectories. However, there is a more satisfying alternative: to proceed to smooth differential equations directly (from which the above equations may then be reobtained as a limiting case when necessary).

4. A SMOOTH MODEL

If the ends of the two bullets in Figure 1 are assumed 'softly elastic' in the longitudinal direction, smooth Hamiltonian forces are implied. This leads to the following continuous model as the simplest possibility:

$$\begin{aligned}
\dot{s}_1 &= p_1 \\
\dot{p}_1 &= \frac{\alpha}{(s_1-0.25)^2} - K_1 - \frac{\gamma}{(0.8-s_1)^2} \\
\dot{s}_2 &= \frac{p_2}{m_2} \\
\dot{p}_2 &= \frac{\delta}{s_2^2} - K_2 \; .
\end{aligned} \tag{4a}$$

Here s_1, s_2 are the positions of the horizontal and the vertical bullet, respectively, with s_1 running from 0.25 (left) to 0.8 (right) and s_2 from zero (top) to some maximal value determined either by the bottom (lower wall of the horizontal tube) or before that by the point of impact with the horizontal bullet.

Every reflecting boundary has its corresponding own force term in Eq.(4a). These terms become large only as the corresponding distance becomes small. Those corresponding to the ends of the tubes are already fully specified in the equation as written (with α, γ, δ small), while the interactional ones (K_1, acting on the horizontal bullet, and K_2, acting on the vertical one) have yet to be entered. K_2 is easier to derive first. As shown in Figure 2, the lower boundary applying to the

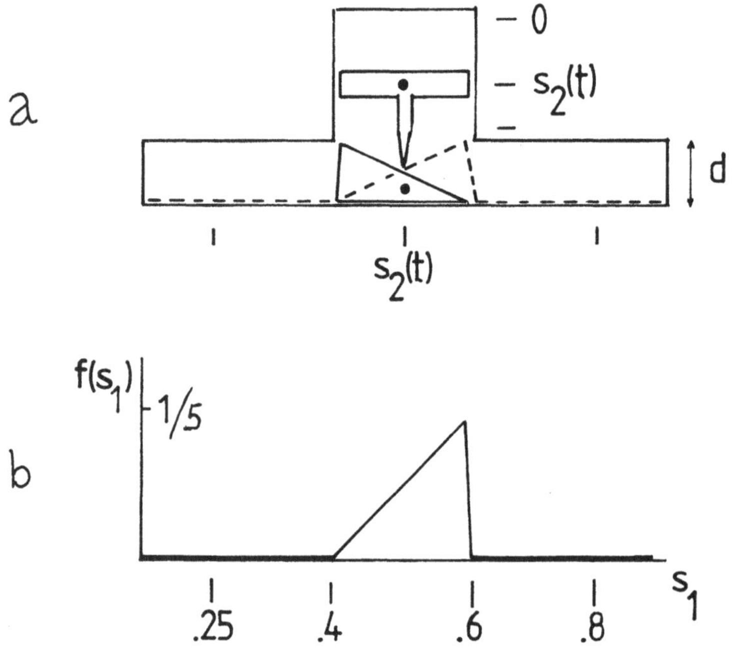

Fig.2 a: An equivalent (somewhat simpler) version to the system of Figure 1. Dashed line = effective lower boundary (s_2-max) of the vertical tube as a function of the position of the horizontal bullet in its own tube (s_1). d = diameter of the horizontal tube.

b: the same function, normalized and redrawn

path of the vertical bullet is simply a function of the actual position s_1 of the horizontal bullet. This function is to be zero outside a central range, and of two different slopes in the central region (Fig.2b). An approximating smooth function is as follows:

$$f(s_1) = \frac{1}{2} \cdot \{ s_1-0.4 + \sqrt{(s_1-0.4)^2+\varepsilon} - 101(s_1 - \frac{302}{505} + \sqrt{(s_1 - \frac{302}{505})^2+\varepsilon})$$

$$+ 100(s_1-0.6 + \sqrt{(s_1-0.6)^2+\varepsilon}) \} . \qquad (4b)$$

(Note that $302/505 = 0.6-(0.6-0.4)/101$; other possible slopes may be constructed by analogy.) In accordance with the postulates of Figure 2b, the present function starts to have appreciable values at $s_1 = 0.4$, then rises with slope unity, then falls with slope 100 down to near zero values at $s_1 = 0.6$. The smaller ε , the sharper the transitions between the four segments.

With $f(s_1)$ given, K_2 now takes on the same form as the rightmost term (containing γ) in Eq.(4a), namely

$$K_2 = \frac{\beta}{(k(0.4-f(s_1))-s_2)^2} \quad . \tag{4c}$$

Here γ has been replaced by β , and the threshold 0.8 by the threshold $k(0.4-f(s_1))$, with k a constant. (Note that $0.4k = s_2\text{-max} = 2d$.)

With K_2 thus given, K_1 is automatically fixed. It simply is

$$K_1 = K_2 \cdot kf' \quad , \tag{4d}$$

where f' is the derivative of $f(s_1)$ by s_1 . Specifically,

$$f' = \frac{1}{2} \cdot \left\{ \frac{s_1 - 0.4}{\sqrt{(s_1-0.4)^2+\varepsilon}} - 101 \frac{s_1 - \frac{302}{505}}{\sqrt{(s_1-\frac{302}{505})^2+\varepsilon}} + 100 \frac{s_1 - 0.6}{\sqrt{(s_1-0.6)^2+\varepsilon}} \right\} . \tag{4e}$$

The whole system of equations thus obtained (Eqs. 4a-e) can now be said to have been derived in the usual way from a Hamiltonian H (using the rule $\dot{s}_i = \frac{\partial H}{\partial p_i}$, $\dot{p}_i = -\frac{\partial H}{\partial s_i}$). Specifically, the underlying Hamiltonian is

$$H = \frac{p_1^2}{2} + \frac{p_2^2}{2m_2} + \frac{\alpha}{s_1-0.25} + \frac{\beta}{k(0.4-f(s_1))-s_2} + \frac{\gamma}{0.8-s_1} + \frac{\delta}{s_2} \quad . \tag{5}$$

Note that Eq.(5) is not much more complicated than Eq.(2).

Picking arbitrary initial conditions for the four variables of Eq.(4a) automatically fixes H . When the system of Eq.(4a-e) is being integrated - for example numerically -, H must not change during the integration. This provides a convenient test for the relative absence of algebraic errors in the process of writing down and programming and solving Eq.(4).

5. NUMERICAL RESULTS

Figure 5 shows a numerical simulation of Eq.(4). Both time plots start from the same initial condition, the second applying to negative time. It was obtained by multiplying the right-hand sides of Eq.(4) by -1.

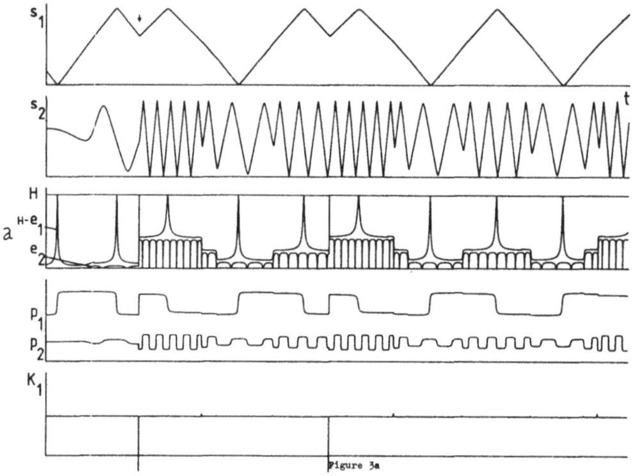

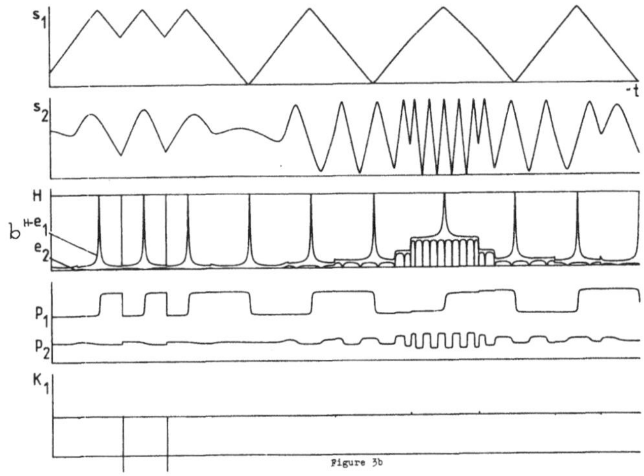

Fig.3 Transient macroscopic behavior in the system of Figure 1. Numerical simulation of Eqs.(4a-e), performed on a HP 9845B desk-top computer with peripherals using a standard Runge-Kutta-Merson integration routine (maximal step error 10^{-4}). a: Positive-time plot. b: Negative-time plot. Parameters: $\alpha = \gamma = 0.005$, $m_2 = 1$, $\delta = \beta = 0.0002$, $\varepsilon = 10^{-7}$, $k = 0.1$. Initial conditions: $s_1(0) = 0.35$, $p_1(0) = -1.8$, $s_2(0) = 0.015$, $p_2(0) = -0.01$. Axes: 0.25...0.8 for s_1 ; 0.04...0 for s_2 ; 0...35 for H, e_1, e_2 ; -9...3 for p_1 ; -3...9 for p_2 , $-2.5\cdot10^4 \ldots 2.5\cdot10^4$ for K_1 ; 0...3 for t and 0...-3 for -t .

The first variable (s_1) is on top. Maxima of s_1 correspond to hits at the right-hand wall of Figure 1. The second trace displays s_2. Its hull curve shows 'indentations' whenever s_1 is in the critical middle range. Next comes a display of H (= constant), along with two sub-energies ($e_1 = p_1^2/2$, $e_2 = p_2^2/2$), the former 'hanging down' from H. In between them, there is a small 'corridor' which contains the 3 potential energies (which are not shown). The next trace shows p_1 and p_2 themselves for completeness. Finally, the interactional force $K_1(t)$ is also displayed.

The initial condition was chosen such that $e_1(0)$ is considerably larger than $e_2(0)$ (compare the beginning of the third trace of Fig.3) - in accordance with the question posed at the end of Section 2. One sees that there indeed are more hits initially to the right-hand side of the horizontal tube. One also sees that the initial excess of e_1 over e_2 decays more or less slowly (with some superimposed 'noise') in both time directions.

This initial behavior is characteristic. On many trials analogous results were obtained. By increasing both k and the right-hand slope of $f(s_1)$, the duration of the 'transient regime' can be increased. Of course, all results apply only statistically: Due to the smoothness of $f(s_1)$ and the potentials chosen (ε as well as α, β, γ, δ are all very small), there are 2 narrow bands in the range of s_2 (middle and bottom) in which the effective slope of $f(s_1)$ has a magnitude close to unity. If s_1 happens to hit s_2 there, the excess energy of e_1 may drop by about 50 per cent in a single collision. An 'accident' like this occurred right at the first major collision in Figure 3a (see the arrow in the top trace).

To sum up, the system of Eq.(4) behaves as if there were two macroscopic regimes: a 'final' attractor, and a 'transient' (or quasi) attractor. Both of these regimes do not normally depend on details of the initial conditions, and both are symmetric under time reversal.

Needless to say, neither of the two regimes is a genuine attractor. The 'transient' one even forms part of the other, being revisited more or less soon. This 'revisiting time' may, compared to larger Hamiltonian systems be rather short. By adding bullets to both sides of the horizontal tube and the top of the vertical tube (after appropriate elongation), however, a derived system with much longer recurrence times can be obtained. In this case, a pseudo-open initial condition would be one in which the upper reservoir of bullets is (initially) 'colder' than the lower one.

6. DISCUSSION

A simple Hamiltonian system determined by Eq.(5) was proposed that presents a peculiar type of 'transients' which preferentially visit the same region of state space in both directions of time. It is tempting to think that macroscopic behavior in statistical mechanical systems may be reducible in part to properties of the present system. For example, one might close the horizontal tube in Figure 1 into a circle and then call the observed direction-preferring transient behavior (which again applies no matter what the initial direction of movement, and what the direction of time) 'convective.' In other words, of the two major types of 'far from equilibrium' behavior to be found in macroscopic physical-chemical systems, 'convection' (facilitated directed motion) and 'catalysis' (facilitated formation of molecular complexes), one can apparently be reproduced in a two-degrees-of-freedom Hamiltonian system already. If this is true, the other type (of chemical evolution) can perhaps also be modeled in a relatively simple setting.

The system of Eq.(4) can be studied further. It is, for example, possible to plot a 2-dimensional Poincaré cross-section through the flow of Fig.3. The pattern of motion to be seen in this map can then be compared with the 'macroscopic' spatial distribution of energy found in Figure 3.

Another open problem is to what extent spatial interpretations of the observed behavior are essential. There might exist other, more subtle (but equally 'quasi-attracting' in both directions of time) types of macroscopic behavior.

A final remark concerns a related 'limit to observation' problem. Every would-be Maxwellian demon (like the vertical bullet in Figure 1) 'spoils' its own observations by perturbing the target. Recently the principle was stated that looking into a Hamiltonian physical system 'from the inside' necessarily leads to radically different results as compared to the usual 'higher world' approach (Rössler, 1981). The system of Eq.(4) was presented above entirely from the higher-level vantage point. Once a precise definition of 'macroscopic subsystem becomes possible, however, the notion of 'measurement from within' (namely, by such a subsystem) will suddenly have quantitative implications.

I thank Dietrich Hoffmann and John Kozak for discussions. This work was partially supported by the "Fonds der chemischen Industrie."

7. REFERENCES

1. Abraham, R. and J.E. Marsden, Foundations of Mechanics, 2nd ed., Reading, Mass., Benjamin/Cummings (1978).
2. Arnold, V.I. and A. Avez, Ergodic Problems of Classical Mechanics. New York, Bemjamin (1968).
3. Birkhoff, G.D., On the periodic motion of dynamical systems, Acta Math. 50, 359-379 (1927).
4. Farquhar, I.E., Ergodicity and related topics, In: Irreversibility in the Many-Body Problem (L.M. Garrido, J.Biel and J. Rae, eds.) pp. 29-104. N.Y. London, Plenum (1972).
5. Feynman, R.P., R.B. Leighton, and M. Sands, The Feynman Lectures on Physics, vol. 1, p. 46-7. Reading, Mass., Addison-Wesley (1965).
6. Greene, J.M., R.S. McKay, F. Vivaldi and M.J. Feigenbaum, Universal behavior in families of area-preserving maps, Physica 3D, 468-486 (1981).
7. Hadamard, J., Les surfaces à curbures opposés et leurs lignes géodésiques, Journ. de Math. (5) 4, 27-73 (1898).
8. Hénon, M. and C. Heiles, The applicability of the third integral of motion: some numerical experiments, Astron. J. 69, 73-79 (1964).
9. Moser, J., Stable and Random Motions in Dynamical Systems. Princeton, N.J., Princeton University Press (1973).
10. Poincaré, H., Sur le problème des trois corps et les équations de la dynamique, Acta Math. 13, 1-271 (1890).
11. Prigogine, I. and I. Stengers, Dialog mit der Natur (Dialogue with Nature). Munchen Zurich, Piper Verlag (1981).
12. Rössler, O.E., Chaos, In: Structural Stability in Physics (W. Güttinger and H. Eikemeier, eds.) pp. 290-309. Berlin-Heidelberg-N.Y., Springer-Verlag (1979).
13. Rössler, O.E., Chaos and bijections across dimensions, In: New Approaches to Nonlinear Problems in Dynamics (P.J. Holmes, ed.), pp. 477-486, Philadelphia, SIAM (1980).
14. Rössler, O.E., Chaos and chemistry, In: Nonlinear Phenomena in Chemical Dynamics (C. Vidal and A. Pacault, eds.), pp. 79-87. Berlin Heidelberg N.Y., Springer-Verlag (1981).
15. Sinai, Ya.G., Dynamical systems with elastic reflections, Russian Math. Surveys 25, 137-189 (1970).

QUANTUM DYNAMICS

B.A. Huberman

Xerox Palo Alto Research Center
3333 Coyote Hill Road
Palo Alto, California 94304, U.S.A.

I think that it is appropriate for a conference devoted to Nonlinear Dynamics and Chaos to include a set of lectures on the problem of quantum dynamics. For although the interest of the past few years has been mainly focused on classical chaotic dynamics, the quantum problem still poses both intriguing questions and tantalizing applications, problems which have not been the focus of much activity until now. As I will try to show in what follows, quantum dynamics still presents unresolved problems; problems whose resolution in a crisp and unambiguous way might lead to insights into the way the classical limit obtains, as well as a deeper understanding of quantum ergodicity.

From the point of view of pure theory, there is, in principle, no quantum dynamics problem. We all know what to do in order to learn about the time evolution of any quantum system - solve the time dependent Schrödinger equation, or equivalently, assume that the state vectors are time independent and work in the Heisenberg representation. But knowing the fundamental equations of quantum theory does not necessarily mean that we can always solve them and therefore predict the time evolution of any system once we know the pertinent Hamiltonian. In the particular case where the Hamiltonian itself is both time dependent and cannot be treated by perturbation theory, to obtain information about global properties presents formidable obstacles, and answers to simple questions such as energy growth or decay are very hard to obtain. These questions are by no means purely academic; a large amount of laser photochemistry research depends on their resolution [1], and the enormous progress that has been made in microelectronics over the past few years is now presenting challenging problems as electron mean free paths inside very small structures become larger than device sizes.[1]

At a more fundamental level, there remains the old question of the quantum behavior of systems which in the classical limit are known to be chaotic. Although for stationary Hamiltonians there exists a fairly large body of knowledge,[3,4] the situation in the time dependent case is less clear. The few results that we have are based on numerical calculations, and although a recurrence theorem now exists for the case of a time periodic Hamiltonian,[5] its proof does not allow for an a priori prediction as to whether a quantum system will satisfy the conditions that would make it recur infinitely often.

In what follows, I will present some results for the case when the Hamiltonian is time periodic, the problem which has been more thoroughly studied in the past few years.[5-11]

TIME PERIODIC HAMILTONIANS

In the particular case where the Hamiltonian is time periodic, one can make some concrete statements concerning the time evolution of a quantum system. This is also the case that has been most thoroughly studied and for which there are a number of good results both analytic and in the form of computer experiments. We start by first considering the following Hamiltonian

$$H=H_o+V(t) \tag{1}$$

where $V(t)=V(t+T)$, T is an arbitrary period, and for the sake of simplicity H_o is assumed to describe a bounded system.

If we now write the wave function in terms of the complete orthonormal set of eigenstates of H_o, $\{u_m(r)\}$, as $\Psi(r,t)=\sum a_m(t)u_m(r)$, it is easy to show that the coefficients $a_m(t)$ make up a vector a(t) which satisfies

$$i\hbar\dot{a}(t)=H(t)a(t) \tag{2}$$

and we can therefore write

$$\| \Psi(t+\tau)-\Psi(t)\|^2 = a(t+\tau)-a(t)|^2 \tag{3}$$

where we have defined $\|\Psi(t)\|^2 \equiv \int dr|\Psi(r,t)|^2$.

Since $H(t)=H(t+T)$ the wave function will satisfy a Floquet theorem, i.e., the vector a(t) is of the form

$$a(t) = \sum_{k=1}^{\infty} \alpha_k \exp[iE_k t/\hbar]\,\Phi_k(t) \tag{4}$$

with $\Phi_k(t+T)=\Phi_k(t)$, $\Phi^+_k(t)\Phi_{k'}(t)=\delta_{kk'}$ for all t. The set $\{E_k\}$ is called the quasienergy spectrum, which incidentally is hard to calculate in most cases. This result, which was first enunciated by Shirley,[9,10] applies to any periodic Hamiltonian.

With this in mind, it is possible to prove the following theorem.[5] Consider any bounded quantum system described by a Hamiltonian H_0 having a discrete spectrum, and subjected to a non-resonant (i.e., the whole system has a discrete quasienergy spectrum) time periodic potential V, with $V(t)=V(t+T)$ and T an arbitrary period, and such that $\|V\|$ is bounded. Given <u>any</u> initial configuration of the system, both the wave function and the energy will return arbitrarily close to their initial values infinitely often. More generally, if we define an almost periodic function, f(t), to be a continuous, bounded function such that for any $\epsilon>0$ there exists a relatively dense set $\{\tau_\epsilon\}$ and for <u>each</u> τ_ϵ in the set, we have $|f(t+\tau_\epsilon)-f(t)|<\epsilon$ for all t, the theorem states that both the normed wave function and the energy are almost periodic functions of time. I should add that the applicability of this theorem to a particular quantum system requires knowledge about the quasienergy spectrum, information that it is hard to obtain a priori from the properties of the Hamiltonian. Nevertheless, the examples that I will show below will demonstrate that for a fairly wide class of problem, recurrence is widespread, a fact that indicates that even the case of resonances (i.e., a continuous quasienergy spectrum) might not be that important in practical situations.

We will first consider the problem of a harmonic oscillator in the field of an arbitrarily large (but not infite) periodic electromagnetic field, E(t), and ask about the behavior of the energy as a function of time. The Hamiltonian of such system is given by $H=H_0+\alpha xE(t)$ with

$$H_0=p^2/2m + \frac{m\omega^2}{2}\,x^2 \qquad (5)$$

The time evolution of the total energy of the system is given by

$$dE/dt = \langle dH/dt\rangle = \alpha\dot{E}(t)\langle x\rangle. \qquad (6)$$

which in turn implies that all that is needed is to calculate the expectation value of the position operator $\langle x\rangle$. Using the Ehrenfest theorem we can write the following equations of motion for $\langle x\rangle$ and $\langle p\rangle$.

$$d\langle x\rangle/dt = \langle p\rangle/m \tag{7}$$

$$d\langle p\rangle/dt = -\langle\nabla V\rangle = -m\omega^2\langle x\rangle - \alpha\dot{E}(t) \tag{8}$$

Since the derivative of the potential is linear in x, we obtain a simple differential equation for $\langle x\rangle$, namely

$$d^2\langle x\rangle / dt^2 + \omega^2\langle x\rangle = -\alpha\,\dot{E}(t)/m. \tag{9}$$

which together with Eq. (6), determines the behavior of the energy. Solving these equations, we obtain[11]

$$E(t)=E_0 + \frac{\alpha^2}{\omega m}\int_0^t dt'\int_0^{t'} dt''\left[\dot{E}(t')\,E(t'')\sin\omega(t'-t'')\right]$$

$$+\alpha c_1\int_0^t dt'\dot{E}(t')\sin\omega t' + \alpha c_2\int_0^t dt' E(t')\cos\omega t' \tag{10}$$

where E_0 is the initial energy and c_1 and c_2 depend on the nature of the initial state.

In order to make contact with quantum maps that we'll study below, we next consider the case where the external field consists of a periodic string of electromagnetic pulses, i.e. $E(t) = \sum_{n=1}^{\infty}\delta(t-nT)$ with T the time between pulses. If the initial state of the oscillator is a pure state, i.e., $\langle x\rangle=\langle p\rangle= 0$, Eq.(10) then gives

$$E^{(n)}=E_0 + \frac{\alpha^2}{2m}\,\frac{\sin^2(n\omega T/2)}{\sin^2(\omega T/2)} \tag{11}$$

which is obviously an almost periodic function in the sense defined above. Concerning this result, some remarks are in order: 1) Notice that it is only when the period between pulses is commensurate with $4\pi n\omega_0$ that the energy recurs periodically. Otherwise $E^{(n)}$ displays quasiperiodic behavior. 2) For short times such that $n\omega T < 1$ Eq.(11) gives $E\simeq E_0+\alpha^2n^2/2m$, i.e., quadratic growth of the energy with time. 3) If, on the other hand, the initial state is not a pure one, Eq.(10) gives

$$E(n)=F(n,c_1,c_2) \tag{12}$$

with F a function which for short times $nT\omega \ll 1$ gives linear growth of the energy with time, instead of the quadratic behavior generated by Eq.(11).

QUANTUM MAPS

In cases where the Ehrenfest theorem does not lead to a linear differential equation for the expectation value of observables, it is more convenient to resort to quantum maps. Basically they are obtained by taking the time dependent potential as a string of delta function pulses, a process which leads to a recursion relation which can then be iterated on a computer. We will first deal with a problem of considerable interest to the physics of microelectronics, namely an electron of mass m and charge e in an infinite square well potential of length L and which is acted upon by a set of electromagnetic pulses of strength E.[5,11] The Hamiltonian of the system is given by

$$H = \frac{p^2}{2m} - eEx \sum_{n=-\infty}^{+\infty} \delta(t/T-n) \tag{13}$$

In order to construct a quantum map, we expand the wave function in terms of the eigenstates of H_o; i.e., we write

$$\Psi(x,t) = \sqrt{2/L} \sum_{n-1}^{\infty} a_n(t)\sin(n\pi x/L) \tag{14}$$

We now notice that in between pulses the a's evolve as $a_n[(N+1)T^-]=a_n(NT^+)e^{-iE_nT/\hbar}=a_n(NT^+)e^{-in2\tau/2}$, where $E_n=\pi^2\hbar^2n^2/2mL^2$ is the nth eigenvalue of H_o and $\tau=\pi^2\hbar T/mL^2$. Furthermore, each pulse induces a change in the wave function which is given by $\Psi[x,(N+1)T^+] = \Psi[x,(N+1)T^-]e^{ikx}$ with $k \equiv eET/\hbar$. We can therefore construct the following map relating the values of a_n just after the (N^{th}) kick to the values of a_n just after the Nth kick;

$$a_n(N+1)=\frac{4i\alpha}{\pi}\sum_{r=1}^{\infty} a_r(N)e^{-ir^2\tau/2}((-1)^{r+n}e^{i\alpha\pi}-1)\frac{nr}{[r^2-(a-n)^2][r^2-(\alpha+n)^2]} \tag{15}$$

where $\alpha = kL/\pi$.

Iterating the map with any given set of initial conditions allows for a calculation of the discrete time evolution of the energy, which is given by

$$E(t) = \sum_{n=1}^{\infty} E_n \left| a_n(t) \right|^2 \qquad (16)$$

and subject to the condition $\sum_{n=1}^{\infty} \left| a_n \right|^2 = 1$. Figure 1 shows a typical computer run on double precision for the electron initially in the ground state and k=3.5; τ =1.432.

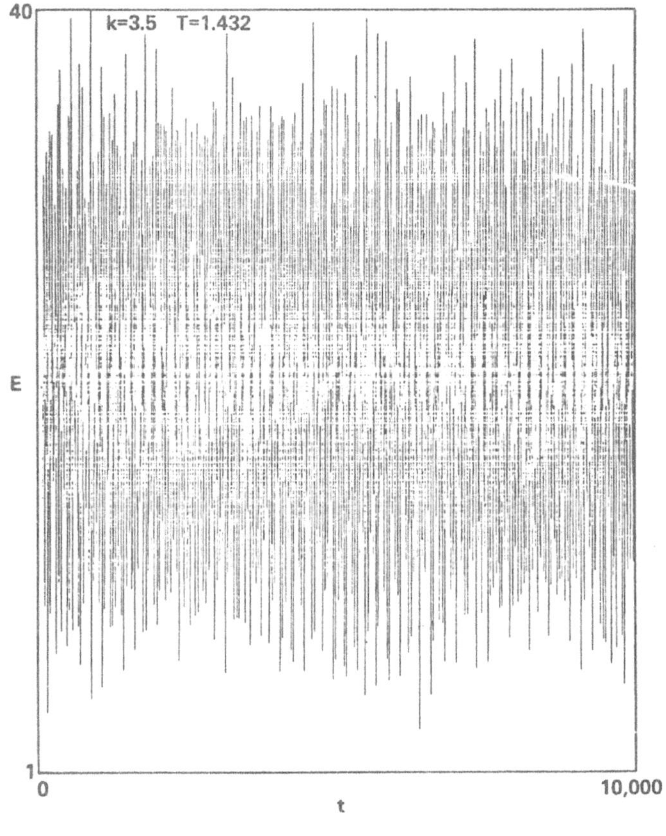

Figure 1

A total of 100 states were used in the calculation, and the normalization of the function was checked at each iteration to within 16 significant digits. (This enormous precision is needed if reliable information is required, for a poor normalization results in a lack of phase conspiracy so as to make the wave function collapse on its initial configuration). As can be seen from the figure, the system reassembles itself many times, and the total variation of its energy is bounded, a fact which could not have been guessed on the basis of a few iterations. Furthermore, it should be stressed that almost periodic behavior does not allow for any concrete prediction concerning the initial behavior for any arbitrary initial state; depending on the particular choice of such initial state, the energy can either grow, decay or stay constant for very many iterations.

One particularly interesting type of recurrence can be studied with the aid of Eq. (15), which can be alternatively written as

$$a_{m+1}=\hat{M}(\tau,kL)\ a_m \tag{17}$$

It is easy to check that for $\tau=2\pi$, and regardless of the value of kL

$$\hat{M}^2(2\pi,kl)=e^{ikL}\ I \tag{18}$$

with I the identify matrix. This in turn implies (Eq.(17)) that $a_{m+1}=a_m e^{ikL}$, and using Eq.(16) we thus obtain the peculiar result that

$$\langle E\rangle_{N+2}=\langle E\rangle_n \tag{19}$$

for all n and any value of kL. Therefore, regardless of how strong the pulses are, the energy will recur every two iterations. This peculiar quantum effect implies that for $\tau=2\pi$, after one pulse one can obtain an enormous spread in the wave function, only to be followed by a process in which it reconstructs itself after the next pulse. This effect, which is also obtained in the case of the harmonic oscillator (Eq.(11)), was first studied in the context of the quantum rotor by the authors of [6].

The last problem which I will mention is the one that still poses many intriguing questions vis a vis the issue quantum chaotic bahavior: the quantum version of a periodically kicked rotor. Its classical counterpart is one of the paradigms of chaotic behavior in non-dissipative dynamical systems and as such it has been thoroughly studied.[12]

Therefore, it has become a testing ground for ideas on quantum ergodicity and on the issue of the semiclassical behavior of non-trivial quantum dynamical systems. Moreover, its quantum version provides quantitative measures of the behavior of the energy as a function of time, and it was expected to become an interesting example of quantum chaotic behavior.

The original numerical investigation of this problem was carried out by Casati and coworkers,[6] who claimed to have observed stochastic behavior in the energy for short initial times at values of the parameter corresponding to classical chaos. Later Hogg and Huberman [5] performed computer experiments for very long times and showed almost periodic behavior for the energy, an indication that the periodically kicked rotor might satisfy the conditions of their theorem and is therefore not chaotic. More recently, Fishman and collaborators have claimed the existence of a discrete quasienergy spectrum for this system by mapping it into a soluble localization problem.[13] However, in spite of all these results, there exist definite problems in obtaining a chaotic classical limit from the quantum maps.

The Hamiltonian for the periodically kicked rotor is

$$H=P^2/2I - \omega_0^2 I\cos(\theta)\sum \delta(t/T-n) \tag{20}$$

where θ is the angle, p is the angular momentum, I is the moment of inertia and the delta functions are understood to be the limit of very narrow Gaussian pulses.

Following the technique of Casati et al.,[6] we expand Ψ in terms of the eigenstates of $H_0=P_\theta^2/2I$ as $\Psi(\theta,t)=(2\pi)^{-1}\sum a_n(t)e^{in\theta}$ to obtain the map

$$a_n(t+T^+)= \sum a_r(t)b_{n-r}(k)e^{-ir^2\tau/2} \tag{21}$$

where $k=\omega_0^2 IT/\hbar$, $\tau=\hbar T/I$ and $b_s(k)=i^sJ_s(k)$ with J_s the ordinary Bessel function of the first kind and order s. Using this map we [5] have computed the energy $E(t)=\sum (n^2/2I)\hbar^2 |a_n(t)|^2$ for several values of k and τ checking the normalization condition $\sum|a_n|^2 = 1$ to 16 digits at every iteration. A typical result is shown in Figure 2 where we show its time evolution in time units of number of pulses for the case k=2.871, t=2.532 and with the initial configuration in the ground state. A total number of 201 states were used. Once again, and in analogy with the problem of the electron in a quantum well, we see that the excursions in the energy are bounded and recur many times. These results were obtained for parameter values such that $k\tau > 1$, a situation which in the classical limit leads to erratic wandering in phase space. Once again the fact that the energy is an almost periodic function of time prevents one from making definite statements concerning its growth or decay, since the choice of inital conditions is completely arbitrary. (For example, one could start the system with a wave function corresponding to the value of the energy at say, 247 iterations and would only observe energy decay).

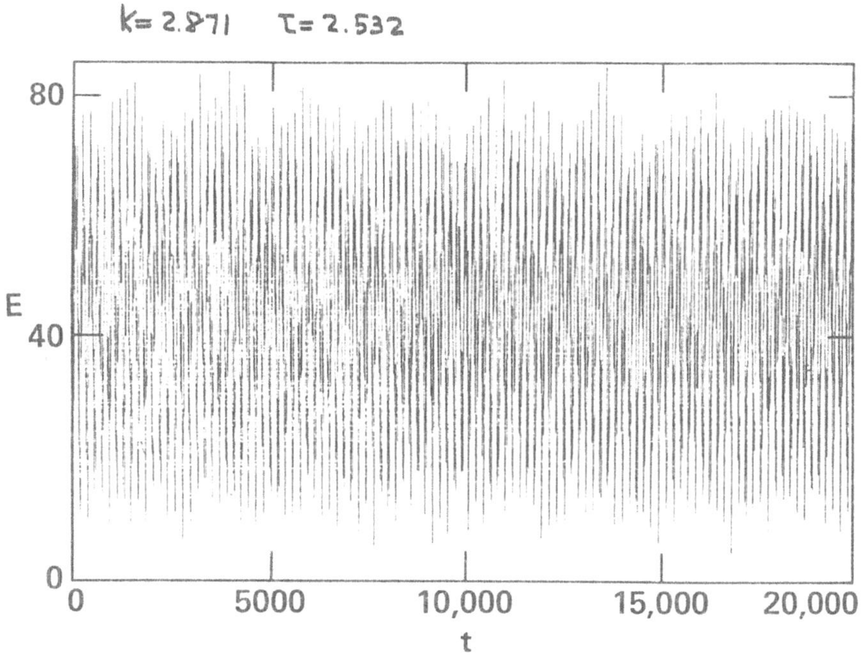

FIGURE 2

An interesting problem is posed by the presence of resonances i.e., special values of the parameters for which one should see unbounded energy growth. In this particular case of the kicked rotor, Izrailev and Shepelyanskii[6] have shown that the quasienergy spectrum is indeed continuous, and that for dense set of values of the kicking period, the energy grows quadratically with time. Nevertheless, the existence of this set, (which is of measure zero) will not, in practice, prevent the system from reassembling itself infinitely often. This stems from the fact that even a minute departure from resonance (say one part in ten to the eight) will eventually produce such a dephasing of the wave function so as to set in motion the mechanism of recurrence. This is illustrated in Figure 3, where we display the value of the energy of the rotor for a resonance condition i.e., k=0.5; $t=8\pi/5$. (A total of 301 states were used, and normalization was preserved throughout the run). Nevertheless, it can be clearly seen that the lack of infinite precision in the numerical value of π makes the energy decrease after 5000 pulses. Beyond this point, it behaves in an almost periodic fashion. This behavior is to be expected under real experimental conditions, as it is unlikely that frequency stability of an external source such as a laser could better the numerical experiment.

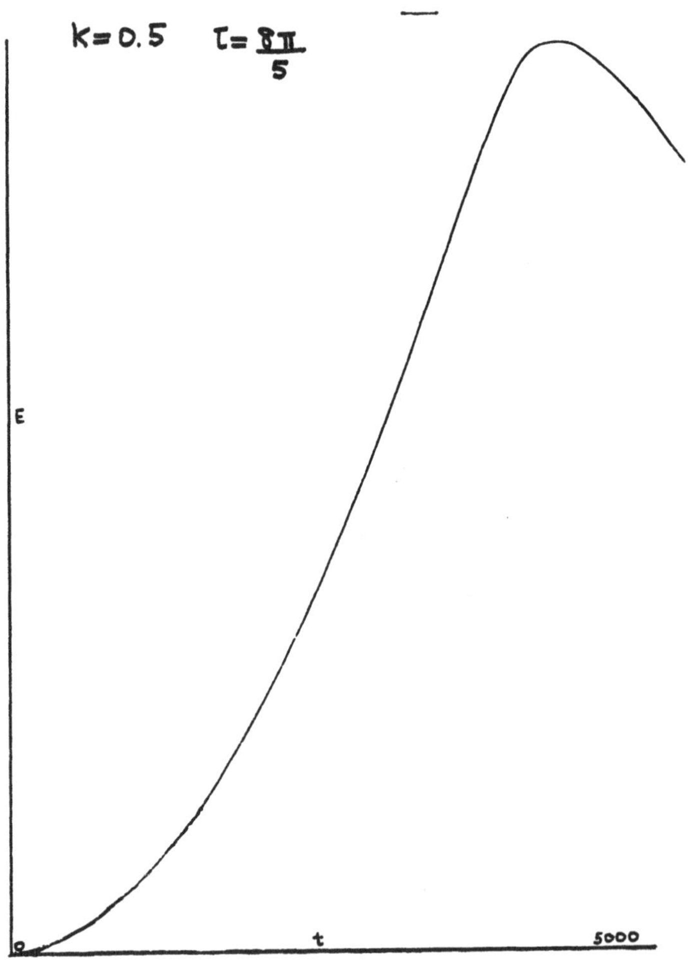

FIGURE 3

These lectures illustrate the fact that in dealing with quantum dynamical systems whose classical counterparts are non-integrable, one finds qualitatively new behavior and peculiar quantum effects which are absent in classical systems. Although the examples which I used were fairly simple, we believe that the absence of quantum chaos in systems with periodic Hamiltonians will be pervasive in more complex systems, such as molecular dynamics and many electron systems. This in turn poses questions concerning the classical limit of quantum maps and the role of damping in quantum systems which are mixing in the classical limit. Once again, although the classical dissipative problem is well characterized through the existence of strange attractors, little is known about the respective quantum problem.

ACKNOWLEDGMENTS

I wish to thank T. Hogg for the very many interesting discussions we've had on this subject and for his help in obtaining the results reported here.

REFERENCES

1. A. Ben Shaul, Y. Haas, K.L. Kompa and R.D. Levine, Lasers and Chemical Change (Springer, Berlin 1981).
2. VLSI Electronics, Microstructure Science, edited by N.G. Einspruch (Academic, N.Y. 1981).
3. J. Von Neumann, Z. Phys. 57, 30 (1929), P. Bocchieri and A. Loinger, Phys. Rev. 107, 337 (1957) and L. Rosenfeld, in Ergodic Theories, edited by P. Caldirola (Academic, N.Y. 1960).
4. M.C. Gutzwiller, Phys. Rev. Lett. 45, 150 (1980); I. Percival, Adv. Chem. Phys. 36, 1 (1977), D.W. Noid, M.L. Koszykowski and R. Marcus, Ann. Rev. Phys. Chem. 32, 267 (1981), and R. Kosloff and S. Rice, J. Chem. Phys. 74, 1340 (1981).
5. T. Hogg and B.A. Huberman, Phys. Rev. Lett. 48, 711 (1982).
6. G. Casati, B.V. Chirikov, F.M. Izraelev and J.Ford, in "Stochastic Behavior in Classical and Quantum Dynamical Systems", edited by G. Casati and J. Ford (Springer, Berlin 1979), F.M. Izrailev and D.L. Shepelyanskii, Theor. Math. Phys. 43, 553 (1980), and B.V. Chirikov, F.M. Izrailev and D.L. Shepelyanskii (preprint 1982).
8. M.V. Berry, N.L. Balasz, M. Tabor and V. Voros, Ann. Phys. (N.Y.) 122, 26 (1979), and J. Korsch and M.V. Berry, Physica (Utrecht) 3D, 627 (1981), and G.P. Berman and G.M. Zaslarsky, Physica 91A, 450 (1978).
9. J.H. Shirley, Phys. Rev. B138, 979 (1965).
10. See also Ya. B. Zeldovich, Sov. Phys. JETP 24, 1006 (1967); V.I. Ritus, Sov. Phys. JETP 24, 1041 (1967); and F. Gesztesy and H. Milter, J. Phys. A14, 179 (1981).
11. T. Hogg and B.A. Huberman (preprint 1982).
12. B.V. Chirikov, Repts. Prog. Phys. 52, 263 (1979).
13. S. Fishman, D.R. Grempel and R.E. Prange, Phys. Rev. Lett. 49, 509 (1982).

A UNIVERSAL TRANSITION FROM QUASI-PERIODICITY TO CHAOS - ABSTRACT

Eric D. Siggia

Laboratory of Atomic and Solid State Physics
Cornell University
Ithaca, New York 14853 U.S.A.

A common route to chaos in dissipative systems proceeds from periodic to quasi-periodic flow (with two independent frequencies). Then, in the absence of rotational symmetry, the system generally mode locks before becoming turbulent. Beyond these qualitative features, the numerous experiments that have examined this regime differ in detail. Dynamical system theory had made the occurrence of the above transitions plausible but has provided no non-trivial quantitative and model independent information.

This situation, on the theoretical side, has recently changed with a proposal on how to modify the experiments so as to make the transition to chaos occur in a quantitatively universal manner.[1,2] The essence of our proposal follows from K.A.M. theory which is the weak coupling limit of the strong coupling problem relevant to the turbulent transition. In addition to the Rayleigh number, the experimenter must control a second parameter so as to maintain the frequency ratio in the quasi-periodic state at a fixed irrational value. The golden mean, $(\sqrt{5}-1)/2$, is the optimal ratio experimentally.

The universality, which is restricted to the low frequencies in the time series, is obtained under the above circumstances because the transition to chaos is continuous. In particular the singular low frequency structure in the spectrum develops continuously as $R \to R_T$ from below or as the frequency ratio approaches $(\sqrt{5}-1)/2$ at R_T. These assertions are rigorously established by a renormalization group analysis that resembles the one developed by Feigenbaum to account for the universal features of period doubling.

Our renormalization group acts on a space of mappings that includes all analytic circle homeomorphisms. It is defined for all rotation numbers, ρ, whether their continued fraction representation is periodic or not. In the former case we conjecture that a fixed point will exist and we have found one numerically for $\rho = 1/1/1\ldots$ and $2/2/2/\ldots$. In the non-periodic case there is numerical evidence that our renormalization transformation has an ergodic attractor. Low frequency spectra of time series are singular and universal in either case but only scale when there is a fixed point.

We have found numerically that any two sufficiently smooth circle homeomorphisms with a single cubic inflection point and the same rotation number are C^1 conjugate. By contrast the conjugacy from a cubic critical map to a pure rotation ($\theta \to \theta + \rho$) is only C^0. The point of inflection corresponds to the onset of turbulence. A power spectrum $|\tilde{s}(\omega)|^2$ of a time series $s(t)$ obtained at the transition has an envelope which scales as ω^2 as the frequency ω tends to zero.

We again stress that all the low frequency complex amplitudes obtained from either a fluid experiment or a forced nonlinear oscillator at the quasi-periodic to turbulent transition are universal. At present, once one establishes their universality, the theoretical numbers are most easily derived numerically by iterating a map such as

$$\phi' = \phi + \omega_0 - \frac{a}{2\pi} \sin(2\pi\phi)$$

for $a = 1$ (corresponding to $R = R_T$) and adjusting ω_0 to achieve the desired rotation number.

The theory summarized above carries over in a quantitative way to dynamical systems in higher dimensions provided there is sufficiently strong contraction onto the 2-torus. We have also considered in a preliminary way situations closer to the model analysed by Ruelle-Takens in which the 2-torus is obtained from mode locking on a 3-torus. Qualitatively different transitions to turbulence are clearly possible and it remains to be seen whether any quantitative and universal predictions can be made.

Models of flow on a 3-torus are also obviously relevant to the quasi-periodic potential problem. The unlocked states are extended while the mode locked regions of parameter space correspond to spectral gaps. Renormalization group techniques will doubtlessly contribute to our understanding of both these problems.

REFERENCES

1. D. Rand, S. Ostlund, J. Sethna, E. Siggia, Phys. Rev. Lett. 49, 132 (1982) and Physica D, submitted.
2. S. Shenker, Physica D, to be published and M. Feigenbaum, L. Kadanoff, S. Shenker, preprint.

SELF-GENERATED DIFFUSION AND UNIVERSAL CRITICAL PROPERTIES IN CHAOTIC SYSTEMS

T. Geisel and J. Nierwetberg

Institut für Theoretische Physik
Universität Regensburg
D-8400 Regensburg, W.Germany

ABSTRACT

This paper reviews the universal critical properties exhibited by some 1d discrete dynamical systems: period-doubling systems and systems generating diffusive motion. While the period-doubling bifurcations have the universal asymptotic bifurcation rate $\delta=4.6692...$, the tangent bifurcations present within the chaotic region do not follow this rate. We show that the tangent bifurcations giving rise to a fine structure of periodic windows have bifurcation rates γ_k which can be calculated analytically. They converge to a universal constant $\gamma=2.94805...$ We have found that a class of dynamical systems show the onset of a diffusive motion in addition to period-doubling. The diffusion is self-generated and does not rely on the presence of random external forces. The onset of diffusion has strong analogies with a phase-transition. The diffusion coefficient is the order parameter and has a universal critical exponent. The dependence on random external fluctuations is also universal and can be expressed in terms of a universal scaling function which is calculated analytically.

1. INTRODUCTION

Chaos in the sense of irregular motion in deterministic systems is observed in a variety of physical systems. The routes leading to the chaotic state when a parameter is changed have been given particular attention. In the last few years it became known that some of these routes have universal critical properties very similar to those observed in phase transitions. The most complicated scaling properties were found in period-doubling systems[1-3]; scaling properties of simpler nature were reported for intermittent chaos[4,5] and for the onset of self-generated diffusion[6].

On the period-doubling route the chaotic state is reached from a periodic regime by a successive doubling of the period when an external parameter is varied. The period 2^{∞} marks the onset of the chaotic regime. The critical parameter values where period-doubling occurs converge geometrically at an asymptotic rate[3] $\delta=4.66920...$ This bifurcation rate has been shown to be a universal constant[1,2]. The period-doubling route has been found in a large variety of physical systems including driven anharmonic oscillators[7] and Rayleigh-Bénard fluids[8]. The second case, the case of intermittency consists of seemingly periodic episodes which are interrupted by chaotic bursts. In a simple model[4,5] the inverse average duration of the almost periodic episodes plays the role of an order parameter with universal critical behavior. The third case, the onset of self-generated diffusion is a transition from a locally bounded motion to an unbounded random walk. This motion is purely deterministic, i.e. does not require random external forces. The diffusion coefficient as an order parameter has universal critical properties[6].

This paper deals with universal scaling properties of period-doubling systems and of systems generating diffusion. In Section 2 we will briefly introduce discrete dynamical systems and review period-doubling bifurcations and their universal properties. Section 3 deals with the fine structure of the chaotic regime, where tangent bifurcations give rise to narrow periodic windows. These tangent bifurcations have bifurcation rates γ_k, which are investigated analytically in Section 4. It is shown that they converge to a universal constant $\gamma=2.94805...$ Section 5 presents an example of self-generated diffusion. In Section 6 we show that the onset of diffusion in strongly dissipative systems has universal critical properties and we derive a universal scaling function.

2. DISCRETE DYNAMICAL SYSTEMS AND PERIOD-DOUBLING

The properties of nonlinear dissipative dynamical systems may conveniently be studied in one-dimensional models of the form

$$x_{t+1} = f(x_t,\mu) \qquad (2.1)$$

where f is a map of an interval into itself, t is a discrete time and μ an externally controllable parameter. Given the starting point x_0 Eq. (2.1) generates a time series of points x_1, x_2, x_3 ... which is called the orbit or trajectory of x_0. Many of the dynamical characteristics known from higher dimensional continuous dissipative systems like the period-doubling route to chaos, intermittency and self-generated diffusion are reflected in such simple discrete models.

Since we are mainly interested in universal properties of these dynamical systems we will consider classes of maps rather than specific maps. In particular we focus on a universality class of maps[1] which is characterized by a parabolic maximum point at $x=x_c$, which is usually called the critical point of f. A well known example of this class is the 'logistic map', i.e. the dynamical system

$$x_{t+1} = \mu(x_t - x_t^2) \tag{2.2}$$

where μ has values between $\mu=1$ and $\mu=4$.

In the following we give a brief survey of the most important properties of the dynamical systems belonging to this universality class. We choose the logistic map as an example, whenever we need to be specific. More details may be found in a review article by May.[9]

A point x* is said to be a periodic point of period p if

$$x^* = f^p(x^*,\mu) \tag{2.3}$$

where f^p denotes the p-fold iterate of f. Equivalently one might say that x* is a fixed point of f^p or that the orbit of x* is periodic with period p. The iterates of x* together with x* form a p-cycle

$$\begin{aligned} x_{i+1} &= f(x_i,\mu) \quad i=1,\dots,p-1 \\ x_1 &:= x^* = f(x_p,\mu) \end{aligned} \tag{2.4}$$

A periodic point x* of period p is said to be (locally) stable if

$$\left|\frac{d}{dx} f^p(x,\mu)\big|_{x^*}\right| = \eta<1 \tag{2.5}$$

which means that a small neighbourhood of length ε shrinks by a factor η when mapped by f^p. Equivalently: an orbit starting in the vicinity of x* converges geometrically towards x*. On the other hand

$$\left|\frac{d}{dx} f^p(x,\mu)\big|_{x^*}\right| = \eta>1 \tag{2.6}$$

means that a small neighbourhood of x* increases by a factor η when mapped by f^p and trajectories starting near x* escape at a rate η. In this situation x* is said to be (locally) unstable.

Figure 2.1 shows typical asymptotic orbits of the logistic map. To generate this figure we computed the iterates of x_c for the logistic map for each of 500 increments of μ between $\mu=3.0$ and $\mu=4.0$. We then plotted the iterates for t=501 to t=900 as points versus the parameter μ. For $\mu<3$ the system has a single stable fixed point of f. As the parameter value is increased the system undergoes a first period-doubling bifurcation at $\mu=\mu_1=3$ which means that the fixed point of f be-

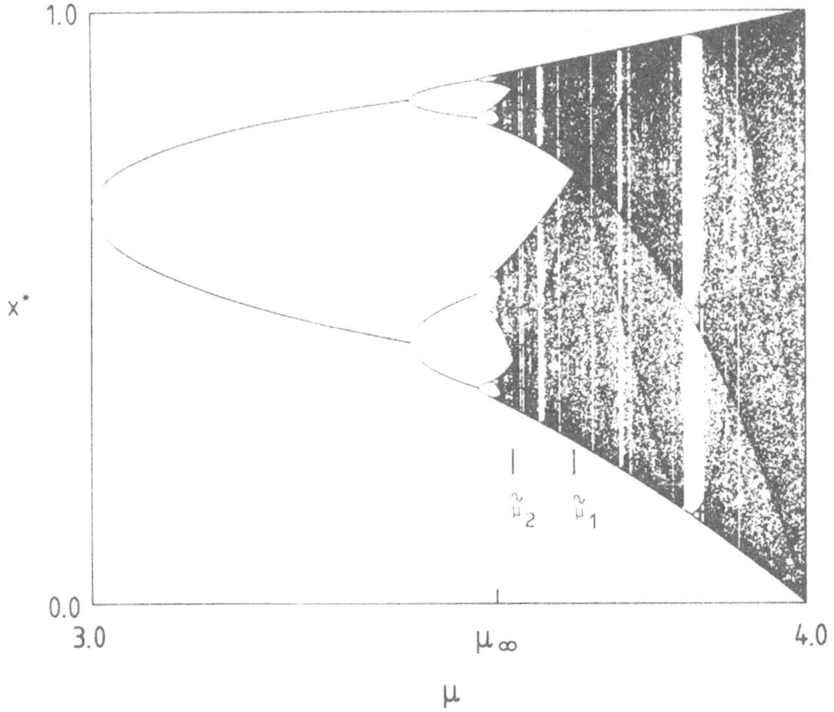

Fig. 2.1 Computer experiment showing the asymptotic orbits x* as a function of μ for the logistic map.

comes unstable while two new and stable fixed points of f^2 appear, which together form a periodic orbit of period two. The system now oscillates between two states, which can be seen in Fig. 2.1. At $\mu=\bar{\mu}_1=1+\sqrt{5}$ this two-cycle enjoys maximum stability. More generally: A periodic point x* of period p is said to be superstable, i.e. has maximum stability, if

$$\frac{d}{dx} f^p(x,\mu)\,|_{x*} = 0 \qquad (2.7)$$

This means that the critical point x_c belongs to the p-cycle of x*. For $\mu=\mu_2$ the two-cycle bifurcates and becomes unstable to give rise to a stable orbit of period four, which in turn becomes superstable for $\mu=\bar{\mu}_2$ and so on.

Generally for $\mu_k<\mu<\mu_{k+1}$ a typical asymptotic orbit of the system (2.1) is a stable 2^k-cycle which appears through a period-doubling bifurcation for $\mu=\mu_k$ out of a 2^{k-1}-cycle. For $\mu=\bar{\mu}_k$ this period 2^k is superstable and it bifurcates for $\mu=\mu_{k+1}$ to create a stable orbit of period 2^{k+1}. The doubling mechanism is sketched in Fig. 2.2 which shows a section of the 2^k-fold iterate f^{2^k} of f and its fixed points near the critical point x_c. For $\mu<\mu_k$ there is only one stable fixed point of the

2^{k-1}-fold iterate $f^{2^{k-1}}$ of f which of course is a stable fixed point of f^{2^k} as well. When μ exceeds μ_k two new and stable fixed points of f^{2^k} are born while the previous one is now unstable. An equivalent process happens for all of the 2^{k-1} fixed points of $f^{2^{k-1}}$ such that altogether 2^k new fixed points appear which form a stable 2^k-cycle.

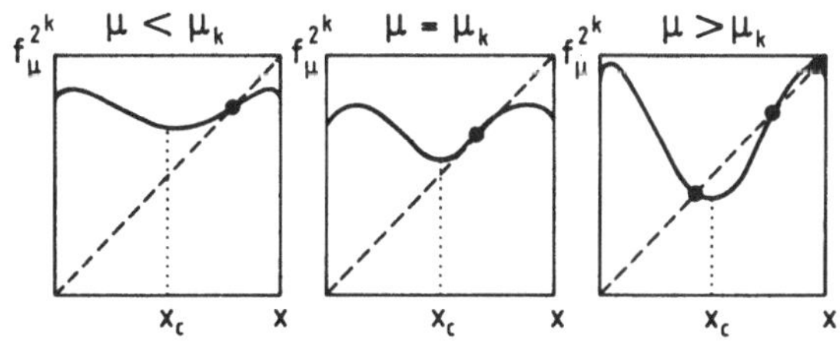

Fig. 2.2 Mechanism of period-doubling bifurcations for a section of the 2^k-fold iterated map.

The parameter sequences μ_k and $\bar{\mu}_k$ converge geometrically towards a limit μ_∞.[3] The rate of convergence δ was shown to have the same value for all dynamical systems belonging to the universality class[1,2] and hence δ is a universal number for the entire class.

$$\lim_{k\to\infty} \frac{\mu_k-\mu_{k-1}}{\mu_{k+1}-\mu_k} = \lim_{k\to\infty} \frac{\bar{\mu}_k-\bar{\mu}_{k-1}}{\bar{\mu}_{k+1}-\bar{\mu}_k} = \delta = 4.66920.... \tag{2.8}$$

From this equation one can derive the following asymptotic relations

$$\begin{aligned} \mu_\infty-\mu_k &= A\,\delta^{-k} \quad A = \text{const} > 0, \quad k\to\infty \\ \mu_\infty-\bar{\mu}_k &= \bar{A}\,\delta^{-k} \quad \bar{A} = \text{const} > 0, \quad k\to\infty \end{aligned} \tag{2.9}$$

When the parameter μ exceeds μ_∞ one finds different types of dynamical behavior: on one hand there exists a set of paramter values of non-zero Lebesgue measure where the system shows chaotic motion. This means that a typical trajectory of Eq. (2.1) is aperiodic and initially neighbouring orbits diverge exponentially in time. From Fig. 2.1 it can be seen that there seemingly exist parameter values above μ_∞ where the orbit densely fills entire subintervals in x. More precisely: one finds a parameter sequence $\tilde{\mu}_k > \mu_\infty$ such that for $\tilde{\mu}_{k+1} < \mu < \tilde{\mu}_k$ the dynamics of Eq. (2.1) is characterized by the existence of 2^k disjoint subintervals which are periodically visited, whereas the motion inside the subintervals is chaotic. These subintervals are often called chaotic bands. The sequence $\tilde{\mu}_k$ then gives the parameter values where 2^k bands merge to 2^{k-1} bands.

These band merging parameters will be of particular importance in the next two sections. The $\tilde{\mu}_k$ converge towards μ_∞ from above and satisfy a relation analogous to (2.9):

$$\tilde{\mu}_k - \mu_\infty = \tilde{A}\,\delta^{-k} \quad \tilde{A} = \text{const} > 0, \quad k\to\infty \qquad (2.10)$$

if we looked very carefully at Fig. 2.1, we would find that the entire parameter regime above μ_∞ is crisscrossed by an infinity of narrow parameter intervals where the asymptotic motion of (2.1) is periodic. We call these intervals periodic windows. In fact, although the set of μ-values where one finds chaos has non-vanishing Lebesgue measure, it contains no intervals. A periodic window occurs after a stable periodic orbit has been created by a so-called tangent bifurcation, a process sketched in Fig. 2.3 which shows a section of the p-fold iterate f^p of f. When μ exceeds the parameter μ_t p of the 2^p-1 extrema of f^p cross the bisector at the same time thus giving rise to 2p new fixed points of f^p which form two p-cycles, one stable and one unstable. Fig. 2.3 shows this process for the extremum at $x=x_c$. We call a stable p-cycle which appeared through a tangent bifurcation a fundamental periodic orbit of period p.

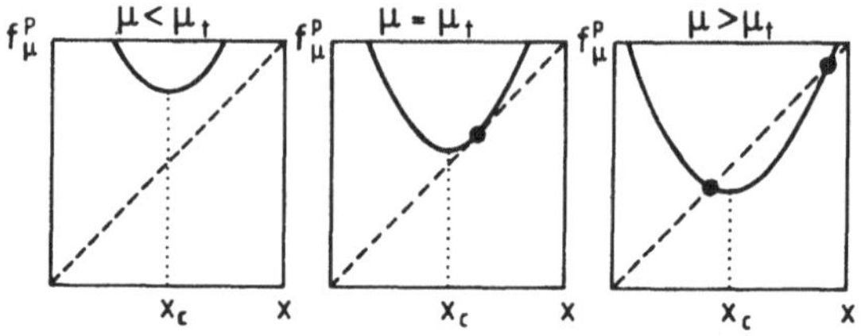

Fig. 2.3 Mechanism of tangent bifurcations for a section of the p-fold iterated map.

Each fundamental periodic orbit is followed by a cascade of period-doubling bifurcations of its own which are followed by chaotic bands. These period doublings and band mergings again follow Feigenbaum's law. Fig. 2.4 shows a magnification of Fig. 2.1 for $3.82<\mu<3.87$ containing the periodic window of period 3.

The universality can be understood in terms of a renormalization transformation. It is based on the idea to transform the 2^k-fold iterate of f such that it looks very much like f and belongs to the same universality class. If we restrict ourselves to maps which are symmetric with respect to the origin this transformation is simply a rescaling by a factor of α^k. Feigenbaum showed that in the limit of large k the scaling factor also becomes a universal number[1] which has the value $\alpha=-2.5029...$ and that the function

$$f^*(x) := \lim_{k\to\infty} \alpha^k f^{2^k}(\alpha^{-k}x, \mu_\infty) \qquad (2.11)$$

is a universal function for the entire class of dynamical models.

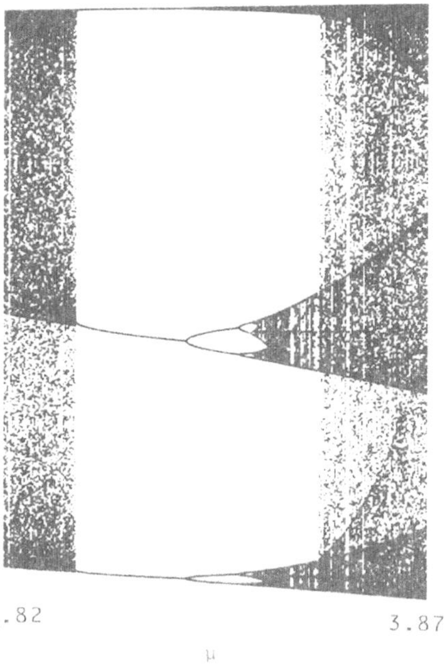

Fig. 2.4

Magnification of Fig. 2.1 for $3.82<\mu<3.87$.

3. FINE STRUCTURE OF THE CHAOTIC REGIME IN PERIOD-DOUBLING SYSTEMS

As mentioned before there is an infinity of periodic windows within the chaotic regime. We have studied the question as to where these periodic windows are located.[10] Previously, mostly qualitative statements were made like in a paper by Metropolis, Stein and Stein[11], which studied the order in which the periods are arranged. We have now investigated quantitatively where fundamental periods occur, where they accumulate and with which bifurcation rates they accumulate. We will first present the phenomena together with numerical results. A theoretical analysis will be given in the next section.

Let us consider in general a region where 2^k bands are present, i.e. $\tilde{\mu}_{k+1}<\mu<\tilde{\mu}_k$. We found that inside this region there exists a sequence of parameter values $\mu_{k,q}$ where the system (2.1) has a superstable fundamtental orbit of period $q\cdot 2^k$ $(q=3,4,5,...)$. This sequence converges towards $\tilde{\mu}_k$ and satisfies the following scaling law:

$$\tilde{\mu}_k - \mu_{k,q} = C_k\, \gamma_k^{-q} \qquad C_k = \mathrm{const} > 0, \quad q\to\infty \qquad (3.1)$$

In the limit of large k the bifurcation rate γ_k approaches a value γ which will turn out to be a universal number for the entire universality class.

$$\gamma := \lim_{k\to\infty} \gamma_k = 2.94805\ldots \qquad (3.2)$$

The validity of Eq. (3.1) is demonstrated numerically for the logistic map in Fig. 3.1. The distance from the accumulation point is shown on a logarithmic scale as a function of q. In agreement with the geometric convergence of Eq. (3.1) the points lie close to straight lines even for small q. The slopes are determined by $-\ln \gamma_k$ and approach a constant $-\ln \gamma$ with increasing k. One observes that also the vertical spacing approaches a constant $\ln \delta$. The prefactor C_k of Eq. (3.1) may therefore be written

$$C_k = C\,\delta^{-k} \qquad C = \mathrm{const} > 0, \quad k\to\infty \qquad (3.3)$$

This can be easily understood in the renormalization picture:

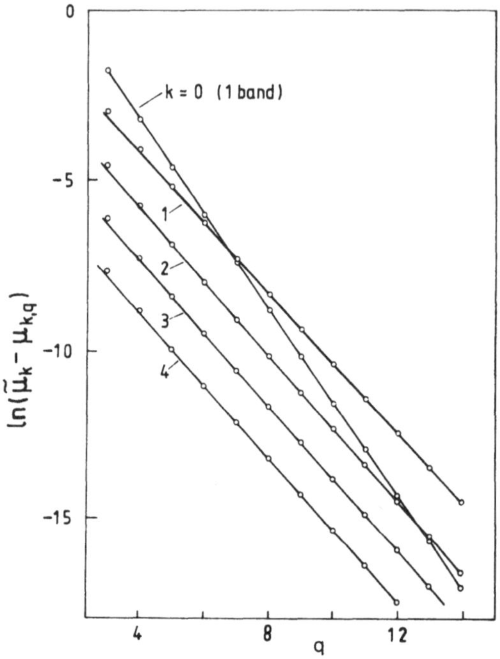

Fig. 3.1 Numerical test of Eqs. (3.1,4) for the logistic map for different numbers of bands 2^k. The slope of the lines approaches $-\ln \gamma$, their vertical spacing $\ln \delta$.

if for $\mu_{k,q}$ we have a superstable $q\cdot 2^k$-cycle the renormalization leads to a $q\cdot 2^{k-1}$-cycle. The parameter distance between these two periods must asymptotically scale with Feigenbaum's δ as well as the sequence $\tilde{\mu}_k$ (Eq. (2.10)). We may thus write

$$\tilde{\mu}_k - \mu_{k,q} = C\,\delta^{-k}\,\gamma^{-q} \qquad k,q\to\infty \qquad (3.4)$$

$$\mu_{k,q} - \mu_\infty = (\tilde{A} - C\gamma^{-q})\,\delta^{-k} \qquad k,q\to\infty \qquad (3.5)$$

The latter is a two-fold scaling law involving the two universal numbers δ and γ. For fixed q the doubling $q\cdot 2^{k-1} \to q\cdot 2^k$ of fundamental orbits when going from a 2^{k-1} to a 2^k-band region follows the rate δ. On the other hand for fixed k incrementation by 2^k, i.e. $q\cdot 2^k \to (q+1)\,2^k$ follows the rate γ on a smaller scale.

Table 3.1 Numerical and analytic values of γ_k in the 2^k-band region for various maps $f(x,\mu)$.

2^k	$\mu(x-x^2)$ numer.	$\mu(x-x^2)$ anal.	$xe^{\mu(1-x)}$ numer.	$\mu x(1-x^2)$ numer.
1	4.0000	4	...	2.5981
2	2.8177	2.81761	3.3605	2.9646
4	2.9634	2.96332	2.8943	2.9472
8	2.9463	2.94625	2.9549	2.9481
16	2.9483	2.94827	2.9472	2.9481
32	2.9481	2.94802	2.9482	2.9481

We have obtained similar figures as Fig. (3.1) for other maps of the same class and have collected the values of γ_k in table 3.1. One observes that in the one and two-band region the values are still different but approach the same value 2.9481 with a rapid convergence.

Before we present the theoretical analysis we note that there are still other sequences obeying scaling laws analogous to those above: not only is $\tilde{\mu}_k$ an accumulation point for fundamental periods $q\cdot 2^k$ from below, but also from above there is an accumulation of fundamental periods $(2q-1)\cdot 2^{k-1}$ $(q=2,3,4,\ldots)$. The rate of accumulation is again γ_k. Moreover there are parameter sequences where chaotic behavior prevails, which have analogous scaling relations. More precisely, we have looked for parameters where an iterate of the critical point x_c ends on an unstable periodic point. There, according to a theorem of Misiurewicz the system shows chaotic behavior with the existence of an ergodic absolutely continuous invariant measure.[12] Let us denote by $\mu'_{k,q}$ the parameter where the $q\cdot 2^k$-th iterate $(q>3)$ of x_c falls on a given unstable cycle e.g. of period 2^k. This parameter sequence satisfies the following asymptotic relations

$$\tilde{\mu}_k-\mu'_{k,q} = C'\delta^{-k}\gamma^{-q} \quad C' = \text{const} > 0, \quad k,q\to\infty \tag{3.6}$$

or equivalently

$$\mu'_{k,q}-\mu_\infty = (\tilde{A}-C'\gamma^{-q})\delta^{-k} \qquad k,q\to\infty \tag{3.7}$$

Again similar sequences of chaotic points also accumulate at $\tilde{\mu}_k$ from above.

The band merging parameters $\tilde{\mu}_k$ thus are accumulation points for both types of behavior: fundamental periodic orbits and chaotic behavior. The sequences above and below $\tilde{\mu}_k$ have a unique accumulation rate

$\gamma_k \to \gamma$. Applying an analysis which we describe in the following section we have derived an analytic expression for the γ_k:

$$\gamma_k = \frac{d}{dx} f^{2^k}(x,\tilde{\mu}_k)|_{x^*_{k-1}} = [\frac{d}{dx} f^{2^{k-1}}(x,\tilde{\mu}_k)|_{x^*_{k-1}}]^2 \qquad (3.8)$$

where x^*_{k-1} denotes a point on the unstable 2^{k-1}-cycle for $\mu=\tilde{\mu}_k$. Therefore if the band merging parameter $\tilde{\mu}_k$ is known, one can analytically calculate γ_k for any finite k. Table 3.1 includes these analytic results together with the numerically determined values for the logistic map. Going to larger values of k we obtained the limit $\gamma=2.94805...$

The universality of γ follows from Eq. (3.8). Assuming that $f(x,\mu)$ has its maximum at $x_c=0$ we can introduce a topologically conjugated map

$$G(x) := \alpha^k f^{2^k}(\alpha^{-k}x,\tilde{\mu}_k) \qquad (3.9)$$

which leaves γ_k unchanged when substituted for f^{2^k} in Eq. (3.8). We know from Eq. (2.11) that in the limit $k\to\infty$ $G(x)$ becomes the universal function $f^*(x)$ and therefore $\gamma=\lim \gamma_k$ is a universal constant.

4. THEORETICAL ANALYSIS OF THE FINE STRUCTURE

In order to make the following considerations more transpartent we first give a more detailed description of the properties of chaotic bands.

Let us assume that there exist 2^k bands, i.e. 2^k disjoint intervals which are periodically visited by a trajectory whereas the motion inside these intervals is chaotic. The 2^{k+1} boundaries of the chaotic bands are given by the iterates of x_c, i.e. the set of points

$$\{y_i := f^i(x_c,\mu)\,|\,i=1,\dots,2^{k+1}\} \qquad (4.1)$$

For $\mu=\tilde{\mu}_k$ the 2^k bands merge into 2^{k-1} which means that 2^k of these boundaries given by $f^j(x_c,\mu)$ $(j=2^k+1,\dots,2^{k+1})$ intersect each other pairwise. This means that at $\tilde{\mu}_k$ these iterates f^j recur and therefore form a periodic orbit, more precisely an unstable cycle of period 2^{k-1}. Thus this unstable cycle may be obtained by e.g.

$$x^*_{k-1} = f^{2^{k+1}}(x_c,\tilde{\mu}_k) \qquad (4.2)$$

and its iterates. In Fig. 2.1, e.g. the point x where at $\tilde{\mu}_1$ the two bands merge is the unstable cycle of period 1.

In order to present the main ideas of the theory let us for the sake of transparency restrict our considerations to the one-band-region. Fig. 4.1 shows the critical point x_c and its iterates near the

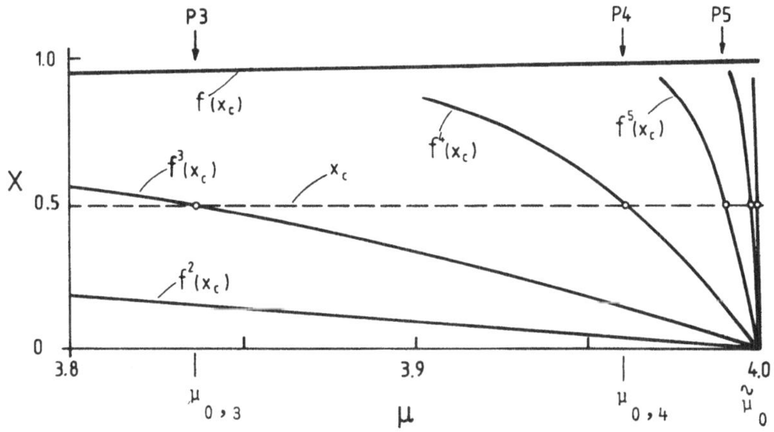

Fig. 4.1 The critical point x_c and its iterates as a function of μ for the logistic map. P3, P4 and P5 denote superstable fundamental periods.

end of the one-band-regime for the logistic map. The first and second iterate of x_c determine the upper and lower bound of the chaotic band respectively. We denote these bounds

$$\begin{aligned} x_{max} &= f(x_c,\mu) \\ x_{min} &= f^2(x_c,\mu) \end{aligned} \tag{4.3}$$

The parameter $\tilde{\mu}_o$ is defined as the point where the band completely fills the interval [0,1]. For the logistic map this is the case for $\mu=\tilde{\mu}_o=4$ where we have (see Eq. (2.2) and Fig. 4.1)

$$x^*_{tr} := 0 = f^q(x_c,\tilde{\mu}_o) \quad (q\geq 2) \tag{4.4}$$

here x^*_{tr} denotes the trivial fixed point of the logistic map. This equation is an example of Eq. (4.2).

Let $\{x^*_i | i=1,2,\ldots,p\}$ be a periodic orbit of period p; to find its parameter of superstability we must solve Eq. (2.7), i.e.

$$\frac{d}{dx} f^p(x,\mu)|_{x^*_j} = 0 \quad j\in\{1,2,\ldots,p\} \tag{4.5}$$

using the fact that x_j belongs to a p-cycle we can rewrite this by means of the chain rule

$$\prod_{i=1}^{p} \frac{d}{dx} f(x,\mu)|_{x^*_i} = 0 \tag{4.6}$$

since f has vanishing slope only for $x=x_c$ we conclude that there exists a j with $1\leq j\leq p$ such that

$$x^*_j = x_c \tag{4.7}$$

Thus it is sufficient to solve

$$x_c = f^p(x_c,\mu) \tag{4.8}$$

After these preliminaries we first show the *existence of a strictly increasing convergent sequence of parameters* $\mu_{o,q}$ *inside the one-band-regime where superstable fundamental orbits of period* q $(q=3,4,5,\dots)$ *occur for the last time below* $\tilde{\mu}_o$.

Let us assume that at $\mu=\mu_{o,p}$ there is the fundamental superstable p-cycle closest to $\mu=\tilde{\mu}_o$, i.e. for $\mu_{o,p}<\mu<\tilde{\mu}_o$ there are no more superstable p-cycles. According to Eq. (4.8)

$$x_c = f^p(x_c,\mu_{o,p})$$

it then follows from Eq. (4.3) that

$$x_{max} = f(x_c,\mu_{o,p}) = f^{p+1}(x_c,\mu_{o,p}) > x_c\ . \tag{4.9}$$

On the other hand we know from Eq. (4.4) that at $\tilde{\mu}_o$

$$x^*_{tr} = 0 = f^{p+1}(x_c,\tilde{\mu}_o) < x_c \tag{4.10}$$

Since $f^{p+1}(x_c,\mu)$ is an analytic function of μ and according to the last two equations has a value $>x_c$ at $\mu_{o,p}$ and a value $<x_c$ at $\tilde{\mu}_o$ there must be a parameter $\mu=\mu_{o,p+1}$ in between, where

$$f^{p+1}(x_c,\mu_{o,p+1}) = x_c \tag{4.11}$$

i.e. there must be a $\mu_{o,p+1}$ where period p+1 is superstable. The argument can be repeated for periods p+2, p+3, ... and thereby establishes the existence of the sequence $\mu_{o,p}$ $(q\geq p)$ provided that $\mu_{o,p}$ as assumed above exists. This is the case since in the one-band region there is exactly one superstable cycle of period 3. This proves the existence of a strictly increasing sequence of parameters $\mu_{o,p}$ $(q\geq 3)$ where a superstable period q occurs for the last time before $\tilde{\mu}_o$.

We now show that these periods q are fundamental periods, i.e. do not arise through a subharmonic bifurcation from period q/2. The statement is of course true for all odd q. It is known that inside the chaotic regime there are no stable fixed points or period-two-cycles. For $q\geq 4$ we can use

$$q/2 < q-1 < q\ . \tag{4.12}$$

If period q at $\mu_{o,q}$ were a subharmonic of period q/2 it would directly follow a parameter interval of a stable period q/2. This cannot be the case since according to the proof above and inequality (4.12) there cannot be a stable period q/2 above $\mu_{o,q-1}$.

The convergence of the sequence $\mu_{o,q}$ simply follows from its monotonicity and boundedness.

Generalizations of the above proofs pertaining to the other parameter sequences mentioned in Sect. 3 are rather straightforward: using the fact that for $\mu_{o,q}$

$$x_{max} = f^{q+1}(x_c,\mu_{o,q}) > x^*$$

where x^* denotes the unstable (non-trivial) fixed point of f and using Eq. (4.4) we can similarily show the existence of a convergent sequence of parameter values $\mu'_{o,q}$ between $\tilde{\mu}_1$ and $\tilde{\mu}_o$ which give the solutions of

$$x^* = f^q(x_c,\mu) \quad (q\geq 4)$$

i.e. where chaotic behavior is found. Furthermore since for $\mu=\mu_{o,3}$

$$x_{min} = f^2(x_c,\mu_{o,3}) = f^5(x_c,\mu_{o,3}) < x_c$$

and

$$x^* = f^5(x_c,\tilde{\mu}_1) > x_c$$

we can inductively show that there is a monotonically decreasing sequence of parameters between $\tilde{\mu}_1$ and $\mu_{o,3}$ where superstable fundamental (2q-1)-cycles occur. All the statements can also be generalized to a parameter region where 2^k bands are present. From the renormalization properties of f it is clear that in this regime one finds a situation analogous to that of Fig. 4.1 if we replace f by f^{2^k}, x^*_{tr} by x^*_{k-1} and $\tilde{\mu}_o$ by $\tilde{\mu}_k$.

We now proceed to the analytic calculation of γ_k. According to Eq. (3.1) we need the distance of $\mu_{k,q}$ (where $f^{q\cdot 2^k}(x_c,\mu)$ intersects x_c) from $\tilde{\mu}_k$. Fig. 4.1 illustrates that in the limit of large q this distance is inversely proportional to the slope $(d/d\mu)f^{q\cdot 2^k}(x_c,\mu)$ at $\tilde{\mu}_k$. Introducing the abbreviation

$$F(x,\mu) = f^{2^k}(x,\mu)$$

we can state this in the form

$$\Delta\mu_{k,q} := \tilde{\mu}_k-\mu_{k,q} \propto \{\frac{d}{d\mu} F^q(x_c,\mu)|_{\tilde{\mu}_k}\}^{-1} \tag{4.13}$$

Using the chain rule, the slope can generally be written

$$\frac{d}{d\mu} F^q(x,\mu) = \frac{\partial}{\partial\mu} F|_{F^{q-1},\mu} + \frac{\partial}{\partial x} F|_{F^{q-1},\mu} \cdot \frac{d}{d\mu} F^{q-1}(x,\mu) \tag{4.14}$$

where we partially dropped the arguments (x,μ). Eq. (4.14) is a recursion relation which may be used to prove the following explicit relation by mathematical induction. As it is straightforward this proof is left out here.

$$\frac{d}{d\mu} F^{q+1}(x,\mu) = \{ \prod_{j=1}^{q} \frac{\partial}{\partial x} F|_{F^j,\mu} \} \frac{d}{d\mu} F(x,\mu)$$

$$+ \sum_{j=1}^{q} \{ \prod_{l=j+1}^{q} \frac{\partial}{\partial x} F|_{F^l,\mu} \} \frac{\partial}{\partial \mu} F|_{F^j,\mu} \tag{4.15}$$

here we use the convention that for $n_1 > n_2$

$$\prod_{j=n_1}^{n_2} a_j := 1$$

Let now $\mu=\tilde{\mu}_k$ and $x=x_m:=F(x_c,\mu)$. There, all the iterates F^j and F^l are equal as according to Eq. (4.2) they end on the unstable point x^*_{k-1} of period 2^{k-1}

$$F^{q+1}(x_c,\tilde{\mu}_k) = F^q(x_m,\tilde{\mu}_k) = x^*_{k-1} \quad (q \geq 1) \tag{4.16}$$

Eq. (4.15) thus simplifies considerably. We abbreviate

$$a := \frac{\partial}{\partial \mu} F|_{x^*_{k-1},\tilde{\mu}_k}$$

$$b := \frac{\partial}{\partial x} F|_{x^*_{k-1},\tilde{\mu}_k}$$

$$d_q := \frac{d}{d\mu} F^q(x_m,\mu)|_{\tilde{\mu}_k} = \frac{d}{d\mu} F^{q+1}(x_c,\mu)|_{\tilde{\mu}_k}$$

Then for $x=x_m$ Eq. (4.15) together with Eq. (4.16) yields

$$d_{q+1} = b^q d_1 + \sum_{j=1}^{q} a b^{q-j}$$

$$= b^{q+2}\left(\frac{d_1}{b^2} + \frac{a}{b^2} \sum_{j=1}^{q} b^{-j}\right) . \tag{4.17}$$

since

$$b = \frac{\partial}{\partial x} F|_{x^*_{k-1},\tilde{\mu}_k} = \frac{d}{dx} f^{2^k}(x,\tilde{\mu}_k)|_{x^*_{k-1}}$$

and due to the fact that x^*_{k-1} lies on the unstable 2^{k-1}-cycle we arrive at:

$$b = \left[\frac{d}{dx} f^{2^{k-1}}(x,\tilde{\mu}_k)|_{x^*_{k-1}}\right]^2 > 1 \tag{4.18}$$

this guarantees the convergence of the sum

$$\sum_{j=1}^{q} b^{-q} = \frac{1}{b} \sum_{j=0}^{q-1} b^{-j} = \frac{1-b^{-q}}{b-1} \xrightarrow[q\to\infty]{} \frac{1}{b-1} \tag{4.19}$$

Thus we finally have:

$$d_q = \frac{d}{d\mu} F^{q+1}(\tilde{x}_c,\mu)|_{\tilde{\mu}_k} \underset{q\to\infty}{\longrightarrow} \left(\frac{d_1}{b^2} + \frac{a}{b^3-b^2}\right) b^{q+1} \tag{4.20}$$

and due to Eq. (4.13)

$$\Delta\mu_{k,q} \propto \frac{1}{d_{q-1}} \sim b^{-q} \tag{4.21}$$

we therefore must identify b with γ_k, i.e.

$$\gamma_k = b = \frac{d}{dx} f^{2^k}(x,\tilde{\mu}_k)|_{x^*_{k-1}} \tag{4.22}$$

5. SELF-GENERATED DIFFUSION - AN EXAMPLE

The remaining two sections will deal with a phenomenon that may occur beyond period-doubling, the onset of a self-generated diffusive motion.[6] It is studied in 1d discrete systems which represent simple models for strongly dissipative physical systems. We expect in particular that the critical properties of these physical systems are the same as those of the 1d models. This is because at the critical point the relevant time scale usually diverges (critical slowing down), and for $t\to\infty$ the dissipation reduces the dimension of phase space. The results presented in the next section are therefore expected to be universal also for more complex dissipative physical systems which are hardly accessible analytically.

As an example we present numerical results for the map

$$x_{t+1} = x_t - \mu \sin(2\pi x_t) \tag{5.1}$$

which is shown in Fig. 5.1 for two different parameters μ. Below μ_c=0.73264 the map exhibits period-doubling and chaotic bands in the interval $-1/2\leq x\leq 1/2$ similar as in Sections 2-4. Above μ_c an orbit originating in this interval can also leave the interval and perform a random walk on the real axis. This is demonstrated in Fig. 5.2 which shows the mean square displacements $\langle(x_t-x_o)^2\rangle$ as a function of time obtained from a computer experiment. The average is taken over 2000 values of x_o in the interval [-1/2,1/2]. The curves increase linearly in time thereby indicating a diffusive process. The diffusion

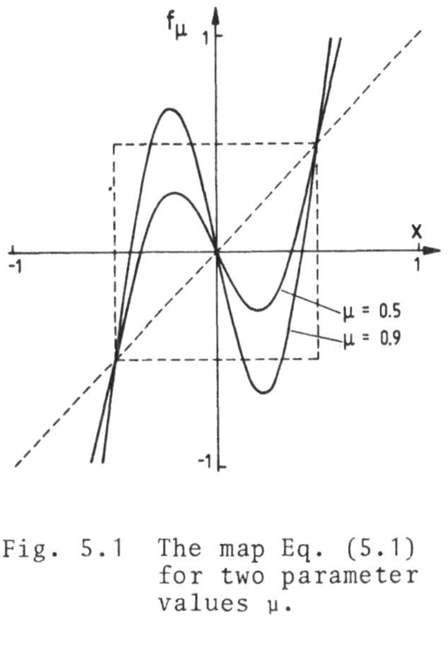

Fig. 5.1 The map Eq. (5.1) for two parameter values μ.

coefficient D equals half the slope. D vanishes below μ_c as can be easily understood from Fig. 5.1. For μ=0.5 all successors of x_0 in [-1/2,1/2] remain in this interval and diffusion cannot occur. It should be emphasized that this diffusion process is generated by a purely deterministic system in the absence of random external fluctuations. This is in contrast to conventional diffusion processes, where in microscopic models (i.e. Langevin equations) one assumes random external forces generating the diffusive motion. The most familiar example is the Brownian motion of a heavy particle which is subject to random collisions with surrounding molecules of a liquid. In the next section we will also investigate the effect of random external fluctuations.

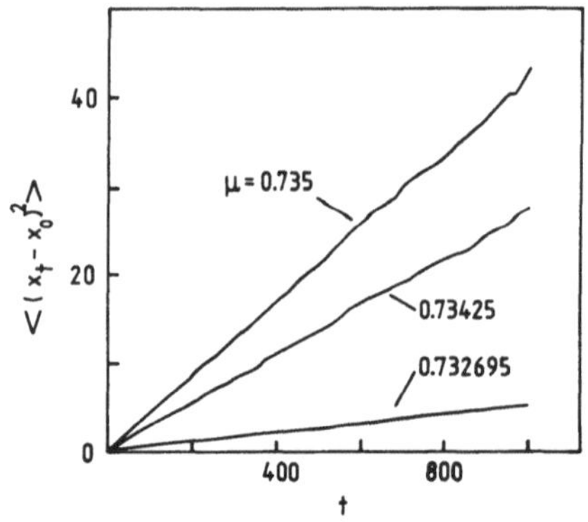

Fig. 5.2 Mean-square displacements as a function of time for Eq. (5.1)

6. UNIVERSAL CRITICAL BEHAVIOR AT THE ONSET OF DIFFUSION

In order to study universal properties we consider general universality classes of models instead of the specific example Eq. (5.1)

$$x_{t+1} = f(x_t) + \sigma\xi_t \tag{6.1}$$

For convenience we immediately consider the presence of external fluctuations $\sigma\xi_t$ of strength $\sigma \ll 1$. The deterministic case is recovered for $\sigma \to 0$. The random variable ξ is assumed to have standard deviation 1, mean zero and a Gaussian distribution $v(\xi)$. The map f(x) is assumed to have the following properties: f(x) is an odd function

$$f(-x) = -f(x) \ , \tag{6.2}$$

f(x)-x is periodic with period 1

$$f(x+n) = n + f(x) \ , \tag{6.3}$$

and f(x) has a relative maximum per period at x_c+n with $-1/2<x_c<0$. The vicinity of the maximum is assumed to be of the form

$$f(x) = a(\mu) - b(\mu)|x-x_c|^z \tag{6.4}$$

where a and b are coefficients depending on a single parameter μ and the exponent z>0 determines the type of maximum and distinguishes different universality classes. Only in the case of external noise we require an additional property

$$|1/2 - f(x)| \ll 1 \Rightarrow |x-x_c| \ll 1 \quad (6.5)$$

Numerical results indicate that the latter condition might be dispensable.

We will make statements about the critical behavior of these universality classes. In the deterministic case ($\sigma=0$), of course, these statements are only valid on the set of those parameters μ for which no stable periodic orbit exists. The study of the external noise case is important because in most physical situations external noise cannot be totally eliminated. The occurrence of irregular motion is then not astonishing, however, for weak external fluctuations ($\sigma\ll1$) the intrinsically generated irregularity is still observable. This case is also of interest because the dependence on the external fluctuations will turn out to exhibit additional scaling properties.

We introduce unit cells of length 1 centered at x=l. We consider the conditional probability $\rho_t(x)dx$ of finding a value between x and x+dx at time t if the initial value x_o was in the 0th cell ($|x_o|<1/2$). Conservation of probability for the step from t to t+1 implies the following relation

$$\rho_{t+1}(y) = \iint\limits_{-\infty}^{+\infty} \rho_t(x)v(\xi)\delta[y-f(x)-\sigma\xi]dxd\xi \quad (6.6)$$

It means that the probability of finding a value y at time t+1 is equal to the sum of probabilities that at time t the system had values x and ξ such that $y=f(x)+\sigma\xi$. From Eq. (6.6) one can derive a Chapman-Kolmogorov equation as shown by Haken and Mayer-Kress.[13] Here instead we carry out some integrations to derive a master equation. We introduce the integrated probability

$$p_t(l) = \int_{l-1/2}^{l+1/2} \rho_t(x)dx \quad (6.7)$$

which is the probability for a transition from cell 0 to cell l in time t. Integration of Eq. (6.6) and neglect of terms of order $\sigma\exp(-1/2\sigma^2)$ leads to[6]

$$p_{t+1}(l)-p_t(l) = -p_t(l)\frac{1}{2}\int_{l-1/2}^{l+1/2} q(x)\{\mathrm{erfc}[\frac{-r(x)}{\sqrt{2}}]+\mathrm{erfc}[\frac{s(x)}{\sqrt{2}}]\}dx$$

$$+p_t(l+1)\frac{1}{2}\int_{l+1/2}^{l+3/2} q(x)\mathrm{erfc}[\frac{-s(x)}{\sqrt{2}}]dx \quad (6.8)$$

+ (next page)

$$+p_t(l-1)\frac{1}{2}\int_{l-3/2}^{l-1/2} q(x)\mathrm{erfc}[\frac{r(x)}{\sqrt{2}}]dx \qquad (6.8)$$

where

$$r(x) = (l-1/2-f(x))/\sigma \qquad (6.9)$$

$$s(x) = (l+1/2-f(x))/\sigma \qquad (6.10)$$

$$q(x) = \rho_t(x)[\int_{l-1/2}^{l+1/2} \rho_t(x')dx']^{-1} \qquad (6.11)$$

Here erfc(x) denotes the complementary error function. Eq. (6.8) is a discrete version of a master equation, i.e. it is a difference equation in time. In the first term one recognizes the flow out of cell l, in the second and third term the flow into cell l. The transition rates are related to that part of f(x) which maps out of a given unit cell. In Fig. 5.1 this is the part lying outside the dashed square. Therefore it is clear that the shape of f(x) close to its maximum is of particular importance. Such master equations are often set up phenomenologically. Here, however, we have derived it from the exact equation (6.6).

For the diffusion coefficient D only the long time behavior of $p_t(l)$ is important. Note that the relevant time scale (residence time in a unit cell) can be made arbitrarily long by letting $D\to 0$ (i.e. in the critical region, the region of interest). In Eq. (6.11) we have assumed that for long times the distribution within a cell, the ratio q(x) becomes independent of t and l, i.e. that within a cell an invariant distribution exists. This can only be verified individually for each system in consideration and may be done numerically. We will show at the end of this section that for the map Eq. (5.1) a distribution is approached which is approximately $q(x)=\pi^{-1}(1/4-x^2)^{-1/2}$ (for $|x|<1/2$).

We can solve the master equation (6.8) for $p_t(l)$ using the symmetry properties Eqs. (6.2-3)

$$p_t(l) = \frac{1}{N}\sum_k \{1-(1-\cos k)\int_{-1/2}^{1/2} q(x)\mathrm{erfc}([1/2-f(x)]/\sqrt{2}\sigma)dx\}^t e^{ikl} \qquad (6.12)$$

where the sum is over N values of k in the first Brillouin zone. This result can be used to compute statistical quantities of interest. From the mean-square displacements the diffusion coefficient follows as

$$D = \frac{1}{2}\int_{-1/2}^{1/2} q(x)\ \mathrm{erfc}\ [\frac{1/2-f(x)}{\sqrt{2}\ \sigma}]dx \qquad (6.13)$$

Near μ_c, which is determined from $a(\mu_c)=1/2$ we can write $a(\mu)=1/2+a'(\mu-\mu_c)$ and $b=b(\mu_c)=\text{const}$. Using Eq. (6.4) we obtain for the deterministic case ($\sigma=0$)

$$D = 2q(x_c)\left(\frac{a'}{b}\right)^{1/z}(\mu-\mu_c)^{1/z} + O(\Delta\mu^{3/z}) \tag{6.14}$$

This means that D grows like an order parameter with a universal critical exponent $1/z$. More generally in the critical region, i.e. for $|\Delta\mu|=|\mu-\mu_c|\ll\mu_c$ and $\sigma\ll 1$ neglecting terms of order $[\sigma^2+\sigma(\mu-\mu_c)]^{1/z}$ the diffusion coefficient can be written

$$D = \sigma^{1/z}\, d\left(\frac{\Delta\mu}{\sigma}\right). \tag{6.15}$$

with

$$d\left(\frac{\Delta\mu}{\sigma}\right) = \frac{q(x_c)}{z}\left(\frac{\sqrt{2}}{b}\right)^{1/z}\int_{-\frac{a'}{\sqrt{2}}\frac{\Delta\mu}{\sigma}}^{\infty} \operatorname{erfc}(u)\left[u+\frac{a'}{\sqrt{2}}\frac{\Delta\mu}{\sigma}\right]^{\frac{1}{z}-1}du \tag{6.16}$$

The function d is a scaling function as it only depends on the ratio of $\Delta\mu$ and σ and remains invariant when both are scaled simultaneously. It is universal (except for nonuniversal prefactors as usually) because for different maps it only depends on the exponent z distinguishing the universality classes. This situation is analogous to the one known e.g. from magnetic phase transitions. D takes the role of the magnetization, μ the role of the temperature and σ the role of the applied magnetic field. For $\mu=\mu_c$ Eq. (6.15) easily yields the critical exponent for the dependence on the external fluctuations

$$D = \sigma^{1/z}\, d(o) \tag{6.17}$$

The universal scaling function d describes the μ and σ dependence in the critical region for any dynamical system belonging to the universality classes Eqs. (6.2-5). Equation (6.16) is an analytic expression which can be computed easily. In order to illustrate our results and to test the accuracy of the scaling behavior we have carried out a computer experiment (Fig. 6.1). For the map

$$f(x) = \mu\sqrt{27}\,\left(2x^3-\tfrac{1}{2}x\right) \qquad (|x|<1/2) \tag{6.18}$$

we have measured the diffusion coefficient D for 3 different noise levels σ and 100 values of $\Delta\mu/\sigma$ between -1 and +1. In the figure D and $\Delta\mu$ are scaled in such a way that according to Eq. (6.15) the scaling function d should result. Indeed the data points for the 3 experiments form a single curve and thereby experimentally show the existence of the scaling function. The line shown in Fig. (6.1) is the analytic result Eq. (6.16) and lies within the region of the experimental data. Note that this agreement is obtained without adjustable parameters.

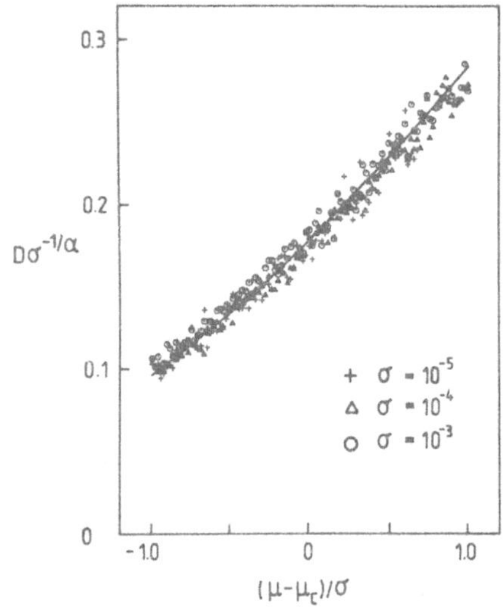

Fig. 6.1 Measurement of the diffusion coefficient D in a computer experiment demonstrating the existence of a scaling function. Full line: Analytic expression for the universal scaling function Eq. (6.16).

A problem which we had postponed so far is the existence of an invariant distribution q(x) in the deterministic case σ=o. This must be investigated for specific maps. Let us consider our example Eq. (5.1) at μ_c. We perform a topological conjugation of this map f

$$u_{t+1} = h^{-1}(f(h(u_t))) = g(u_t) \qquad (6.19)$$

using the transformation

$$x = h(u) = -\frac{1}{2}\cos(\pi u) \qquad (0\leq u\leq 1) \ . \qquad (6.20)$$

The conjugated map g is

$$u_{t+1} = \frac{1}{\pi}\arccos\{\cos(\pi u_t) - 2\mu_c \sin[\pi\cos(\pi u_t)]\} \qquad (6.21)$$

The latter map is a piecewise C^∞function and with the exception of the extrema its slope is everywhere $|g'(u)|>2$ as illustrated in Fig. 6.2

Then, applying a theorem of Lasota and Yorke[14] we conclude that the map g has an invariant distribution $q_g(u)$ and thus it follows that f has an invariant distribution as well, which we denote by q(x).

The topological conjugation also allows us to calculate the distribution function q(x). The map g (see Fig. 6.2) is very close to a piecewise linear map which exactly folds the interval $0\leq u\leq 1$ three times onto itself. The latter map would have a constant distribution as may be shown by solving the Frobenius-Perron equation. Therefore in a good approximation we can use $q_g(u)=1=$const. for the map g. This approximation is compared with numerical results in Fig. 6.3. The condition of conservation of probability under conjugation then gives us an approximation for q(x) (belonging to the map f)

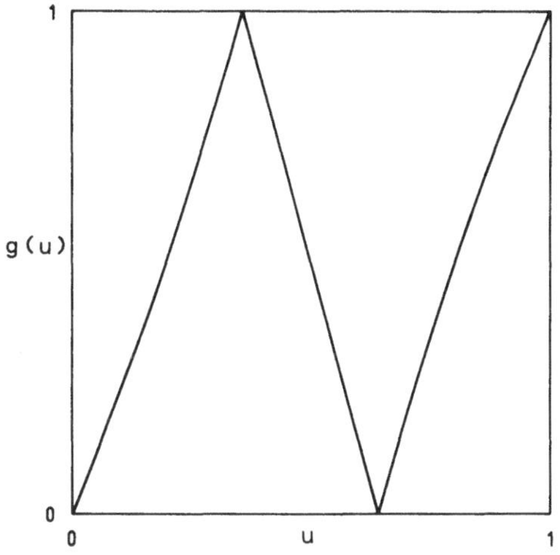

Fig. 6.2 Topologically conjugated map (6.21).

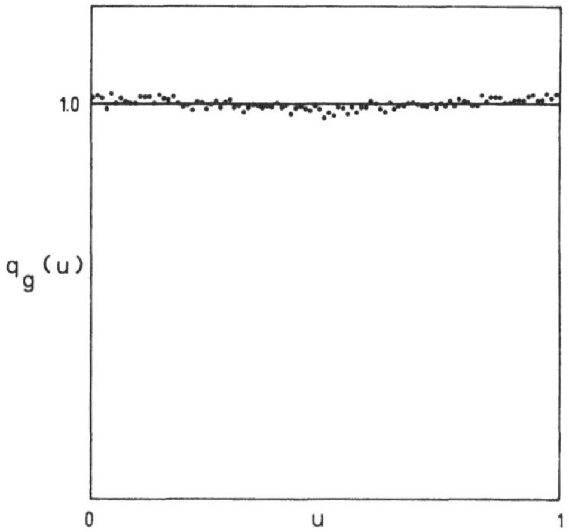

Fig. 6.3 Invariant distribution $q_g(u)$ of the conjugated map (6.21) compared with the approximation $q_g(u)=1$.

$$q(x) = q_g(u)\left|\frac{du}{dx}\right| \tag{6.22}$$

$$\Rightarrow q(x) \approx \left|\frac{du}{dx}\right| = \frac{1}{\pi}(1/4-x^2)^{-1/2}. \tag{6.23}$$

The self-generated diffusion process described here has recently been studied also by Grossmann and Fujisaka[15] and by Schell, Fraser and Kapral[16] in the deterministic case. These authors also discuss other dynamical features exhibited by the map (5.1) like drifting periodic and drifting chaotic orbits.

REFERENCES

1. M.J. Feigenbaum, J. Stat. Phys. 19, 25 (1978) and 21, 669 (1979).
2. P. Coullet and C. Tresser, J. Phys. Coll. (Paris) 39, C5-25 (1978).
3. S. Grossmann and S. Thomae, Z. Naturforsch. 32a, 1353 (1977).
4. P. Manneville and Y. Pomeau, Phys. Lett. 75A, 1 (1979) and Physica 1D, 219 (1980).
5. J.E. Hirsch, B.A. Huberman and D.J. Scalapino, Phys. Rev. 25A, 519 (1982)
6. T. Geisel and J. Nierwetberg, Phys. Rev. Lett. 48, 7 (1982).
7. B.A. Huberman and J.P. Crutchfield, Phys. Rev. Lett. 43, 1743 (1979).
8. A. Libchaber and J. Maurer, J. Phys. Lett. (Paris) 40, 419 (1979).
9. R.M. May, Nature 261, 459 (1976).

10. T. Geisel and J. Nierwetberg, Phys. Rev. Lett. 47, 975 (1981).
11. N. Metropolis, M.L. Stein and P.R. Stein, J. Comb. Theory A15, 25 (1973).
12. M. Misiurewicz, Publ. Math. IHES 53, 17 (1981).
13. H. Haken and G. Mayer-Kress, Z. Phys. B43, 185 (1981).
14. A. Lasota and J.A. Yorke, Trans. Am. Math. Soc. 186, 481 (1973).
15. S. Grossmann and H. Fujisaka, preprint.
16. M. Schell, S. Fraser and R. Kapral, preprint.

SUBHARMONICS AND THE TRANSITION TO CHAOS

Joseph Rudnick

Physics Department
University of California, Santa Cruz
Santa Cruz, California 95064 (U.S.A.)

1. INTRODUCTION

We all know that a sinusoidally driven system will respond both at the frequency of the drive and at harmonics of the driving frequency. Physicists are now becoming accustomed to the notion that a driven system can also respond at subharmonics of the driving frequency. This is not a recent discovery. Over 150 years ago Michael Faraday[1] observed a system responding at one half the frequency of the drive. Electrical engineers have been aware for some time of the possibility of subharmonic generation in certain amplifier circuits.[2] Furthermore there are some nice examples of simple mechanical systems with subharmonic response.[3]

That the possibility of subharmonic response is not a part of the general lore of physics follows from two of its characteristics. First, and most important, response at a subharmonic is not a ubiquitous, or as yet technologically important, phenomenon. Second, the theory of subharmonic response does not develop smoothly from the theory of linear response. An arbitrarily small nonlinearity in a system will lead to response at a harmonic of an arbitrarily weak drive. In the case of a weak nonlinearity one can construct a theory for the harmonic content of the response with the use of perturbative methods, and perturbation calculations are the bread-and-butter work of the theoretical physicist. We all learn to do them in graduate school. You will never, however, see response at a subharmonic arising from a simple perturbation theory; it is a nonperturbative phenomenon. Like most nonperturbative effects it is not a simple manifestation of an all-encompassing generic phenomenon but is an effect that must be anticipated to be investigated theoretically. In this respect subharmonic response is like a soliton or a phase transition.

Subharmonic response especially resembles a phase transition in that it is a threshold phenomenon. You will not see subharmonic response in a driven system unless you drive it hard enough. Of course, if you apply a really strong drive to a system anything can happen, including turbulent, or chaotic, response. Subharmonics play an especially important role in the very interesting dynamical transition from non-chaotic to chaotic behavior in certain systems. One need only think of the transition to chaos via period doubling,[4] the experiments on Couette flow[5] and Rayleigh-Bénard convection[6-8] and the recent variant of Faraday's original investigation[9] to appreciate how essential nonharmonic response is to the study of the preturbulent, or prechaotic state of physical systems.

In what follows we will look at a system having both simple and complex subharmonic response, the parametrically driven pendulum. The behavior of this pendulum will be formulated in terms of a two-dimensional map. Some simple properties of the map will be investigated and others will be pointed to. Then we will consider the spectral signature of period doubling and the structure of the invariant map and its compositions at the accumulation point of the period doubling cascade. This latter discussion supplements reported work performed in collaboration with Michael Nauenberg.[10]

2. SIMPLE SUBHARMONIC RESPONSE: THE PARAMETRICALLY DRIVEN PENDULUM

Imagine a simple pendulum whose point of support is oscillated vertically, as shown in Figure 1.

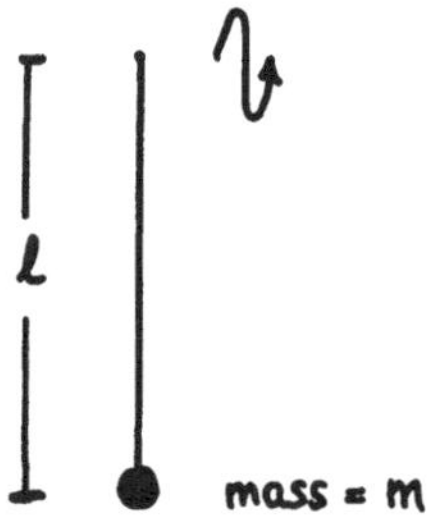

Fig. 1

The elevation of the point of support, $z(t)$, is given by

$$z(t) = -\frac{a}{\omega^2}\cos\omega t \tag{2.1}$$

In the frame of reference moving with the point of support the equation of motion for the angular displacement of the pendulum from the vertical, $\theta(t)$, is

$$\begin{aligned} m\ell\frac{d^2\theta}{dt^2} &= -m\left[g - \frac{d^2z}{dt^2}\right]\sin\theta(t) \\ &= -m[g + a\cos\omega t]\sin\theta(t) \ . \end{aligned} \tag{2.2}$$

If the displacement is small we have $\sin\theta(t) \approx \theta(t)$ and (2.2) becomes

$$\frac{d^2\theta}{dt^2} + \frac{1}{\ell}[g + a\cos\omega t]\,\theta(t) = 0 \; . \qquad (2.3)$$

This method of excitation is called parametric because what we are doing, in effect, is to introduce a time dependence into one of the parameters (the restoring force) in the equation of motion of the pendulum.

One solution to either (2.2) or (2.3) is $\theta(t) = 0$. This solution applies for arbitrary drive amplitude, and it is not too hard to see why. We are not driving the pendulum directly. If we did push it from side to side then it would swing no matter how gently we applied the sideways force. The parametrically driven pendulum, on the other hand, will continue to hang vertically until the drive amplitude passes a threshold. At that point the solution $\theta(t) = 0$ becomes dynamically unstable-small deviations from verticality grow in time.

The onset of instability can be studied by looking at equation (2.3). This equation, the Mathieu equation, is also the Schroedinger equation for a point particle in a one dimensional sinusoidal potential.[11] The two different possibilities for the wave function of such a particle correspond to two different kinds of behavior of the pendulum in the linear regime. The particle can be in a band state with a wave function modulated periodically in space, the periodic modulation corresponding to its crystal momentum. In its corresponding dynamical regime the pendulum will oscillate about its hanging position if perturbed from it, never wandering much further out than its initial conditions place it. The quantum mechanical alternative to a band state is a forbidden state, in the band gap. Here the wave function grows exponentially as one follows it out to infinity. For the pendulum this corresponds to dynamic instability. Small deviations from the vertical grow exponentially in time.

When the periodic potential in the quantum mechanical problem is weak whether or not a particle will find itself in the band gap depends sensitively on how nearly its wavelength matches the fundamental wavelength of the potential. If a half wavelength of the free particle matches the wavelength of the potential then coherent scattering leads to exponential growth of the wave function. In the same way when the fundamental frequency, $\sqrt{\frac{g}{\ell}}$, of the pendulum is a multiple of <u>one half</u> the driving frequency, ω, an arbitrarily small parametric drive will lead to a dynamical instability. When this frequency matching condition is not closely approximated the pendulum when perturbed from verticality simply oscillates about its equilibrium condition. If there is damping these oscillations decay exponentially and the energy fed into the pendulum when there is exponential growth has to surpass the energy dissipated in order for the exponential growth to persist.

The response of the pendulum at one half the driving frequency, which is a subharmonic response, can be understood in an intuitive way. Suppose we wanted to feed energy into the pendulum with maximum efficiency. We would accomplish this by pulling up on the point of support when the pendulum is swinging down because by doing so we are increasing the restoring force and feeding kinetic energy into the pendulum. If we allow the point of support to accelerate downwards when the pendulum swings out we reduce the restoring force and allow it to go further out from the vertical. The pendulum swings down and out twice in each complete period. We are thus pulling it up and allowing it to drop twice in the time it takes the pendulum to complete one full cycle, or going through two complete drive cycles for each cycle of the pendulum.

A strictly sinusoidal drive is easy to envision but difficult to analyze mathematically. A much more tractable model replaces the sinusoidal drive by a series of impulses. That is, one replaces a $\cos\omega t$ in (2.2) or (2.3) by

$$a(t) = V_0 \sum_{n=-\infty}^{\infty} \delta(t - n\tau) \quad . \tag{2.4}$$

The point of support is jerked in a periodic series of impulses. Between these impulses the pendulum swings as if it were not driven. In the quantum mechanical analogue (2.4) corresponds to the Kronig-Penney model.[11]

We solve the equation of motion of the impulsively driven pendulum by matching angular displacements just before and after each impulse and by adding an appropriate increment to $\frac{d\theta}{df}$ just before each impulse to obtain $\frac{d\theta}{df}$ just afterwards. Starting with the linearized equation (2.3) we assume the following form for the solution

$$x(t) = A_n \cos(\omega(t - n\tau) + \delta_n) \qquad n\tau < t < (n+1)\tau \tag{2.5}$$

The matching condition for $\theta(t)$ yields

$$A_{n+1} \cos \delta_{n+1} = A_n \cos(\delta_n + \omega\tau) \tag{2.6}$$

and the condition on $\frac{d\theta(t)}{dt}$ yields

$$A_{n+1} \sin \delta_{n+1} = A_n \sin(\delta_n + \omega\tau) - \frac{A_n V_o}{\omega} \cos(\delta_n + \omega\tau). \tag{2.7}$$

Dividing (2.7) by (2.6) we obtain the following amplitude-independent recursion relation

$$\tan\delta_{n+1} = -\frac{V_o}{\omega} + \frac{\tan\delta_n + \tan\omega\tau}{1 - \tan\delta_n \tan\omega\tau} \quad . \tag{2.8}$$

Let

$$z_n = \tan\delta_n \sin\omega\tau + \cos\omega\tau \tag{2.9}$$

and $\frac{V_o}{\omega} = 2\tan\Delta$. The following recursion relation for z_n then follows from (2.8)

$$z_{n+1} = \frac{2\cos(\omega\tau - \Delta)}{\cos\Delta} - \frac{1}{z_n} \tag{2.10}$$

If we define $a \equiv \cos(\omega\tau - \Delta)/\cos\Delta$ we can express z_n as a continued fraction in terms of z_o

$$z_n = 2a - \cfrac{1}{2a - \cfrac{1}{\ddots - \cfrac{1}{2a - \cfrac{1}{z_o}}}} \tag{2.11}$$

According to the standard theory of continued fractions[12] we can also write

$$z_n = \frac{A_n z_o + B_n}{C_n z_o + D_n} \tag{2.12}$$

where (A_n, B_n) satisfies

$$\begin{pmatrix} A_{n+1} \\ B_{n+1} \end{pmatrix} = \begin{pmatrix} 2a & 1 \\ -1 & 0 \end{pmatrix} \begin{pmatrix} A_n \\ B_n \end{pmatrix} \quad . \tag{2.13}$$

The pair (C_n, D_n) obeys the same recursion relation. The eigenvalues of the matrix in (2.13) are

$$\lambda_\pm = a \pm \sqrt{a^2 - 1} \quad . \tag{2.14}$$

If $|a| < 1$ the λ's are complex numbers of the form $e^{\pm i\phi}$ ($\cos\phi = a$). If $|a| > 1$ the λ's are of the form $e^{\pm k}$ ($\cosh k = a$). In the latter case z_n approaches the fixed point e^k as $n \to \infty$ for almost all z_o's. When $|a| < 1$ the solution for z_n is

$$z_n = \frac{z_0 \sin(n+1)\phi - \sin n\phi}{z_0 \sin n\phi - \sin(n-1)\phi} \tag{2.15}$$

which wanders about, never settling on a fixed point.

The case $|a| > 1$ corresponds to exponential growth of the solutions (2.6) and (2.7). This is because of the relationship

$$\frac{A_n \cos(\delta_n + \omega\tau)}{A_0 \cos\delta_0} = z_n \, z_{n-1} \cdots z_1 \, z_0 \tag{2.16}$$

which follows from (2.6) and (2.9). If $z \to e^k$ then clearly $A_n \sim e^{nk}$.

The transition from periodicity to exponential growth in the impulsively driven undamped pendulum is very much like the "intermittency" transition from chaos to periodicity in the one dimensional map and certain physical systems.[13-16] Both are transitions via tangent bifurcations. In the case at hand tangency occurs when the right hand side of (2.10) plotted as a function of z_n achieves tangency with the 45° line $z_{n+1} = z_n$. This occurs at $z_n = z_{n+1} = \pm 1$. Periodicity in the pendulum corresponds to intermittency in the map while exponential growth corresponds to periodicity in the map. An amusing aspect of the correspondence has to do with the renormalization group treatment of the intermittency transition.[15,16] This approach posits a universal map, or recursion relation, that emerges upon repeated compositions, or iterations, of the map at tangency. In the case at hand the map at tangency when expressed as a recursion relation for $w_{n+1} = z_{n+1} - 1$ in terms of w_n is

$$\begin{aligned} w_{n+1} &= a-1 - \frac{1}{w_n + 1} = 1 - \frac{1}{w_n + 1} \\ &= \frac{w_n}{w_n + 1} \\ &\equiv f(w_n) \end{aligned} \tag{2.17}$$

and is <u>immediately</u> of the form of the universal map. It obeys the functional equation

$$f(f(w)) = 2f(w/2), \tag{2.18}$$

the version of Feigenbaum's[4] fixed point equation that applies to this dynamical transition.

Returning to the parametrically driven pendulum, we can insert two effects into our impulsive system to make it a better model of a real, nonlinear system. First, we will account for the fact that there is always some attenuation in a real physical system (superconductivity and superfluidity aside). Second, we introduce the important nonlinear effect of an amplitude-dependent period. We achieve the former by multiplying the right hand side of (2.6) by $e^{-\mu}$, and the latter by replacing $\omega\tau$ in the recursion relations by $\psi(A_n)$. Our new recursion relations are

$$A_{n+1} = \frac{A_n \cos(\delta_n + \psi(A_n))e^{-\mu}}{\cos \delta_{n+1}} \qquad (2.19)$$

and

$$\tan\delta_{n+1} = \tan(\delta_n + \psi(A_n)) - 2 \tan \Delta \quad . \qquad (2.20)$$

These two recursion relations give us a two dimensional map with some nice properties. If we define an area element $da_n \equiv A_n dA_n d\delta_n$, an appropriate definition if we think of A_n and δ_n as action and angle variables, then $da_{n+1} = e^{2\mu} da_n$. The map is area contracting when there is dissipation and area preserving when there is not. Furthermore the map admits of a stable parametrically excited state. Linear dissipation alone will not stabilize the oscillator against growth since the energy fed into it by the drive increases with the amplitude of the swings at the same qualitative rate as does the energy lost through linear dissipation. The oscillator stabilizes by dephasing due to the dependence of its natural frequency on the amplitude of its swings. In the limit of very large dissipation ($e^{-\mu} \ll 1$) the two recursion relations (2.14) and (2.20) reduce to a one dimensional map with the qualitative features of the single humped map with period doubling. When the dissipation is not small the map's attractors have much of the beautiful structure of those of the two dimensional Hénon map.

I would like to conclude this section by mentioning two examples of parametrically driven systems. The first is the system studied by Faraday.[1] He looked at water in a dish whose bottom was periodically oscillated up and down. In the frame of reference of the bottom this leads to a periodic modulation of the gravitational constant, g. Gravity waves are parametrically excited and one can see ripples ("crispations" according to Faraday) having half the frequency of the drive. The modern version of the experiment[9] replaced Faraday's dish by a channel. This simplifies the mode structure and makes the analysis of experimental results much more straightforward. The experimental results turn out to be quite spectacular. Both subharmonic response and chaos are seen, as in other systems, but the subharmonic spectrum is much richer than had been previously observed. Responses at one sixteenth, one twenty-sixth and one thirty-fourth of the driving frequency indicate, but do not exhaust the range of possibilities. The response seems to be coupled to the mode spectrum if you oscillate the channel at a frequency that is sixteen times the frequency of the lowest lying channel wave you can expect to see response at one sixteenth the driving frequency - but the precise mechanism for this subharmonic response, and in particular the way in which the mode structure influences this response, remains to be worked out.

The attendees of this summer school need not look far (in principle) for an example of the second kind of parametrically excited oscillators. They are edge waves,[17-19] which leave their mark as scallops in the sand at the surf's edge. The theory of edge waves is a theory of parametric excitation.[19,20] Briefly, they are standing waves with a finite wavelength along the shoreline that are driven by waves normally incident to the shore, and they have a natural frequency of one half the frequency of the driving wave train. Whether other subharmonics can show up in edge waves is as yet unsettled. While they may not prove an important example of subharmonic generation and dynamical transitions edge waves at least allow researchers in nonlinear dynamics to point to a beautiful manifestation of the effects they spend their time investigating.[25]

3. THE SPECTRAL SIGNATURE OF PERIOD DOUBLING AND THE STRUCTURE OF THE UNIVERSAL MAP

One signature of period-doubling in a system is the appearance of new lines in its power spectrum at each period-doubling bifurcation. The new lines are at odd multiples of a new fundamental frequency, equal to one half the frequency of the mode that has doubled its period.[21] In the case of the period doubling route to chaos the power spectrum accumulates more and more lines to form a spectral pattern with universal characteristics. Through these characteristics we can identify the transition as being of the type studied by Feigenbaum[4,21] and test in a detailed way for the accuracy of detailed predictions based on his renormalization group analysis.

As with other power spectra this one derives from the behavior of the system's autocorrelation function. Since our model system, the iterated map, is discrete in time its Fourier spectrum is uniquely defined in a frequency interval of finite width. For simplicity we take the time interval to be one. The frequency interval can then be taken to be between $\omega = -\pi$ and $\omega = \pi$. The power spectrum is the Fourier transform of the autocorrelation function

$$c(j) = \lim_{N\to\infty} \frac{1}{N} \sum_{k=1}^{N} x_k \, x_{k+j} \equiv \langle x_0 x_j \rangle \tag{3.1}$$

The Fourier transform $C(\omega)$ is

$$C(\omega) = c(0) + 2 \sum_{j=1}^{\infty} c(j) \cos j\omega \ . \qquad (3.2)$$

This quantity has delta function spikes at $\omega_0 = \frac{2\pi}{n}$ and its harmonics when the motion has a period n. As this system takes the period-doubling route to chaos new spikes appear at odd multiples of $\frac{\omega_0}{2^m}$ with each bifurcation, m increasing by one from bifurcation to bifurcation. One universal feature of this spectrum first noted by Feigenbaum,[21] has to do with the size of newly emerging peaks relative to those already developed or, alternatively, to the sizes of peaks at odd multiples of $\frac{\omega_0}{2^{m+1}}$ as compared to those at $\frac{\omega_0}{2^m}$ when the bifurcation sequence has proceeded well beyond the $m+1^{st}$. What has been found[10] is that if one defines $\phi(k)/\phi(k+1)$ as the ratio between the average of the weights in the peaks at odd multiples of $\omega_0/2^k$ and those at $\omega/2^{k+1}$ the universal ratio $2\beta^{(2)} = 20.9634...$ is approached. This ratio has been tested for and verified in numerical experiments on maps and differential equations with period doubling. Real experiments have produced results that do not strikingly confirm this prediction, but do not rule it out, either.

Before looking at the universal spectrum, though, lets consider the nature of the highly bifurcated orbits. Each bifurcation looks generically like a pitchfork. Here are two:

Each tine of a pitchfork is the handle of a succeeding one. At the onset of periodicity a stable fixed point bifurcates into the two points of a period two orbit. These points start out close to the fixed point and then separate as the next bifurcation is approached. At the second bifurcation the period two orbit becomes a period four orbit in which x_{j+2} nearly equals x_j. This near equality vanishes as the third bifurcation is approached. This development of orbits with increasing periods leaves us with a period 2^n orbit after the nth bifurcation in which $x_{j+2^{n-1}}$ nearly equals x_j since deep in the sequence the tines hardly separate at all. Furthermore $x_{j+2^{n-2}}$ is further away but is also very close to x_j. The period 2^∞ orbit - the attractor - at the accumulation point of the bifurcations will be called the critical attractor. On this attractor x_{j+2^m} comes closer and closer to x_j as m increases. We can, in fact, define an exponent describing the rate of convergence of x_{j+2^m} to x_j as $m \to \infty$. What we consider is the quantity

$$D(j) = \langle (x_{j'} - x_{j'+j})^2 \rangle$$

$$= 2 \langle x_{j'}^2 \rangle - 2 \langle x_{j'} x_{j'+j} \rangle = 2[c(o) - c(j)] \tag{3.3}$$

We find

$$D(2^m) \propto (\beta^{(2)})^{-n} \tag{3.4}$$

with $\beta^{(2)}$ the constant previously mentioned in connection with the power spectrum.

This points to a kind of power law decay of correlations at the transition from periodicity to chaos that is in some respects like the power law decay. Since the time elapsed after 2^n iterations of the map is $T = 2^n$ we can replace n by $\ln T/\ln 2$. This means that

$$\begin{aligned} c(T) &= c(0) + K\beta^{(2)-\ln T/\ln 2} \\ &= c(0) + KT^{-\ln \beta^{(2)}/\ln 2} \end{aligned} \tag{3.5}$$

It is important to note that this power law describes only what happens after time intervals of 2^n. The point x_{2^n+1} is significantly farther away from x_o than is x_{2^n} since x_{2^n+1} is the map of x_{2^n} and is thus nearly equal to the map of x_o since $x_{2^n} \approx x_o$. The true autocorrelation function reflects all the complexity of the critical attractor.

The connection between the scaling of autocorrelations and that of the power series follows from straightforward arguments. If we are interested in the sum of the weights of the peaks in the power spectrum at odd multiples of $\pi/2^k$ (we are here taking $\omega_o = 1$) we want the following sum

$$\sum_{\ell=1}^{2^k} \lim_{N\to\infty} \frac{1}{N} [c(0) + 2 \sum_{j=1}^{N-1} c(j) \cos[\frac{(2\ell - 1)j\pi}{2^k}]] \tag{3.6}$$

Rearranging the order of summation and using trigonometric identities we obtain

$$\begin{aligned} \phi(k) &= \lim_{N\to\infty} \frac{2^k}{N} \sum_{j=1}^{N-1} \cos\frac{\pi j}{2} \cos\frac{\pi j}{4} \dots \cos\frac{\pi j}{2^k} c(j) \\ &= \lim_{N\to\infty} \frac{2^k}{N} \sum_{m=o}^{N/2^k} (-1)^m \langle x_o x_{m2^k} \rangle \end{aligned} \tag{3.7}$$

If we let $N = 2^k M$ we have for our sum

$$\lim_{M\to\infty} \frac{1}{M} \sum_{m=o}^{M} (-1)^m C(m2^k) \tag{3.8}$$

If we let M be even we can add and subtract (2.10) to recast (3.8) into the form

$$\lim_{M\to\infty} \frac{1}{M} \sum_{m=o}^{M} (-1)^m D(m2^k) \tag{3.9}$$

This last result strongly suggests that scaling of D(j) ought to control scaling in the Fourier spectrum. Thc only complication is that we have to consider scaling of $D(m2^k)$ as well as $D(2^k)$. The remainder of this section will consist of a discussion of the universal map and the structure of its compositions with an eye to justifying the simplified assumption that scaling of $D(m2^k)$ is the same as that of $D(2^k)$ and to elucidating the properties of the map at criticality in general.

First let's consider the critical attractor. This is just the closure of the orbit followed by the initial point $x = o$ under iterated applications of Feigenbaum's invariant map, f(x) which satisfies $f(-1) = f(x)$, $f(o) = 1$ and

$$f(f(x)) = -\frac{1}{\alpha} f(\alpha x), \tag{3.10}$$

α being the constant 2.50.. .

Repeated applications of (3.10) allow us to express the images of $x_o = 0$ in a binary form. Using our three conditions on f(x) we have

$$x_1 = f(x_o) = f(o) = 1$$
$$x_2 = f(f(o)) = -\frac{1}{\alpha} f(o) = -\frac{1}{\alpha}$$
$$x_3 = f(-\frac{1}{\alpha}) = f(\frac{1}{\alpha})$$
$$x_4 = f(f(\frac{1}{\alpha})) = -\frac{1}{\alpha} f(1) = \frac{1}{\alpha^2}$$

continuing in this way we can justify the following formula yielding the location of points in the orbit. Define

$$a \equiv -\frac{1}{\alpha}$$
$$b \equiv f(\frac{1}{\alpha}) \tag{3.11}$$

and define multiplication of a and b in terms of composition of the right hand sides above. Finally set the argument of the resulting function to zero, so that abab0 = $-\frac{1}{\alpha} f(-\frac{1}{\alpha^2} f(\frac{o}{\alpha}))$. Then the nth element of the orbit defining the attractor will be represented by a string of a's and b's obtained by writing n as a binary number, replacing the ones by b's and the zeros by a's reversing the order of the strings and appending a zero to the left. Thus $x = abb0 = -\frac{1}{\alpha} f(\frac{1}{\alpha} f(\frac{o}{\alpha})) = -\frac{1}{\alpha} f(\frac{1}{\alpha})$. Any point on the attractor can thus be represented by a symbol string of a's and b's with a well-defined meaning in terms of the invariant function.

Now suppose we look at the effect of applications of $-\frac{1}{\alpha}$ and $f(\frac{1}{\alpha})$ on the distance between two points x_a and x_b where $o \leq x_a, x_b \leq 1$. Since $f(\frac{x}{\alpha})$ has a slope with an absolute value less than one in the interval $-1 < x < 1$ as does $-\frac{x}{\alpha}$ we will have the distance between $f(\frac{x_a}{\alpha})$ and $f(\frac{x_b}{\alpha})$ smaller than that between x_a and x_b and similarly for $a - \frac{x_{a,b}}{\alpha}$. Repeated applications bring them closer and closer together. This means that two points on an attractor represented by symbol strings that start out with the same m symbols will be closer and closer together as m increases. On this basis we know immediately that $x_{j+2k} \approx x_j$ for large k.

Now, consider a composition of the invariant map, shown in Figure 2. As indicated the composition consists of two submaps, each with its own range and each covering the attractor. The central submap is an exact scaled down and flipped over replica of f(x). The side map looks qualitatively like f(x) but does not have precisely the same shape. Adopting a useful nomenclature we call the central map the C map and the side map the S map. If we compose $f^{(2)}(x)$ with itself to obtain $f^{(4)}(x)$ we find the C map and the S map each splitting into two yet smaller maps. The S map will split into a central map having the same extremum which we will call the CS map and a side map, the SS map. Similarly the C map splits into a

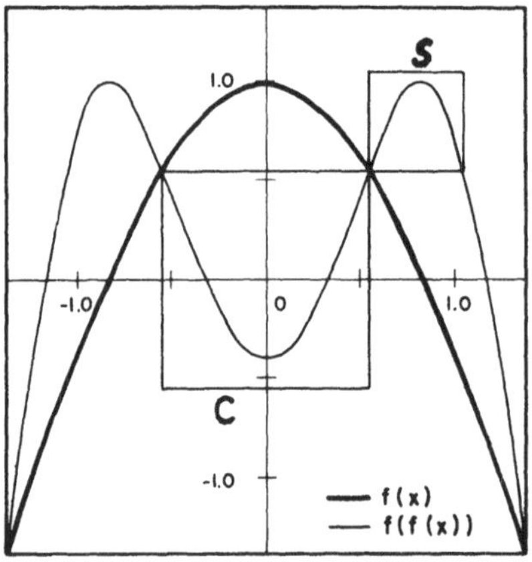

Fig. 2

CC and an SC map. Continuing on $F^{(2^n)}(x)$ will consist of 2^n submaps, each representable by a string of C's and S's. Each of these submaps looks qualitatively like $f(x)$, but only the map representable by a string of n C's is an exact scaled down version. The other $2^n - 1$ submaps are each different, in scale and shape. However it will turn out that they can be lumped into classes, defined by the first few symbols in the string of C's and S's, each member of which has pretty much the same shape.

To establish this we first note that each of the above submaps covers a portion of the critical attractor consisting of the points x_{j+m2^n} with $o < j \leq 2^n$ and $m \geq o$. The portion covered by each submap can be obtained by finding the correspondence between the symbol sequence identifying the submap and the symbol sequence identifying points on the critical attractor.

A consideration of the location of points in the orbit, and of the way in which the submaps develop under repeated compositions of $f^{(2^n)}(x)$ yields the following rule for constructing the correspondence.[22] First, we obtain a binary number for each submap by replacing S's by zeros and C's by ones. Then we add one to this number. This gives us the number j for each submap. To find the expression in terms of a's and b's for the points of the attractor in each submap we add an arbitrary integer times 2^{n+1} to the binary digit j. Then we replace the ones by b's and the zeros by a's and reverse the string so that it reads from right to left instead of left to right. This gives us a simple rule for the expression in terms of a's and b's of all the points belonging to a submap representable by a symbol string ending in an S. Simply replace the endmost S by a C then replace all S's by a's and all C's by b's, flip the symbol string from left to right and append a zero. Any point whose symbol string starts out with these n a's and b's will be in the region covered by that submap. This the submap CSSCS contains all points whose symbol string starts with bbaab.

Now, the shape of the universal map is controlled, in principle at least, by the critical attractor, since the attractor provides us with an infinite set of relations of the form $f(x_i) = y_i (= x_{i+1})$ on the interval $-1 < x_i, y_i \leq 1$. If we are to represent $f(x)$ as, say, a Taylor expansion in x we can determine any number of coefficients. Similarly, the shape of any of the submaps will be controlled by the portion of the critical attractor lying in its range. Thus, the shape of the submap CSSCS is determined by the trajectory bbaab0, bbaabb0, bbaabab0, ... If we are interested in eliminating the overall scale of the map we need only consider ratios of distances between points in the trajectories, that is, ratios like $x_{k+2^n} - x_{k+2\cdot 2^n})/(x_{k+2\cdot 2^n} - x_{k+3\cdot 2^n})$. In fact, it is straightforward to verify that if the ratios $x_{k_i+j_1 2^n} - x_{k_i+j_2 2^n})/(x_{k_i+j_3 2^n} - x_{k_i+j_4 2^n})$ are the same for all $j_1,..j_4$ for two different k_i's then the two submaps associated with those k_i's will have the same shape, if a different scale.

Now, we can show that ratios of the type above are controlled by the first few symbols in the designation of the submap. Let us start with maps whose designation ends in an S. A map like SCCSS .. S will have the critical attractor points b...baa bba X_n where X_n is a point designated by a term in the sequence a, b, ab, bb, ... We take $X_0 = 0$, $X_1 = b$ and so on. The distance $x_{k+j_1 2^n} - x_{k+j_2 2^n}$ is the distance between b...baabaX_{j_1} and b...baabaX_{j_2}. It is also the distance between b....bY_{j_1} and b...bY_{j_2} where Y_{j_i} = aabaX_{j_i}. According to arguments above both Y_{j_1} and Y_{j_2} will be close together, and both will be close to aaba0 $(=\frac{1}{\alpha^2} f(\frac{0}{\alpha}) = \frac{1}{\alpha^2})$. If we let $Y_{j_1} - Y_{j_2} = \Delta$ then the distance between b...bY_{j_1} and b...bY_{j_2} will equal, to a good approximation, Δ times the derivative of $f(\frac{1}{\alpha}....f(\frac{x}{\alpha}))$ evaluated at x = aaba. Since this x does not depend on X_{j_1} and the derivative will cancel out when we take the ratio

$$(b...baabaX_{j_1} - b...baabaX_{j_2})/(b...baabaX_{j_3} - b...baabaX_{j_4})$$

and the ratios determining the shape of the submaps depend only on the first few symbols in the sequence of C's and S's describing the submap.

Since the submaps whose designations end with a C are the images under the invariant map of those ending in an S -- i.e. CSCC is the image of CSCS. The submaps of $f^{(2^m)}(x)$ $(m>>1)$ have ranges that are very small so in that range f(x) is nearly linear, which means that the image of CSCS has the same shape as CSCS.

The net result of all of this is that $D(n2^k)$ scales like $D(2^k)$ since in a given class of submaps the difference between two points on the attractor, say b...aabX_3 and b...aabX_5 depends in the same way on the scale of the submap as does the distance between b...aabX_3 and b...aabX_4. The behavior of $D(2^k)$ as a function of k has been discussed in other articles.[10,23]

The shape of the submaps in the composed invariant map controls the structure of the bands in the highly bifurcated chaotic regime. Our ability to separate the submaps into a finite number of classes allows us to do the same thing with the bands. Furthermore it allows us to correct an improper characterization of the bands and the submaps in an earlier article on scaling at the period-doubling transition.[24] Fortunately the main results in that paper are not sensitive to the structure of the submaps.

Acknowledgements

Much of what is reported here was obtained in collaboration with Michael Nauenberg. His substantial contributions are gratefully acknowledged.

REFERENCES

1. M. Faraday, Philos. Trans. Roy. Soc. London 299, Sec. 103 (1831).
2. C. Hayashi, Nonlinear Oscillations in Physical Systems (McGraw-Hill, New York, 1964).
3. A.B. Pippard, The Physics of Vibration (Cambridge Univ. Press, Cambridge, 1978) Vol 1, Chap. 9.
4. M. Feigenbaum, J. Stat. Phys. 19, 25 (1978) and 21, 669 (1979).
5. M. Gorman and H.L. Swinney, Phys. Rev. Lett. 43, 1871 (1979).
6. J.P. Gollub, S.V. Benson and J. Steinman, Ann. N.Y. Acad. Sci. 357, 22 (1980).
7. A. Libchaber and J. Maurer, J. Phys. (Paris), Colloq. 41, C3-51 (1980).
8. M. Giglio, S. Musazzi and V. Perini, Phys. Rev. Lett. 47, 243 (1981).
9. R. Keolian, L.A. Turkevich, S.J. Putterman, I. Rudnick and J. Rudnick, Phys. Rev, Lett. 47, 1133 (1981).
10. M. Nauenberg and J. Rudnick, Phys. Rev. B24, 493 (1981).
11. N.W. Ashcroft and N.D. Mermin, Solid State Physics (Holt, Rinehart and Winston, 1976).
12. A.I. Khinchin, Continued Fractions (Noordhoff, 1963).
13. P. Manneville and Y. Pomeau, Phys. Lett. 75A, (1979) and Physica 1D, 219 (1980); Y Pomeau and P. Manneville, Commun. Math. Phys. 74, 189 (1980).
14. J.E. Hirsch, B.A. Huberman and D.J. Scalapino, Phys. Rev. A25, 519 (1982).
15. J.E. Hirsch, M. Nauenberg and D.J. Scalapino, Phys. Lett. 87A, 391 (1982).
16. B. Hu and J. Rudnick, Phys. Rev. Lett. 48, 1645 (1982).
17. Eckart, C. Wave Rep. 100, Scripps Inst. of Oceanogr. Univ. of Calif., La Jolla (1951).
18. Ursell, F. Proc. Roy. Soc. London, Ser. A, 214, 79 (1952).
19. R.T. Guza and R.E. Davis, J. Geophys. Research, 79, 1285 (1974).
20. D. Rudnick, private communication.
21. M.J. Feigenbaum, Phys. Lett. 74A, 375 (1979); Commun. Math. Phys. 77, 65 (1980).
22. Filling in the steps here is left as a challenging, but not overwhelming, excercise.
23. P. Collet, J.-P. Eckmann and L. Thomas, Commun, Math. Phys. 81, 261 (1981).
24. B.A. Huberman and J. Rudnick, Phys. Rev. Lett. 45, 154 (1980).
25. See, also, the "Amateur Scientist" section of Scientific American, 247 (1982).

LOW DIMENSIONAL DYNAMICS AND THE PERIOD DOUBLING SCENARIO

Mitchell Feigenbaum

Laboratory of Atomic and Solid State Physics
Cornell University
Ithaca, New York 14853 (U.S.A.)

1. INTRODUCTION

During the last years, a strategy has been developed to quantitatively determine the nature of the onset of chaotic motion in deterministic systems. This strategy incorporates two main ingredients. The first is a serious consequence of dissipation--in general only dissipative systems are to be considered, although special cases of conservative dynamics can be treated along the lines of the second ingredient. The special role of dissipation is that although the systems considered generally possess a large or infinite number of degrees of freedoms, the onset regime is characterized by the excitation of only a few of them. Thus, there exists an effective model of the system in this regime much simpler than the basic local description. The other main ingredient is that a scaling phenomenon for these systems, near the transition to chaos, is governed by a renormalization group theory, so that any low modal model in the same universality class serves as an appropriate model to determine quantitative properties of the actual system. Thus, a general class of one dimensional discrete-time models serves to determine the onset behavior of some real fluid configurations. In these lectures only the period doubling route will be seriously considered--the theory will be sketched, experimentally measurable phenomena worked out and the results of measurement discussed. Other lecturers will discuss the quasiperiodicity transition from this viewpoint.

2. MINIMAL MODELS AND MAPPINGS

Without energy input, a dissipative system relaxes to its static "ground" state. The systems we have in mind are described through a finite or infinite set of coupled first order differential equations--these may be chemical rate equations, equations of motion of a system of coupled nonlinear oscillators, or the equations satisfied by the amplitudes of the spatial Fourier decomposition of the field equations of a continuum. These equations read

$$\dot{\underset{\sim}{x}} = \underset{\sim}{f}_\lambda(\underset{\sim}{x}) \tag{1}$$

in N dimensions (N degrees of freedom), with λ a set of parameters which we take, for example as just an external energy source. Time is explicitly absent in f: should the system be inhomogeneous in time, t is elevated to the position of a degree of a freedom by introducing a new trajectory parameter and writing a differential equation for t. The determination of the "ground" state is simply $\dot{\underset{\sim}{x}} = 0$ which determines in general an isolated set of singular points satisfying $\underset{\sim}{f}_\lambda(\underset{\sim}{x}) = 0$: such an $\underset{\sim}{x}$ is the specification of a static configuration. For no external source, at least one of these configurations should be stable: (1) is expanded about the static solution, linearly in all perturbations, and this linear system subjected to a usual stability (eigenvalue) analysis. These eigenvalues are functions of λ and as λ increases either one of them, or a complex conjugate pair produces a growing mode. Typically, the latter case occurs, and an oscillatory instability is excited and saturates because of the nonlinearities. This is called a Hopf bifurcation and obviously requires at least two first order degrees of freedom. However, whatever the actual dimensionality of the system, at this stage of excitation, the system has only two effective degrees of freedom: it is simply an oscillator. Of course, any of the original specified coordinates (as opposed to the eigenmode) generally exhibits an oscillatory motion which is a projection of the full orbit in N dimensions upon the chosen coordinate axis. Experimentally, the leading eigenvalues can be determined by slightly perturbing the system and observing its manner of relaxation.

As λ is further increased, the asymptotic motion of this system at some new bifurcation point will again expose more degrees of freedom. This presents a much harder problem than the first such bifurcation, as now (1) must be linearly expanded about the time dependent true solution: in general the saturated oscillatory solution is not analytically available, and now the linear system whose stability is to be analyzed has time dependent coefficients. Nevertheless, a secondary Hopf bifurcation might be determined to occur, producing quasiperiodic motion (motion on a 2 torus). Evidently, three rectilinear degrees of freedom are necessary to obtain this motion. Indeed, this is the minimal dimension necessary for a system to execute a complex motion: for a smooth differential system of equations, in order to be nonintersecting, motion on a 2D surface can be no more complicated than a closed loop.

As soon as a third dimension is available, a simple closed loop can be given a half twist and the two halves folded to lie one just atop the other producing a motion that is approximately two copies of the previous one, and closing after twice the original period. This is the behavior of a period doubling (subharmonic) bifurcation, which can recur as λ further increases and lead into an aperiodic compact trajectory.

In order to facilitate the description of an orbit, it is useful to construct a map. Basically we have an oscillatory motion with each cycle slightly different from the next. This motion can be identified by tagging each such cycle at some determined point on it. Then an orbit of n distinct cycles is described as a set of n points. To accomplish this, all we need do is determine where each cycle punctures an N-1 dimensional hypersurface. Such a surface of section must be transverse to the motion at each cycle. What is required is a dynamical clock (of irregular beats) so that each clock beat occurs every one cycle of variable duration. If x_N is a typical coordinate for the motion, then its time record will exhibit all n approximate cycles, and will have n maxima. These moments of maximum value will serve as the required dynamical clock beats. However, they occur when $\dot{x}_N = 0$, which by (1) is whenever the trajectory punctures the surface $f_N(x) = 0$ in the same sense of orientation. This then identifies a useful surface of section, and moreover one naturally suited to experimental triggering.

Once a surface has been specified, we proceed to replace the continuous dynamics by a discrete one that determines the coordinates at the next point of puncture in terms of the present one. That is,

$$\underset{\sim}{x}_{t+1} = \underset{\sim}{F}_{\lambda}(\underset{\sim}{x}_t) \tag{2}$$

with t discrete, $\underset{\sim}{x}_t$ the N-1 coordinates on the surface of section at the tth crossing, and the map F is called the Poincaré map induced by the differential flow. Of course, x_{t+1} is just the finite evolution of the initial condition x_t under (1), and so F is a smooth map. Also, (1) can be reversed in time, so that F also has a smooth inverse (i.e. F is a diffeamorphism on the surface of section.) At this point, we have replaced the original differential system by a discrete mapping one: equation (2) is the framework in which we continue. Observe though that (1) in three dimensions reduces to (2) in 2D, which is then a minimal context in which to observe complex motions of an arbitrary dissipative system which is just bifurcating into the regime of chaotic behavior. However, if the system (1) was N dimensional, the invertible map F is N-1 dimensional. Should a 2D map be sufficient, it is only because the rest of the dimensions are transiently excited and asymptotically not relevant. That is, the asymptotic behavior of (2) can be equivalent to a 2D F: (2) must be iterated enough times so that the smaller eigenvalues no longer matter compared to the larger ones. That is, the asymptotically high iterates of F become noninvertible. In particular, if just a 2D F suffices in this sense, and about some orbit one eigenvalue is larger than the other, then for asymptotic behavior a 1D may suffices, and this effective map need not be invertible. We have thus arrived at the following picture: A dissipative system of the form (1) can have asymptotic motions equivalent to that of (2) in one dimension. Experimentally, after transients have decayed, any typical coordinate of the system, with a suitable surface of section, will reveal a 1D F of (2) when x_{t+1} is plotted against x_t. Indeed the simplest way to do this is to use the maxima of this particular x as the clock beats, and plot a given maximum value of x against the previous one. In this 1D regime, the points will then fall on the graph of F. (In present parlance this reduced Poincaré map of the asymptotic motion is called a return map.)

3. PERIOD DOUBLING

Once it is known that 1D discrete dynamics pertain to a given motion, it remains to know how such systems behave. We shall only discuss 1D period doubling, and that only schematically, since many detailed accounts are in the literature.

The dynamical equation we consider is

$$x_{t+1} = f_\lambda(x_t) \tag{3}$$

where f_λ is a one parameter family of maps on an interval resembling the prototypic family

$$f_\lambda(x) = 4\lambda x(1 - x) \tag{4}$$

Periodic orbits of (3) are finite sets of points satisfying

$$x_{t+n} = x_t \qquad x_{t+m} \neq x_t \qquad m < n. \tag{5}$$

To compute x_{t+n} (3) must be iterated n times:

$$x_{t+n} = f^n(x_t) \tag{6}$$

where f^n is the nth iterate of f (f composed with itself n times). Then, the n distinct points x_i in the orbit are each fixed points of f^n:

$$x_i = f^n(x_i) \qquad i = 1,\ldots,n. \tag{7}$$

(It is easy to see that f must be noninvertible to possess an orbit with $n > 1$.) Now f_λ might not possess a real n-orbit for all λ; in general there is a minimal λ for which such an orbit exists. Also, to be asymptotically interesting, the orbit must be stable, so that the system relaxes to it if perturbed away from it. By a simple stability analysis, if x_i is a point of the orbit, then every n iterates, a deviation from x_i moves geometrically with

$$\xi_{nt} \sim [Df^n(x_i)]^t \, \xi_0. \tag{8}$$

Should

$$|Df^n(x_i)| < 1 \tag{9}$$

then the orbit is stable. We shall call the value of this dervative the stability, μ, of the orbit. (If $|\mu| > 1$ the orbit is unstable.)

Assuming that at some λ an n-cycle is stable, then since $|\mu| < 1$ and μ is a continuous function of λ, there is a range of λ over which it remains stable. Instability occurs at the end points of the λ interval above $\mu \pm 1$. Should $\mu \to -1$ as we increase λ, then at the bifurcation point

$$\mu_n = Df^n(x_i) = -1, \mu_{2n} = Df^{2n}(x_i) = +1$$

In a general context, for maps like (4) μ_n will decrease below -1 as λ increases while each x_i throws off a pair of points $x_i\pm$ such that

$$f^n(x_i\pm) = x_i\mp, \; f^{2n}(x_i\pm) = x_i\pm$$

which form a 2n cycle and for which $\mu_{2n} < 1$ and so stable. This is a period doubling bifurcation. (The original n cycle is still present, but unstable: at the bifurcation value of λ each x_i is a triply degenerate fixed point of f^{2n}.) As λ further increases, μ_{2n} monotonically decreases until at -1, the 2n cycle again bifurcates, throwing off a 4n cycle. This phenomenon recurs until at λ_∞ the orbit is no longer periodic.

From what has been said, there is a succession of adjacent λ intervals I_n in which a 2^n cycle is stable, and for which the stability of the stable orbit varies from +1 to - 1 as λ increases through I_n. To compare λ values in different I_n's, the values at identical stability are always considered. Thus, for each n we consider

$$f_\lambda^{2^n}(x) = x \tag{10}$$

$$Df_\lambda^{2^n}(x) = \mu \tag{11}$$

which determine both an $x_n(\mu)$ and $\lambda_n(\mu)$. It is then true that

$$\lambda_\infty - \lambda_n(\mu) \sim \delta^{-n} \qquad \delta = 4.6692016... \tag{12}$$

$$x_n(\mu) - x_c \sim \alpha^{-n} \qquad \alpha = -2.5029078... \tag{13}$$

where x_c is the extremal (critical) point of f (which by a coordinate transformation can be moved to the origin--as we assume from now on), and $x_n(\mu)$ is the point on the orbit nearest to x_c. Both (12) and (13) are directly suitable to experimental investigation. Superfically, $\mu = -1$, the bifurcation value, would seem most suitable. However, it is most difficult to accurately locate these values. By observing the relaxation of perturbations, however, one could determine the values at $\mu = 0$, since here relaxation changes over from monotone to alternating and the orbit is least sensitive to other perturbations.

From (11) and (13) one is led to the correct conclusion that

$$\lim_{n\to\infty} \alpha^n f^{2^n}_{\lambda_n(\mu)} (x/\alpha^n) = g_\mu(x) \tag{14}$$

exists, and similarly to the numbers of (12) and (13) is a universal one parameter family of functions. This fact is very important because it asserts that the fine structure of asymptotically long orbits is universal. We will see in the next lecture that this fact determines a universal transition power spectrum.

With our choice $x_c = 0$ [assumed in (14)], it is easy to see that at the fixed point of g_μ nearest to the origin

$$g_\mu(x_\mu) = x_\mu,\; g_\mu'(x_\mu) = \mu \tag{15}$$

In particular, there is a special (unstable) value of $\mu = \mu^* \sim -1.6$ at which the convergence of (12) is much faster, so that

$$\lambda_n(\mu^*) \sim \lambda_\infty.$$

Calling $g_{\mu^*} \equiv g$, it is indeed true that

$$\lim_{n\to\infty} \alpha^n f^{2^n}_{\lambda_\infty}(x/\alpha^n) = g(x) \tag{16}$$

so that λ_∞ is special in that it is the unique isolated value of λ for which the magnified high iterates of the fixed map f_{λ_∞} themselves converge. From (16), computing at finite n and then taking the limit, it is easy to verify that g obeys

$$g(x) = \alpha g(g(x/\alpha)). \tag{17}$$

Normalizing to g(0) = 1, (17) is a functional (infinite dimensional) equation that determines both g and α.

To finish off the theory of period doubling, define the operator T by

$$(Tf)(x) = \alpha f(f(x/\alpha)) \tag{18}$$

so that g is the fixed point of T. Equation (14) now reads

$$\lim_{n\to\infty} T^n f_{\lambda_n(\mu)} = g_\mu . \tag{19}$$

Since $\lambda_n \to \lambda_\infty$, to compute g_μ we need to know the asymptotic behavior of

$$T^n f_\lambda = T^n f_{\lambda\infty + (\lambda - \lambda\infty)}$$

But,

$$\begin{aligned} T^n f_{\lambda\infty + \mu} &= T^n(f_{\lambda\infty} + \mu\partial_\lambda f_{\lambda\infty} + \ldots) \\ &= T^n f_{\lambda\infty} + \mu\ DT^n[f_{\lambda\infty}] \cdot \partial_\lambda f_{\lambda\infty} + \ldots \\ &\sim g + \mu\ DT^n[f_{\lambda\infty}] \cdot \partial_\lambda f_{\lambda\infty} + \ldots \end{aligned} \tag{20}$$

where DT^n is the derivative of T^n. Since

$$DT^n[f] = DT[T^{n-1}f]\ldots DT[Tf] \cdot DT[f],$$

taking the fact that DT[g] has a unique growing eigenvalue, which as we shall see is δ of (12), (20) becomes

$$T^n f_\lambda \sim g + (\lambda - \lambda\infty)\delta^n \psi(x)\ c(f) \tag{21}$$

where ψ is the eigenvector of DT at δ and c(f) is a constant for a given family f_λ. Substituting (12), we now have

$$T^n f_{\lambda_n(\mu)} \sim g - k_\mu \psi + \ldots \tag{22}$$

where the higher order nonlinear terms are sizable in (22), but the form of (12) together with the identification of δ is now clear in order that the limit in (19) exist and not generally be g. In order to compute g_μ, the following procedure is necessary.

Define

$$\lim_{n\to\infty} T^n f_{\lambda_{n+r}}(\mu) = g_{\mu,r};\ g_{\mu,0} = g_\mu$$

By (21) we then have

$$T^n f_{\lambda_{n+r}(\mu)} \sim g - \delta^{-r}\psi(x)k_\mu + \ldots$$

for which, at sufficiently large r, we need retain only the leading term. k_μ is determined from the requirement that

$$T^r(g - \delta^{-r}\psi k_\mu)$$

have slope μ at its fixed point nearest x = 0 for large enough r. Having solved this nonlinear problem of evaluating k_μ we then have

$$g_{\mu,r} \sim g - \delta^{-r}\psi k_\mu \qquad (23)$$

where all ingredients are now available from the theory of T. Finally, it is easy to see that

$$Tg_{\mu,r} = g_{\mu,r-1}$$

so that the exact application of T r times produces g_μ from (23)

4. TRAJECTORY SCALING

Recall that at a bifurcation point the period of an orbit doubles in the fashion that after the original period n, the cycle just fails to close on itself, requiring another n periods to exactly close. Thus the double period cycle resembles the parent very closely, the distinction being a set of n small errors to close near each point of the parent. If we can determine how these errors are related at different stages of period doubling then, of course, we can determine the entire asymptotic orbit through a recursive procedure. A different way to say this is that a 2^n cycle is really a very noisy 2 cycle, a less noisy 2^2 cycle, and so forth, since these errors must certainly decrease from level to level. This noisiness is determined by the hierachy of errors, and so these will determine the nature of the power spectrum. In particular, the higher order and smaller errors have longer periodicities, and so determine the low frequency part of the power spectrum (which, of course, modulates the other higher frequency lines.) Since (14) determines these sufficiently high order errors, it follows that the "noisy" aspects of the spectrum should be universal. Let us work out the details.

At λ_n let $x_c = 0$ be a point in the 2^n cycle. At the bifurcation point Λ_n, the point on the orbit which was at x_c is now some small distance away, and throws off a pair of points belonging to the new stable 2^{n+1} cycle. At λ_{n+1} μ is again 0 and one of these two points has moved to x_c. It requires 2^n steps to visit the other adjacent point which had split at Λ_n. Accordingly, the "error" produced in increasing λ_n to λ_{n+1} is just

$$d_n \equiv f^{2^n}_{\lambda_{n+1}}(x_c) - x_c = f^{2^n}_{\lambda_{n+1}}(0). \tag{25}$$

By (23), suppressing the μ index (μ is here 0),

$$\alpha^n f^{2^n}_{\lambda_{n+1}}(0) \sim g_1(0)$$

so that

$$d_n \sim \alpha^{-n} \tag{26}$$

and so decreases geometrically at each stage of doubling. Now it might have been the case that each error along the orbit also scales with α. Had this been the case, the Fourier analysis would have worked out trivially. Unfortunately, it turns out that the scaling varies along the trajectory. This is very easy to see.

First, let us write down the general error formula. Starting with $x_0^{(n)} = 0$ and $x_r^{(n)}$ its rth iterate at λ_{n+1},

$$d_r^{(n)} = x^{(n)}_{r+2^n} - x_r^{(n)} = f^{2^n}_{\lambda_{n+1}}(x_r^{(n)}) - x_r^{(n)}. \tag{27}$$

Clearly, $d_0^{(n)}$ is just (25), there are 2^n such errors at $r = 0,1,\ldots 2^n-1$ and

$$d^{(n)}_{r+2^n} = -d^{(n)}_r, \quad d^{(n)}_{r+2^{n+1}} = d^{(n)}_r \tag{28}$$

Consider $d_1^{(n)}$.

$$\begin{aligned} d_1^{(n)} &= f^{2^n}_{\lambda_{n+1}}(x_i^{(n)}) - x_1^{(n)} = f^{2^n}_{\lambda_{n+1}}(f_{\lambda_{n+1}}(0)) - f_{\lambda_{n+1}}(0) \\ &= f_{\lambda_{n+1}}(f^{2^n}_{\lambda_{n+1}}(0)) - f_{\lambda_{n+1}}(0) = f_{\lambda_{n+1}}(d_n) - f_{\lambda_{n+1}}(0). \end{aligned} \tag{29}$$

Since f_λ has a critical point at $x = 0$, and d_n is very small by (26), we see by expanding (29) that

$$d_1^{(n)} \sim d_n^2 \sim (\alpha^2)^{-n}.$$

Thus the splitting of the point visited next after x_c scales with α^2. However, applying the same manipulations to $d_2^{(n)}$ as we did in (29), since $f_{\lambda_{n+1}}$ has a nonzero derivative at x_1, $d_2^{(n)}$ must also scale with α^2. Indeed the scaling must remain at α^2 until the orbit passes close to x_c. This reasoning in reverse, since x_c is the image along the orbit of a point which is also not critical, implies that for the time steps just before landing at x_c, the scaling must be α. Thus the crude picture is that half the errors scale like α, the other half like α^2. By (28) a trivial abrupt change occurs at $r = 2^n$. However, at $r = 2^{n-1}$ and $2^n + 2^{n-1}$ the orbit makes next closest returns to x_c and abrupt changes in scaling also occur. Similarly at odd multiples of 2^{n-2} next closest approaches to x_c occur, and smaller changes in scaling again occur. In this way it can be anticipated that along the entire trajectory, to first approximation there are just the two scalings α and α^2. At the next level of approximation, each of these time intervals is split in half with two intervals of different but closer value, and so forth. All that remains is to make this exact. However, it is very important to observe that while the actual orbit is determined by <u>global</u> and nonuniversal properties of f, the scaling of the errors is determined by only close passages to x_c, and such information <u>is</u> universally furnished by (23). Accordingly the power spectrum itself is not truly universal; rather internal self-similarities are.

Let us precisely define a trajectory scaling function. It is

$$d_r^{(n)} \equiv \sigma_n\left(\frac{r-1}{2^{n+1}}\right) d_r^{(n-1)} \tag{30}$$

It is crucial to the applications of these results to continuous time systems that errors at different stages of doubling are compared at the same (discrete) time r. (Other scalings, comparing at proportionate times, exist but are of no interest.) By (27),

$$\sigma_n\left(\frac{r-1}{2^{n+1}}\right) = \left(f_{\lambda_{n+1}}^{2^n}(x_r^{(n)}) - x_r^{(n)}\right)/\left(f_{\lambda_n}^{2^{n-1}}(x_r^{(n-1)}) - x_r^{(n-1)}\right). \tag{31}$$

As immediate consequences of (28)

$$\sigma_n(t+1) = \sigma_n(t), \quad \sigma_n\left(t + \frac{1}{2}\right) = -\sigma_n(t). \tag{32}$$

Also,

$$\sigma_n(0) = d_1^{(n)}/d_1^{(n-1)} \sim \alpha^{-2}.$$

Writing

$$\lim_{n\to\infty} \sigma_n(t) = \sigma(t) \tag{33}$$

this result is $\sigma(0) = \alpha^{-2}$.

Also, by (25),

$$\sigma_n\left(-\frac{1}{2^{n+1}}\right) = d_0^{(n)}/d_0^{(n-1)} \sim \alpha^{-1}$$

or, $\sigma(1-\varepsilon) = \sigma(-\varepsilon) = -\alpha^{-1}$ so that σ has a strong discontinuity at t = 0. Also, by (32)

$$\sigma\left(\frac{1}{2} - \varepsilon\right) = -\alpha^{-1}, \quad \alpha\left(\frac{1}{2} + \varepsilon\right) = -\alpha^{-2}.$$

Since the next few iterates after a discontinuity have the same value, when the limit (33) is taken, we obtain for σ a function that is constant except at locations of discontinuities. From the previous discussion, these discontinuities occur at values of t for which t has a finite binary expansion (the dyadic rationals); the longer the expansion, the less close the approach to the critical point, and so the smaller the discontinuity. Thus, σ has derivative zero almost everywhere, jumping at a subset of the rationals.

Let us demonstrate how these discontinuities are calculated. Consider $r = 2^{n-s}$. Then

$$\sigma(2^{-s-1}-\varepsilon) = \lim_{n\to\infty}\left(f_{\lambda_{n+1}}^{2n}(x_{2^{n-s}}^{(n)}) - x_{2^{n-s}}^{(n)}\right)/(n \to n-1,\ s \to s-1). \quad (34)$$

But $x_{2^{n-s}}^{(n)} = f_{\lambda_{n+1}}^{2^{n-s}}(0)$, so that the numerator is

$$f_{\lambda_{n+1}}^{2^{n-s}}(f_{\lambda_{n+1}}^{2^n}(0)) - f_{\lambda_{n+1}}^{2^{n-s}}(0)$$

which by (23) is asymptotically

$$f_{\lambda_{n+1}}^{2^{n-s}}(\alpha^{-n}g_1(0)) - f_{\lambda_{n+1}}^{2^{n-s}}(0)$$

which again by (23) is asymptotically

$$\alpha^{-n+s}[g_{s+1}(\alpha^{-s}g_1(0)) - g_{s+1}(0)].$$

Substituting in (34),

$$\sigma(2^{-s-1}-\varepsilon)=(g_{s+1}(\alpha^{-s}g_1(0))-g_{s+1}(0))/(g_s(\alpha^{-s+1}g_1(0))-g_s(0)). \quad (35)$$

To obtain $\sigma(2^{-s-1}+\varepsilon)$ we set $r = 2^{n-s} + 1$, and the analogous numerator to (34) is

$$f(f^{2^{n-s}}(f^{2^n}(0)) - f(f^{2^{n-s}}(0)).$$

Since the argument of each f is of order α^{-n+s}, the f's are expanded about x_c and only the quadratic terms matter. Thus at

$2^{-s-1}+\varepsilon$, all we need do is replace each g_{s+1}, g_s by its square. Finally, to obtain an arbitrary discontinuity, consider $r = 2^{n-s_1} + 2^{n-s_2} + \ldots$, and obtain the discontinuities of σ at $t = 2^{-s_1-1} + 2^{-s_2-1} + \ldots$ by formulas analogous to (35) but employing compositions of various of the g_s. Since these are all universal, it now follows that $\sigma(t)$ is a universal scaling function. We shall now show that the power spectrum is determined just by σ.

5. POWER SPECTRUM

Although we can Fourier analyze this discrete dynamics of a map, at this point we want to return to the actual system (1) of physical interest. So long as we know that some 1D map suffices to determine the behavior in the regime of interest, since 1D maps possess all the universality we have discussed, of course so too does the continuous system. We now want to extend the trajectory scaling to the original N dimensional continuous system.

Should the system be driven periodically in time, then the unit interval of the map is just the drive period, and period doubling results in periods 2^n times the drive. In the more general case, when the clock beats are the dynamical ones of the surface of section, then as we vary λ a periodic orbit has a period $T(\lambda)$. When the orbit bifurcates, then it is still true that the new orbit has exactly double the period--but only at the bifurcation point. As λ increases, the new orbit at λ_{n+1} will only approximately have twice the period of that at λ_n. However, since $\lambda_n \to \lambda_\infty$ very rapidly, asymptotically the doubling is exact, and generally, denoting the period at λ_n by Tn, $2^{-n}T_n \to T_0{}^*$ and converges to this value geometrically at the usual rate δ. We will analyze here just the asymptotic behavior when $T_n \sim 2^n T_0{}^*$.

The first item to generalize is the trajectory splitting, which at the clock beats is just the map splitting of (27). So, we define

$$\underset{\sim}{d}_n(t) \equiv \underset{\sim}{x}_n(t + T_n) - \underset{\sim}{x}_n(t) \tag{36}$$

when x_n has periodicity T_{n+1}. $d^{(n)}$ is of course an ND vector. At the map times $\approx rT_0{}^*$, d_n is a vector in the surface of section. Since, however, the map F^{2^n} has, per assumption, just one significantly nonzero eigenvalue, this splitting vector must lie along it, and so d_n and d_{n+1} are parallel at this point. Thus, at the map times, (30) must still hold with the same scalar function σ connecting the vector d's. Next, apart from the "infrequent" discontinuities, σ is constant. Thus, in general, from one map time to the next, we expect σ to retain its constant map-set value. It is precisely for this reason that the scaling law we posited in (30) related the same time along the respective trajectories. We now elevate (30) to the continuous case as

$$\underset{\sim}{d}_n(t) \equiv \sigma_n(t/T_{n+1})\ \underset{\sim}{d}_{n-1}(t). \tag{37}$$

In fact, it is a simple exercise to demonstrate that a smooth flow protects the extension of (30) to (37) for a scalar σ with $\dot{\sigma} = 0$ almost everywhere. The serious content of (37) is that the same σ as in the 1D map suffices for the continuous ND problem, without ever having to determine the "underlying" 1D map, nor the eigendirection in the full space along which the motion occurs.

At this point, (37) is a deep experimentally testable precition: since $\sigma_n \to \sigma$, (37) says that for sufficiently large n, measuring the trajectory splittings d_{n-1} and d_n will produce σ for the quotient of each coordinate (test of 1D) and that d_{n+1} can now be computed. While this is already a strong scaling prediction, a still stronger fact exists--namely that σ is on a prior known function taking on, in lowest approximation, the values α^{-1} and α^{-2}. Moreover, if $d_{n-1}(t)$ is known, (36) can be solved to obtain $x_n(t)$ after d_n is obtained from (37). Thus, if the global trajectory $x_n(t)$ is measured at a sufficient level of period doubling, iterating (37) determines x(t) for all future levels. Thus, the parametric dependence of x(t) is known throughout the period doubling transition regime.

At this point we have discussed the deepest consequences of our theory. We conclude by actually determining the form of the power spectrum. By definition,

$$\hat{x}_p^{(n)} = \frac{1}{2^{n+1}} \int_0^{2^{n+1}} dt\ x_n(t)\ e^{2\pi i pt/2^{n+1}} \tag{37}$$

where we have taken $T_0{*}$ to be unity. Breaking (37) into two halves,

$$\hat{x}_{2p+1}^{(n)} = \frac{1}{2^n} \int_0^{2^n} dt\ \left[\frac{x_n(t) - x_n(t+2^n)}{2}\right]\ e^{2\pi i \frac{2p+1}{2^{n+1}} t}. \tag{38}$$

The crucial consequence of (38) is that the spectral components at the odd multiples of the fundamental $1/2^{n+1}$ are determined precisely by d_n: each period doubling introduces a new set of odd multiples of the new subharmonic fundamental, and these are determined just by the trajectory splitting function.

It is convenient to interpolate these spectral components. We do so by continuing (38) with its half period integration (to give a smoother interpolation) to arbitrary ω:

$$\hat{x}_n(\omega) \equiv \frac{1}{2^n} \int_0^{2^n} dt\, d_n(t)\, e^{i\omega t} \tag{39}$$

where d_n is now the bracketed term in (38). Obviously,

$$\hat{x}^{(n)}_{2p+1} = \hat{x}_n\left(\pi \frac{2p+1}{2^n}\right).$$

(By substituting in (39) the inverse transform of $d_n(t)$ determined by (38), this interpolation is explicitly obtained.) We now use (37) to compute $\hat{x}_{n+1}(\omega)$. By (39),

$$\hat{x}_n(\omega) = \frac{1}{2^n} \int_0^{2^n} dt\, d_n(t)\, e^{i\omega t}$$

$$\sim \frac{1}{2^n} \int^{2^n} dt\, \sigma(t/2^{n+1})\, d_{n-1}(t)\, e^{i\omega t}. \tag{40}$$

At this point we use the lowest order approximation for σ:

$$\sigma(t) = \alpha^{-2} \qquad 0 < t < 1/4$$

$$\sigma(t) = -\alpha^{-1} \qquad 1/4 < t < 1/2$$

so that splitting the integral, (40) gives

$$\hat{x}_n(\omega) \simeq \frac{1}{2^n} \int_0^{2^{n-1}} dt(\alpha^{-2} d_{n-1}(t) - \alpha^{-1} d_{n-1}(t+2^{n-1}) e^{i2^{n-1}\omega}) e^{i\omega t}.$$

By (36), $d_{n-1}(t+2^{n-1}) = -\, d_{n-1}(t)$, so that we have

$$\hat{x}_n(\omega) \simeq \frac{1}{2\alpha}\left(\frac{1}{\alpha} + e^{i2^{n-1}\omega}\right) \frac{1}{2^{n-1}} \int_0^{2^{n-1}} dt\, d_{n-1}(t)\, e^{i\omega t}$$

which by (39) becomes, simply,

$$\hat{x}_{n+1}(\omega) \simeq \frac{1}{2\alpha}\left(\frac{1}{\alpha} + e^{i2^n\omega}\right) \hat{x}_n(\omega). \tag{41}$$

Thus, to a good approximation, (37) implies a similar scaling formula for the interpolations of the successive odd spectral elements. The iteration of (37) is precisely the construction of a nonuniform Cantor set; (41) determines the same construction for the Fourier transform.

We conclude these lectures with a brief account of the import of (41). First, the new spectral lines of x_{n+1} are at the frequencies $\omega = \frac{\pi}{2}(2p+1/2^n)$; substituted into (41) we have

$$\hat{x}^{(n+1)}_{2p+1} \simeq \frac{1}{2\alpha}\left(\frac{1}{\alpha} + e^{i(\pi/2)(2p+1)}\right) = \frac{1}{2\alpha}\left(\frac{1}{\alpha} + i(-1)^p\right)\hat{x}_n\left(\frac{\pi}{2}\,\frac{2p+1}{2^n}\right) \tag{42}$$

or,

$$|\hat{x}_{n+1}| \simeq \frac{1}{2}\sqrt{\frac{1}{\alpha^2} + \frac{1}{\alpha^4}}\;|\hat{x}_n| \simeq \frac{1}{4.6}\,|\hat{x}_n|$$

where $20 \log_{10} 4.6 = 13.4$ db. (43)

(Observe, however, that (42) connects the new spectral elements in terms of the interpolation of the old ones.) While (43) gives an idea of the successive drop of spectral amplitudes, it is important to realize that the full set of new spectral lines has increasingly large variance as n increase, so that a careful test for agreement between experiment and theory is complicated. To see this consider $\omega = \pi(2p+1/2^n)$ in (41):

$$\hat{x}_{n+1} \simeq \frac{1}{2\alpha}\left(\frac{1}{\alpha} - 1\right) x_n \simeq -\frac{1}{3.6}\,\hat{x}_n$$

$$20 \log_{10} 3.6 = 11 \text{ db}.$$

Thus at the next level after which a spectral line appears, the interpolation drops 11 db.

Next, consider $\omega = 2\pi(2p+1/2^n)$ in (41):

$$\hat{x}_{n+1} \simeq \frac{1}{2\alpha}\left(\frac{1}{\alpha} + 1\right) x_n \simeq -\frac{1}{8.3} x_n$$

$$20 \log_{10} 8.3 = 18 \text{ db}.$$

Thus at all levels two or more after a spectral line has appeared, the interpolation drops 18 db per doubling at these frequencies. The consequence of these two results is that the interpolations after n levels of doubling has a variance of ~ 7n db, so that the true frequencies at the new subharmonics, as given by (42) results in a similar variance of these lines. The upshot of this discussion is that (37) is the proper test of the theory, with (41) giving a good picture of what the spectrum looks like.

STRANGE ATTRACTORS IN FLUID DYNAMICS

John Guckenheimer*

Department of Mathematics
University of California
Santa Cruz, California 95064

In 1971, Ruelle and Takens (1971) published a provocative paper about the nature of turbulence. Their work motivated experimental physicists to study the transition to aperiodic flow in traditional laboratory experiments. Some of these experiments, beginning with those of Gollub and Swinney (1975) have produced results which appear to corroborate the basic ideas of Ruelle and Takens. However, there are fluid dynamicists who remain skeptical of these strange attractor models of turbulence. This paper is a discussion of these issues. I propose means by which the Ruelle-Takens theories can be tested directly and offer suggestions for alternative models of the transition to aperiodic flow in situations where the strange attractor models appear inappropriate.

The strange attractor models are based upon systems of nonlinear ordinary differential equations. They have two essential features which are definite reductions of the physical situations being studied: (1) they have a finite number of degrees of freedom and (2) they are deterministic. The Ruelle-Takens hypothesis is that fluid flow in any particular experiment can be described by a "typical" finite dimensional, deterministic system of differential equations. As experimental parameters are varied, the transitions which one observes should correspond to "typical" changes in the properties of solutions to a family of differential equations. As far as the theory has been developed thus far, finite dimensional should be interpreted as meaning low dimensional. In particular, experience about the "typical" behavior expected from systems of differential equations with aperiodic solutions is based largely on systems whose state space has dimension three.

Fluid systems have an infinite number of degrees of freedom, and the jump from three to infinity is a long one. It is therefore necessary to investigate the circumstances under which one expects a

*Research partially supported by the National Science Foundation

system with few degrees of freedom could describe the characteristics of the time dependence in a fluid flow. An initial attempt to address this question begins with examination of the linearized fluid equations about an equilibrium representing the basic steady fluid state in an experiment with small forcing. Low dimensional models can be expected to have some validity in situations for which the spectrum associated with these linearized equations lies well to the left of the imaginary axis apart from a few eigenvalues which become unstable as the forcing parameter is increased. There is then hope that the fluid motion can be described in terms of equations expressing the (nonlinear) interaction of these unstable modes.

The spectra of linearized equations depend strongly upon the mathematical formulation of the problem. The onset of fluid convection provides a good example. In a three dimensional layer which is infinite in horizontal directions, there is infinite degeneracy in the eigenspace of zero at the onset of convection corresponding to an indeterminacy in the orientation of convective rolls (Busse 1981). Moreover, the spectrum for this problem is not discrete: eigenvalues depend continuously on the wave number of convective rolls. Lateral boundary conditions in the mathematical problem destroy these degeneracies, but in large aspect ratio containers there are likely to be many modes which are near instability at the onset of convection. Experimental observation (Gollub et. al. 1982) indicates that these modes typically have a complicated spatial structure that is not readily amenable to further analysis of higher instabilities. In small aspect ratio containers, the situation is much more hopeful for finding linearized equations with few modes near the threshold of instability. Experimental results here (Gollub & Benson 1981, Libchaber & Maurer 1982) do correspond nicely with predictions of the Ruelle-Takens hypothesis about the characteristics of instabilities.

Eckmann (1981) has reviewed the "routes to chaos" which have been observed and analyzed in low dimensional systems of differential equations. The principal results from the bifurcation theory of dynamical systems characterize several different mechanisms by which aperiodic trajectories appear in a system as parameters are varied. Notable here are two predictions which have been repeatedly verified in a wide variety of small aspect ratio fluid systems: (1) Ruelle-Takens' (1971) assertion that one should not expect to find quasi-periodic flow with many independent frequencies, and (2) cascades of period-doubling bifurcation whose quantitative features approximate those observed by Feigenbaum (1978) in the iteration of one dimensional mappings.

Neither bifurcation theory nor experiments observing the routes to chaos address clearly the question of whether the fluid flow continues to be described by a system with few degrees of freedom beyond the onset of chaos. This is the issue considered here. The heart of the matter revolves around whether the aperiodicity of the motion can be ascribed to the nonlinear interaction of the few unstable modes or whether other mechanisms play an essential role. It is quite possible that the answer to this question will be system dependent and that different alternatives will hold in different situations. Let me try to portray two different scenarios for how the transition to chaotic flow might produce flows that are not readily describable by low dimensional dynamical systems.

The first scenario with many degrees of freedom is based upon an interpretation of the observations by Gollub et. al. (1982) of convection in a container with a large aspect ratio. As noted above, there are many modes of the system which are approaching instability at the onset of convection. At physically realized Rayleigh numbers above onset, it is plausible to think of there being many modes for the trivial solution which are actually unstable. With an experimental protocol in which the Rayleigh number is slowly increased from below its critical value, the mode(s) represented in the initial convective motion may be determined by small fluctuations within the system and consequently unpredictable. There are likely to be many non-trivial equilibria for the supercritical flow. Their stability will be determined by nonlinear interactions. At small amplitudes near the critical Rayleigh number, a model for such a system comes from perturbing the normal form of an equilibrium with many zero eigenvalues (Guckenheimer 1980). The dynamics of such systems have not been studied, but already in systems of codimension three one expects to find chaotic solutions (Arneodo et. al. 1981).

The second scenario for the transition to chaos with many degrees of freedom entails spatially localized regions of dissipation within the fluid. There has been considerable speculation about the singularity structure of the Euler equations and its relationship with intermittent concentrations of vorticity in fluid flow (Chorin 1981). The stretching and crinkling of vortex lines provides a spatially localized mechanism for the dissipation of energy in a flow. Insofar as this type of mechanism is associated with the onset of aperiodicity in an experiment, it seems plausible that the flow itself will be affected by the small scales where the dissipation takes place. In these situations, the onset of chaos could appear to be as random as the thermal fluctuations within the fluid itself. Greenside et. al. (1981) present arguments for the stochastic nature of the onset of chaotic motion in convection.

Can these different possibilities be distinguished? Theoretically, the answer to this question is probably negative. Nonetheless, there are practical things which can be done with time series data that shed light on the distinctions which we have drawn above. The issues involved are ones of dimensionality and determinism. I have outlined elsewhere (Guckenheimer 1982) statistical procedures aimed at testing whether a given aperiodic time series can be modelled by a solution of a discrete dynamical system of a given order. The rest of this paper is a further discussion of these methods.

The starting point for this time series analysis is a more quantitative formulation of the strange attractor hypothesis:

SA Hypothesis:

If $\xi(t)$ is a discrete time series, then there is

(1) a d-dimensional manifold M^d

(2) a discrete dynamical system $f: M \to M$

(3) a function $h: M \to \mathbb{R}$

(4) $L_1 \varepsilon \mathbb{R}$, $L_2 \varepsilon \mathbb{R}$

(5) $x \varepsilon M$

such that $\xi(t) = h(f^t(x))$, $\|Df\| < L_1$, and $\|Dh\| < L_2$. The goal is a set of procedures to assess the statistical validity of this hypothesis. Because all physical systems are subject to some "noise", the SA Hypothesis cannot be expected to be exactly true for any physical data. Therefore, the hypothesis should be modified to allow random perturbations of a deterministic system in (2). This creates an estimation problem: if one assumes that a set of data was generated by a stochastic dynamical system lying within a specified class of random perturbations of discrete dynamical systems $f: M \to M$, then one would like to estimate the variance of an optimal model for the data. In terms of power spectra, one would like to make a reasonable estimate of the energy in the continuous part of the spectrum attributable to a low dimensional strange attractor. For data produced by random perturbations of low dimensional dynamical systems, this appears to be a feasible task.

The strategy I have employed in developing these statistical procedures relies upon the reconstruction of a strange attractor from a time series $\xi(t)$ produced by a model satisfying the SA Hypothesis. The basis of this reconstruction procedure is apparently due to Ruelle and is discussed formally by Takens (1981). A k-dimensional data state is a k-vector $\Xi(t) = (\xi(t),\cdots,\xi(t,k-1))$ formed from k successive values in the time series. With reasonable genericity hypothesis on the function h in the SA Hypothesis, Takens (1981) proves that the mapping $\Psi: M \to \mathbb{R}^k$ given by $\Psi(x) = (h(x),\cdots,h(f^{k-1}(x)))$ is injective if $k > 2d$. If $\xi(t) = h(f^t(x))$, then $\Xi(t) = \Psi(f^t(x))$ and the k-dimensional data state determines a unique point in the state space M. Therefore, data produced by an SA model can be analyzed effectively in terms of data states. The question of choosing an efficient value k for the data state dimension is an important one which is not discussed here.

Assume now that the time series is transformed into a time series of data states, the question of dimensionality of a good model for the data having been resolved. One can then proceed to the estimation problem described above. For a given dimension of data states, what is the variance needed in a stochastic dynamical system to represent the time series $\xi(t)$? In the first instance, this is a question of short term predictability for the data. In a strange attractor, sensitive dependence to initial conditions destroys the possibility of making long time predictions about the specific location of an approximately known initial condition, but for moderate periods of time determined by the rate of separation of trajectories (the Liapunov exponents) prediction is possible.

In practical terms, assessing the predictability of a time series is not so simple. Suppose that one has an ensemble E of initial conditions x lying in a small region of a strange attractor. The separation of trajectories along unstable directions in the strange attractor will lead to a progressive increase in the diameter of the sets $f^t(E)$. This growth in the diameter of an ensemble of trajectories will be indistinguishable from that seen in random perturbations provided that the variance of the random perturbation is smaller than the diameter of E. Thus the detection of randomness by direct comparison of the separation of data states which are close to one another requires sufficient data for there to be data states whose separation is much smaller than the variance in the randomness one is trying to find. With less data than this more sophisticated methods are necessary.

In strange attractors, the separation of trajectories occurs in an orderly fashion governed by the unstable manifolds within the system (Ruelle 1979). The procedures used in Guckenheimer (1982) assess this separation in attractors with one unstable direction by comparing the separation of data states with a function of the form $s_{a,b}(t) = a_t \exp bt$. Here a_t should be determined by the approximate location of a trajectory in the state space and therefore will be assumed constant for an ensemble of nearby trajectories. The constant b is the Liapunov exponent for the attractor and a_t reflects nonuniformity in the separation of trajectories in different regions of the attractor. If $\Xi(t_0), \Xi(t_1), \Xi(t_2)$ are three nearby data states whose separation is of the form $s_{a,b}(t)$, then the ratios $R_2(\tau) = \dfrac{\xi(t_0+k+\tau) - \xi(t_1+k+\tau)}{\xi(t_2+k+\tau) - \xi(t_1+k+\tau)}$ should be approximately constant in τ (Guckenheimer 1982). The deviations of $R_2(\tau)$ from a constant function of τ provide a comparison of the separation of trajectories with functions of the form $s_{a,b}(t)$.

For simple models of the form $f_t(x) = ax(1-x) + \lambda e_t$, where $x \varepsilon [0,1]$ and e_t is a Gaussian random variable, I have examined the extent to which $R_2(\tau)$ correlates with the amplitude λ of the random perturbation of the underlying deterministic function. Table 1 below presents the data from this numerical experiment. I do not expect higher dimensional models will behave as simply, but the results from this one dimensional iteration are certainly encouraging! The averaging procedure used in constructing Table 1 appeared to work better than computing the variance of $R_2(\tau)$.

Table 1

λ	$\overline{\left\|\log\left\|\frac{R_2(\tau+1)}{R_2(\tau)}\right\|\right\|}$
0	$.694 \times 10^{-3}$
10^{-7}	$.930 \times 10^{-3}$
10^{-6}	$.339 \times 10^{-2}$
10^{-5}	$.990 \times 10^{-2}$
10^{-4}	$.405 \times 10^{-1}$
10^{-3}	$.505 \times 10^{-1}$
10^{-2}	$.616 \times 10^{-1}$
10^{-1}	$.925 \times 10^{-1}$

$\xi(t+1) = 3.942x(t)(1-x(t) + e_t \; ; \; 0 \leq t < 5000$

e_t is a normally distributed random variable

$R_2(\;) = \frac{\xi(t_0+\tau) - \xi(t_1+\tau)}{\xi(t_2+\tau) - \xi(t_1+\tau)}$ with t_0, t_1, t_2 selected

so that $\xi(t_i) - (t_j) < 10^{-5}$ and $0 \leq \tau \leq 10$

$(\overline{\;})$ denotes the average of $(\;)$

The above results indicate that it is possible to estimate the variance in a random perturbation of a strange attractor. I am optimistic that these methods can be successfully applied to experimental data from fluid experiments in the regime just beyond the onset of chaos in order to test the strange attractor hypothesis as formulated above. If the hypothesis fails, then one is left with the possibilities that (1) the state of the system is characterized by a variable of large dimension, but the evolution of the system is predictable on short time scales, or (2) there is no reasonable description of the system which allows short term prediction. This question is much like that of trying to predict the weather by finding past analogues for the current atmospheric state and predicting that the weather will reproduce the past evolution over short time scales. Lorenz found that there was insufficient data in our knowledge of the past weather to find such analogues, but in the laboratory it should be feasible to collect data for which good analogues are present.

"Chaos" in fluids need not be regarded as formless and without structure but rather can be compared with a variety of models in a way which hopefully will lead to a deeper understanding of these systems.

REFERENCES

1. A. Arneodo, P. Coullet, E.A. Spiegel and C. Tresser, Bifurcations and Chaos, preprint (1981).
2. F.H. Busse, Transition to Turbulence in Rayleigh-Benard Convection in Hydrodynamic Instabilities and the Transition to Turbulence, ed. H.L. Swinney and J.P. Gollub, Springer-Verlag (1981).
3. A.J. Chorin, Estimates of Intermittency, Spectra, and Blow-up in Developed Turbulence, Comm. Pure Appl. Math. 34, 853-866 (1981).
4. J.P. Eckmann, Roads to Turbulence in Dissipative Dynamical Systems, Rev. Mod. Phys. 53, 643-654 (1981).
5. M.J. Feigenbaum, Quantitative universality for a class of nonlinear transformations, J. Stat. Phys. 19, 25-52 (1978).
6. H.S. Greenside, G. Ahlers, P.C. Hohenberg, and R.W. Walden, A simple stochastic model for the onset of turbulence in Rayleigh-Benard convection, preprint (1981).
7. J. Gollub and H. Swinney, Onset of Turbulence in a Rotating Fluid, Physical Review Letters 35, 927-930 (1975).
8. J.P. Gollub and S.V. Benson, Many routes to turbulent convection, J. Fluid Mech. 100, 449-470.
9. J.P. Gollub, A.R. McCarriar, and J.F. Steinman, Convective Pattern Evolution and Secondary Instabilities, J. Fluid Mech., to appear.
10. J. Guckenheimer, Patterns of Bifurcation, New Approaches to Nonlinear Problems in Dynamics, SIAM (1980).
11. J. Guckenheimer, Noise in Chaotic Systems, Nature 298, 358-361 (1982).
12. A. Libchaber and J. Maurer, A Rayleigh-Benard experiment: helium in a small box, Geilo NATO School, April (1981).
13. D. Ruelle, Ergodic Theory of Differentiable Dynamical Systems, Publ. I.H.E.S. 50, 27-58 (1979).
14. D. Ruelle and F. Takens, On the nature of turbulence, Comm. Math. Phys. 20, 167-192 (1971).
15. F. Takens, Detecting strange attractors in turbulence. Dynamical Systems and Turbulence, Warwick 1980, ed. D.A. Rand and L.S. Young, Springer Lecture Notes in Mathematics 898, 366-381 (1981).

EXPERIMENTAL ASPECTS OF THE PERIOD DOUBLING SCENARIO

A. Libchaber

Groupe de Physique des Solides de l'Ecole Normale Supérieure
24 rue Lhomond, 75231 Paris Cedex 05, France

Rayleigh-Benard experiments in Helium[1], Water[2] and Mercury[3] have shown that, in small and moderate Prandtl number fluids, the period doubling scenario is one of the possible routes to chaos. One usually analyses the experimental results in terms of a one dimensional mapping. In such a model[4], the mapping of the interval as a dynamical system, there is a strict hierarchy in the order of appearance of bifurcations as a function of the control parameter R. As R increases, a cascade of period doubling bifurcations (pitchfork bifurcations) evolves to its limit, for R_∞. Beyond this value, a chaotic states sets in, in which two remarkable phenomena occur. First, a reverse bifurcation sequence[5] appears which shows period halving with increasing R. Second, within this region, a universal sequence[6] of periodic states (tangent bifurcations) occurs. We will review in the first part of this paper the results of our Rayleigh-Benard experiments in Helium and Mercury, showing that the whole scenario is experimentally observed.

But all the experiments indicate also that a somewhat different scenario is very frequent, where the period doubling cascade is interrupted. We will show, in the second part of this paper, that within the framework of a 2D map this behaviour can be explained[7]. The two-dimensional mapping used is the Henon[8] one.

In larger dimensional systems, the scenario can become even more complex. For example, Franceschini[9] has shown that, in a seven mode truncation of the Navier-Stokes equations, after two period doubling bifurcations, a transition to a strange attractor occurs.

For a theoretical presentation of the period doubling scenario, we refer to the existing literature[6]. In this short review, we will not detail the experimental situation but refer to reference 1 for Helium and reference 3 for Mercury. In this last reference, the importance of an applied magnetic field is stressed. The experiments were performed in small aspect ratio cells with up to 6 convective rolls present.

1. THE PERIOD DOUBLING SCENARIO

a. The cascade of pitchfork bifurcations

In all the experiments up to four period doubling bifurcations were observed. One can then compute the universal Feigenbaum number δ and the ratio of the successive subharmonics amplitude μ. The results are shown on table 1 for the last bifurcation.

	δ	μ
Water[2]	4.3	3 to 4
Mercury[3]	4.4 ± .1	5
Theory[4]	4.669	4.58

Given the very large reduction in amplitude of the subharmonics, as the cascade evolves, it is difficult to go beyond the fifth bifurcation. To give some concrete feeling, let us look at our results in Mercury shown on Fig.1 for the Fourier spectrum and on Fig.2 for the direct time recordings.

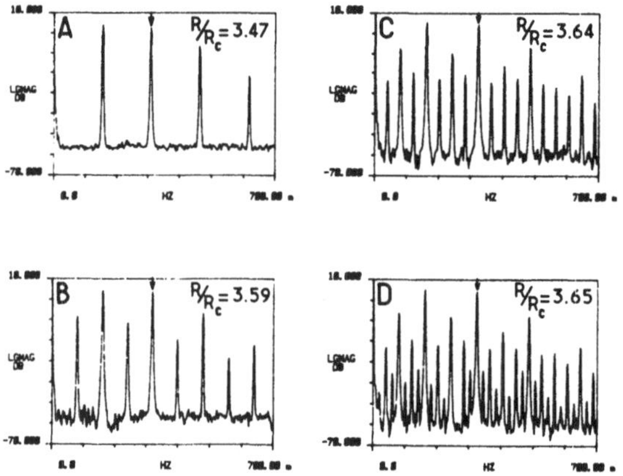

Fig.1 : The Fourier spectrum of the period doubling cascade in mercury. Arrows indicate the peak at the frequency f.

The signal is about 70 dB above noise in the Fourier spectrum, and at each bifurcation the reduction of the subharmonics amplitude is between 10 and 14 dB. Thus, after the f/32 bifurcation, the signal becomes barely visible. It thus seems difficult to go very far in the cascade to get a precise measurement of the asymptotic values of δ and μ.

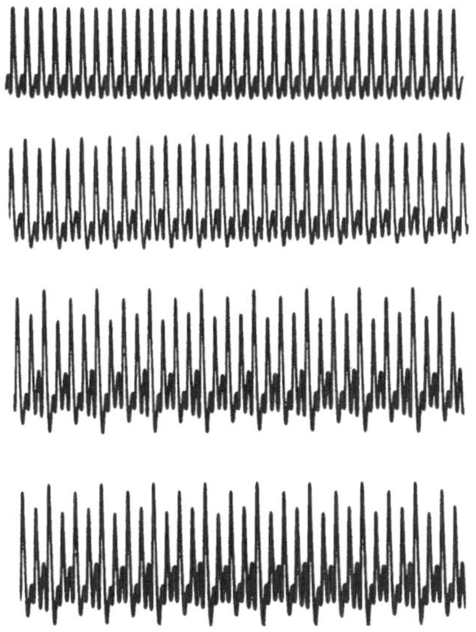

Fig.2 : Direct time recording of temperature for the R/R_C values corresponding to Fig.1.

Up to the asymptotic Rayleigh number R_∞, the noise level is unchanged. As one increases R slightly beyond this point, one observes that the base line of the Fourier spectrum is unaffected but that a low frequency noise modulates the various peaks. This heralds the onset of chaos in the period doubling scenario. It is a very gradual turbulent transition and in this aspect totally different from all the other routes to chaos[10,3], where the noise amplitude grows more abruptly.

b. The reverse bifurcation sequence

Beyond the accumulation point, a mirror image of the cascade[5] exists as one keeps increasing the Rayleigh number, plus noise. This reverse bifurcation sequence is shown on Fig.3, taken from our early Helium experiment[1]. In order to show the effect clearly, we show an enlarged portion of the Fourier spectrum around the peak at the frequency f showing the sattelites peaks at the frequencies $f \pm f/16$ and $f + f/8$. The top spectrum is just beyond R_∞, all the sattelite peaks can be seen modulated by noise. In the middle spectrum, for a slight increase of R, the sattelite $f \pm f/16$ is washed out replaced by noise. For a larger increase of R, the sattelite $f + f/8$ disappears also.

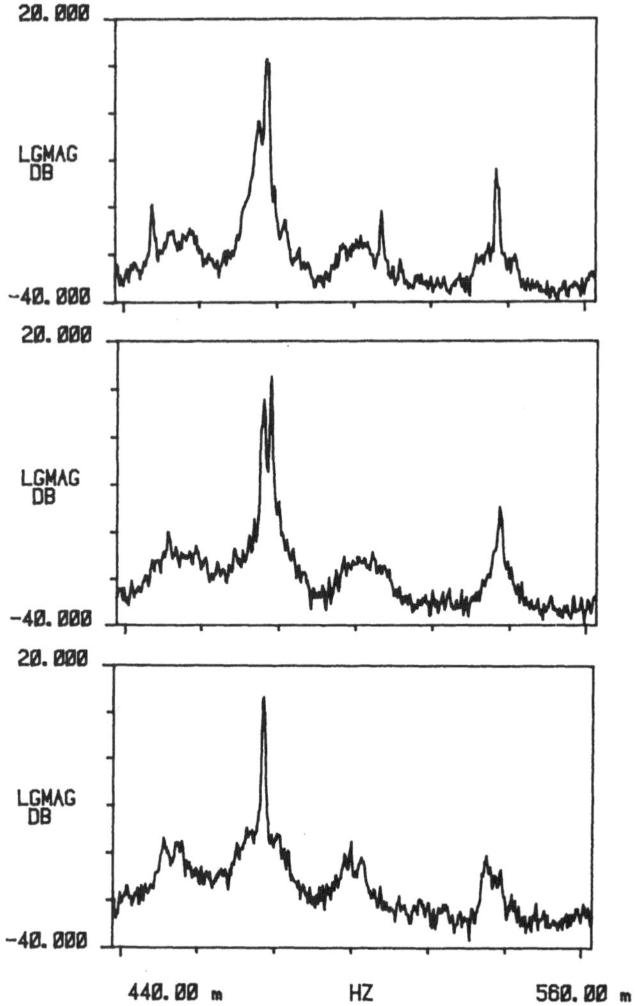

Fig.3 : The reverse bifurcation sequence in liquid Helium. The Rayleigh number increases from the top to the bottom spectra by a small amount beyond R_∞.

There is a close analogy between this route to chaos and equilibrium continuous phase transitions[11], the noisy part of the spectrum being equivalent to the order parameter. Also, as the transition point is approached from either side, the period doubling cascade grows in subharmonics.

c. The Universal sequence of periodic states

A very striking phenomena occurs within the chaotic region. One finds domains in Rayleigh number where the regime becomes periodic again (laminar). Tangent bifurcations[6] are responsible for the onset of those new fundamental periods p. There is a definite ordering of the periodic states in the theory of one dimensional maps. In a recent experiment[3] in Mercury, we have found three of those periodic states, in the chaotic region. The periods are p = 10, 3 and 9, in this order as the Rayleigh number is increased. We show in Fig.4 the Fourier spectra corresponding to period 10 (top figure) and period 9 (bottom figure).

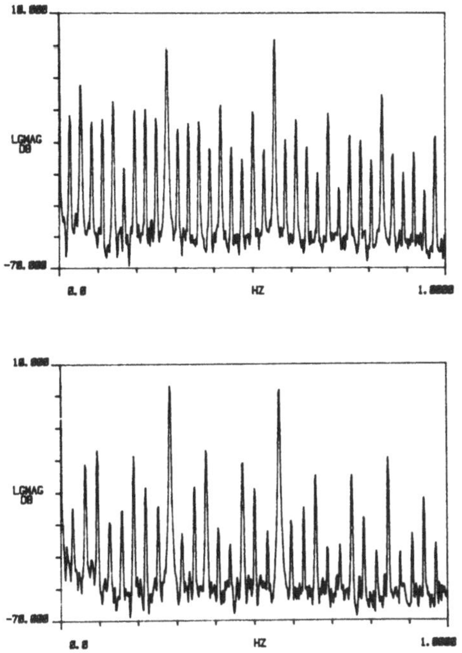

Fig.4 : Fourier spectrum of period 10 (top figure) and period 9 (bottom figure) of the U sequence in the chaotic region.

One leaves those periodic states by a pitchfork cascade of period doubling bifurcations, followed by a reverse bifurcation scheme. This is shown on Fig.5 for the period 3 state of the chaotic region. Period 3 is clearly seen with sharp peaks and also period 3.2. Periods 3.2^2 and 3.2^3 are rounded, we are on the reversed bifurcation sequence.

In conclusion, the three main phenomena of the period doubling scenario appear in those Rayleigh-Benard experiments. From the experimental point of view, two conditions have to be met : very good stability of the temperature regulation and an almost adiabatic increase of the Rayleigh number (temperature difference).

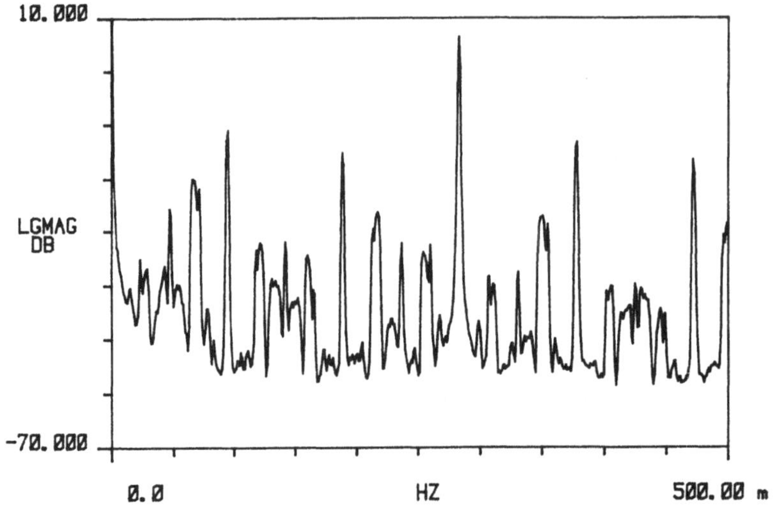

Fig.5 : Fourier spectrum of period 3 of the U sequence, as one leaves it.

2. THE INTERRUPTED CASCADE

In most experimental runs in Helium and in Mercury, we found that the period doubling cascade was interrupted and the next bifurcation was one of the period of the U sequence. This was happening for Rayleigh numbers far from the onset of the chaotic state. This can in no way be explained by a 1D map. Arneodo et al[7] proposed an explanation in terms of 2D maps. 2D dynamical systems, such as the Henon mapping[8], can present for given values of the parameters several stable dynamical states. The expression for the Henon mapping is :

$$\begin{aligned} X_{n+1} &= 1 - aX_n^2 + Y_n \\ Y_{n+1} &= bX_n \end{aligned} \qquad (1)$$

This involves two parameters. The parameter a can be roughly thought of as a constraint parameter, like the Rayleigh number in the convection experiment. The parameter b represents the area contraction rate at each iteration. Negative b values are more physical, because they correspond to orientation preserving maps. $b = -1$ represents the conservative case, and $|b| < 1$ a dissipative one.

For $b = 0$, the mapping reduces to the non invertible one dimensional map :

$$X_{n+1} = 1 - aX_n^2 \qquad (2)$$

At first sight, the parameter b could be related to the Prandtl number of the fluid in the convection experiment. But the Prandtl number is associated to a global dissipation for the convection equations, whereas b acts as a local dissipation for the dynamical behaviour of the few modes involved in the convection experiment. Thus, the connection between b and the Prandtl number is difficult, even qualitatively, to formulate. For a finite b value, there exist several competing dynamical states with distinct basins of attraction, for a given value of a and b. Fig.6 shows in the a,b plane some of the periodic stable states. We have essentially drawn the location of a few bifurcations of the period doubling cascade, the period three, and a few other periods as an illustration. It is clear that there is a large domain of values where, for example, period three coexists with the period doubling cascade. But, this is not a sufficient condition to jump from the cascading state to period three. If we vary adiabatically a control parameter, one must always stay on the continuous branches of a bifurcation tree. Thus, one follows the period doubling cascade to its end. In order to observe an unterminated cascade, a non adiabatic change of the control parameter is necessary. This is clearly what could happen in an experiment, where any realistic change in the Rayleigh number imposes some discontinuous jump. Background noise could have the same effect.

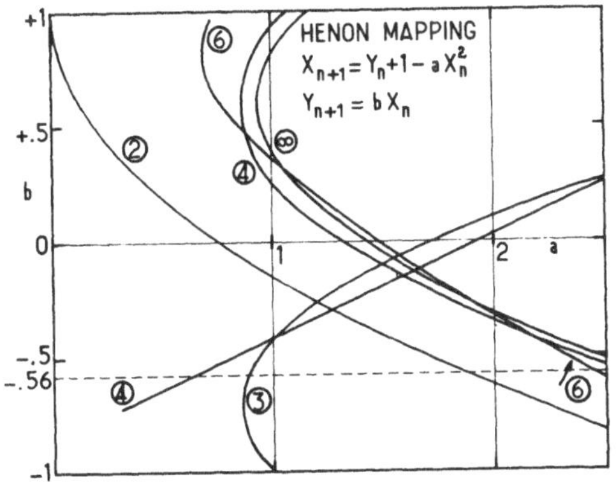

Fig.6 : Henon mapping. Traces of the period doubling cascade 2, 4, ∞. Traces of some periodic attractors 3, 4, 6.

In the Helium and in the Mercury experiments, we found that two common basins of attraction which would interrupt the cascade were period 3 and sometimes period 5. Also very large integers (like p = 27) were found after a few period doubling bifurcations.

Coming back to the Henon map, it is clear that, for very small b values, there is no competing basin of attraction to the period doubling cascade. We believe that such is the case in Mercury experiments with a large DC magnetic field present[3]. There, the dynamic modes are damped, which is equivalent to a small b value.

REFERENCES

1. A. Libchaber and J. Maurer, "Nonlinear phenomena at phase transition and instabilities", T. Riste ed. (Plenum N.Y. 1981) p.259-286.
   J. Maurer, A. Libchaber, J. Physique Lett. 40, 419 (1979).
   A. Libchaber, J. Maurer, J. Physique Colloq. 41, C3-51, (1980).
2. M. Giglio, S. Muzzati, U. Perini, Phys. Rev. Lett. 47, 243 (1981).
3. A. Libchaber, C. Laroche, S. Fauve, J. Physique Lett. 43, 211 (1982).
   A. Libchaber, S. Fauve, C. Laroche, "Two parameters study of the routes to chaos" preprint.
4. J.P. Eckmann, Rev. Mod. Phys. 53, 643 (1981).
   M.J. Feigenbaum, Phys. Lett. 75A, 375 (1979).
   N. Nauenberg, J. Rudnick, Phys. Rev. B24, 493 (1981).
5. S. Grossmann, S. Thomae, S. Naturforsch. 32a, 1353 (1977).
   E.N. Lorenz, Ann. N.Y. Acad. Sci. 357, 282 (1980).
6. N. Metropolis, M.L. Stein, P.R. Stein, J. Comb. Theory A15, 25 (1973).
   P. Collet, J.P. Eckmann, Iterated maps of the interval as dynamical systems (Birkhäuser, Boston, 1980).
7. A. Arneodo, P. Coullet, C. Tresser, A. Libchaber, J. Maurer, D. d'Humières, Physica D, in press.
8. M. Henon, Comm. Math. Phys. 50, 69 (1976).
9. V. Franceschini, Los Alamos preprint.
10. J.P. Gollub, S.V. Benson, J. Fluid Mech. 100, 449 (1980).
11. P. Martin, Melting, Localization and Chaos, R.K. Kalia, P. Vashishta Ed. (North Holland, 1982).

ENTROPY AND SMOOTH DYNAMICS

S.E. Newhouse

Department of Mathematics
University of North Carolina
Chapel Hill, North Carolina 27514 (USA)

1. INTRODUCTION

It is a matter of common experience that simple non-linear differential equations often have many solutions which exhibit complicated and erratic time dependence. For low dimensional systems there has recently been considerable progress made on the problem of describing such motion. This gives one optimism for understanding physical systems whose principal qualities can be described by a small number of modes.

To take a specific case, consider the following Duffing equation which arises in forced oscillations.

$$\ddot{x} + k\dot{x} + \beta x^3 - \alpha x = A\cos 2\pi\omega t \tag{1.1}$$

Here α, and β are positive constants, k is a real constant, and A is thought of as a movable external parameter.

We first consider the corresponding system in $\mathbb{R}^3$ ((x,v,t)-space)

$$\begin{aligned} \dot{x} &= v \\ \dot{v} &= -kv + \alpha x - \beta x^3 + A\cos 2\pi\omega t \\ \dot{t} &= 1 \end{aligned} \tag{1.2}$$

Let $\phi_s(x,v,t)$ be the unique solution at time s of (1.2) which equals (x,v,t) at $s = 0$. Using $\phi_{\frac{1}{\omega}}(x,v,0)$, and identifying the $t=\frac{1}{\omega}$-plane with the $t=0$ plane gives us a one-to-one differentiable map f_A from $\mathbb{R}^2$ to $\mathbb{R}^2$ — the so-called Poincare map or first return map. The orbits of f_A are in one-to-one correspondence with the solutions of (1.2).

Equation (1.1) and related equations have been studied by many authors. Some recent studies are in Holmes [7], Ueda [20], and Chow, Hale, Mallet-Paret [2]. We will describe the structure of the orbits of f_A for various values of the parameters of the equation. For $k = A = 0$ we consider (1.2) in the (x,v)-plane. The system is Hamiltonian. There are three equilibria: two centers and one saddle. Call the centers p_0 and p_1 and the saddle z_0. There are two orbits doubly asymptotic to z_0 and all other orbits are periodic. This is depicted in Figure 1.

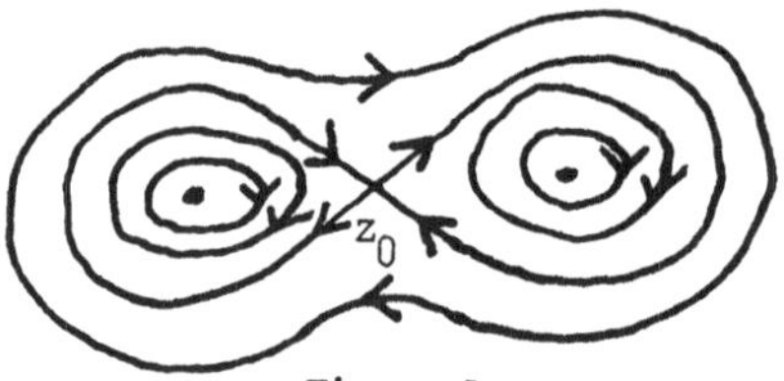

Figure 1

Now suppose k is a fixed positive constant. For $A = 0$, the centers become asymptotically stable, and attract all but three orbits as in figure 2.

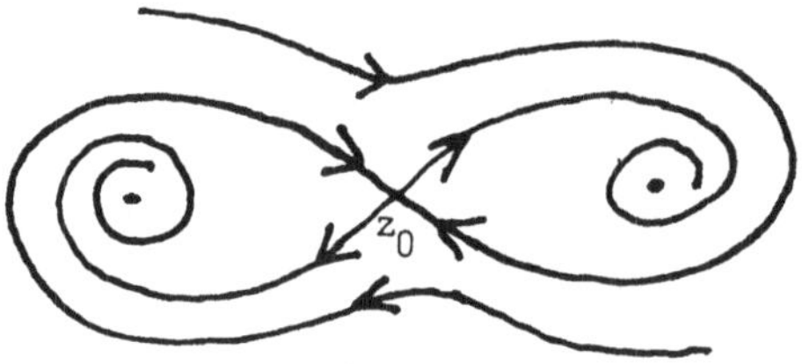

Figure 2

The map f_A is the time $-\frac{1}{\omega}$ map of these flows. For small positive A, the points p_0 and p_1 give rise to asymptotically stable fixed points, say p_{0A} and p_{1A}, of f_A, and z_0 becomes a saddle fixed point z_{0A} for f_A. The qualitative structures of the maps f_0 and f_A (with $k > 0$ fixed) are essentially the same.

As A is increased further, however, interesting changes occur. Denote by f_A^n the iterated map $\underbrace{f_A \circ \dots \circ f_A}_{n=\text{times}}$ for $n > 0$, f^0 = identity, and $f_A^{-n} = (f_A^n)^{-1}$. Let $W^s(z_{0A}, f_A) = \{z \in \mathbb{R}^2 : f_A^n(z) \to z_{0A} \text{ as } n \to \infty\}$ and $W^u(z_{0A}, f) = \{z \in \mathbb{R}^2 : f_A^{-n} z \to z_{0A} \text{ as } n \to \infty\}$. Then, $W^s(z_{0A}, f_A)$ and $W^u(z_{0A}, f_A)$ are smooth curves which only meet at z_{0A} for A small. As A increases, to a certain value $A = A_0$, the left branches of these curves become tangent at some point q as in figure 3.

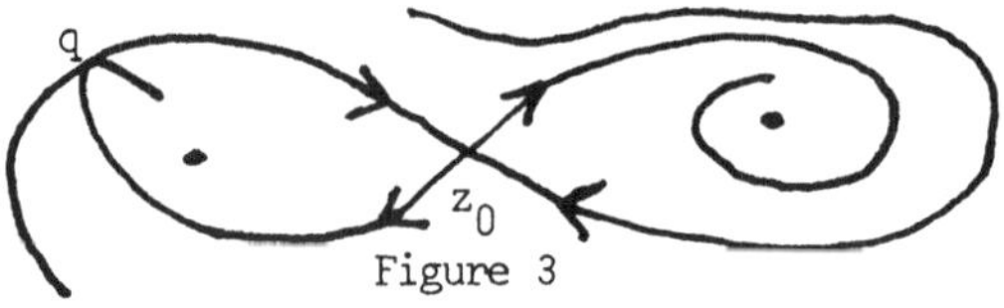

Figure 3

Part of the left branch of $W^u(z_{0A}, f_A)$ then lifts above that of $W^s(z_{0A}, f_A)$ so that there are two transverse intersections of $W^u(z_{0A}, f_A)$ and $W^s(z_{0A}, f_A)$ near q. These intersections are called transverse homoclinic points. Such points imply the existence of compact invariant sets with infinitely many periodic orbits and many orbits which exhibit erratic time dependence (see [18],[11;§2]). In this example, it can actually be shown that for certain values of A less than A_0, f_A has other saddle periodic points with transverse homoclinic points. For certain values of A it seems that the results in [11,§8] can be used to show T_A has infinitely many distinct asymptotically stable periodic orbits. Numerical results indicate that for certain values of A there are large sets of initial points whose orbits exhibit erratic time dependence.

The basic problem is to describe the orbit structure of f_A as A varies. In these maps, the presence of some chaotic orbits usually implies the existence of uncountably many such orbits. Thus, it is natural to use statistical methods in which certain orbits can be neglected and the essence of the structure can be obtained from the remaining orbits. This leads one into the subject of smooth ergodic theory.

2. ENTROPY AND ϕ^u-EQUILIBRIUM STATES

We will discuss certain orbit structure invariants (measure theoretic entropy, ϕ^u-equilibrium states) and relate them to theoretical and observable aspects of the smooth structure. First, we make some comments on the observability of chaotic motion.

Typically, chaotic motion in a mapping f is detected in one of the following ways

(1) the existence of points x whose orbits have highly aperiodic behavior.

(2) the existence of points x and vectors v such that $\liminf_{n\to\infty} \frac{1}{n} \log |T_x f^n(v)| > 0$

(3) the exponential decrease of autocorrelation functions along certain orbits (continuous frequency spectrum).

It is not clear exactly what mathematical formulation of the above properties conforms to their actual detection in experiments. The mere existence of a few points x satisfying (1) and (2) above is very hard to observe. For instance, consider the map $f(x) = 3.83x(1-x)$ for x in the interval $[0,1]$. Its structure is completely understood. There is one attracting periodic point p of period 3. The complement of the set of points forward asymptotic to the orbit of p is a Cantor set with many points satisfying (1) and (2) above. Yet numerical experiments will almost always yield the orbit of p.

A commonly accepted dictum is that (1) and (2) above will be observable if the Lebesgue measure of the set S of points x in (1) and (2) is positive and not too small. Proposition (4.2) below gives a theoretical condition guaranteeing that this Lebesgue measure is positive. Unfortunately, we have nothing to say at present about the very interesting problem of estimating the actual measure of the set S.

Let us recall some notions from ergodic theory. For the time being M will be the closure of an open set in Euclidean space or a compact smooth manifold, and $f:M\to M$ will be a smooth map. All subsets of M are assumed to be Borel sets and all measures are assumed to be Borel probability measures. Let $M(f)$ denote the set of f-invariant measures on M; i.e. $\mu(f^{-1}B) = \mu(B)$ for each subset B. A topology on $M(f)$ is defined by saying the sequence μ_i converges to μ if and only if $\int\phi d\mu_i \to \int \phi d\mu$ for each continuous function $\phi:M\to\mathbb{R}$. This topology makes $M(f)$ compact and metrizable.

Metric entropy: Given a finite partition α of M, and a $\mu\in M(f)$, set $H_\mu(\alpha)= -\sum_{A\in\alpha}\mu(A)\log\mu(A)$. For two partitions α and β, define $\alpha\vee\beta = \{A\cap B: A\in\alpha, B\in\beta\}$. Set $h_\mu(\alpha,f)=\lim_{n\to\infty}\frac{1}{n} H_\mu(\alpha\vee f^{-1}\alpha\vee\ldots\vee f^{-n+1}\alpha)$, and $h_\mu(f) = \sup_\alpha h_\mu(\alpha,f)$. The number $h_\mu(f)$ is the measure theoretic or metric entropy of f and μ. For the basic properties of $h_\mu(f)$, see [3].

The following theorem is due to Oseledec [12]. See also [4], [16].

Theorem 1. Let $f:M\to M$ be a C^1 diffeomorphism of the compact manifold M. There is a subset $\Gamma\subset M$ such that $f(\Gamma) = \Gamma$ with the following properties.

(1.1) $\mu(\Gamma) = 1$ for every $\mu\in M(f)$.

(1.2) For each $x\in\Gamma$ there are numbers $\lambda_1(x) < \lambda_2(x) < \ldots < \lambda_{r(x)}(x)$ and a splitting $T_xM = E_1(x)\oplus\ldots\oplus E_{r(x)}(x)$ such that

(a) $\lim_{n\to\pm\infty}\frac{1}{n}\log|T_x f^n(v)| = \lambda_i(x)$ for $v\in E_i(x) - \{0\}$,

(b) $\lim_{n\to\pm\infty}\frac{1}{n}\log|\det T_x f^n| = \sum_{i=1}^{r(x)}\lambda_i(x)\dim E_i(x)$

(c) $T_xf(E_i(x)) = E_i(f(x))$ for $1\le i\le r(x)$

(1.3) the functions $x\mapsto\lambda_i(x)$, $x\mapsto r(x)$, and $x\mapsto E_i(x)$ are Borel measurable, and $x\to\lambda_i(x)$, $x\to r(x)$ are constant on orbits of f.

The points x in Γ will be called the Lyapunov regular points of f. For $x\in\Gamma$, the numbers $\lambda_i(x)$ are called the characteristic exponents of x. Given a Lyapunov regular point x, let $X^+(x) = \sum_{\lambda_i(x)\ge 0}\lambda_i(x)\dim E_i(x)$.

Pesin [13] has proved that if f is C^2 and $\mu\in M(f)$ is absolutely continuous with respect to the Lebesgue measure on M, then $h_\mu(f) = \int X^+(x)d\mu$ (see [9] for a simpler proof). Ruelle has proved that even for C^1 maps f and any $\mu\in M(f)$, one has $h_\mu(f)\le\int X^+(x)d\mu$ [15].

To understand the orbit structure of f it is natural to try to find measures of maximal entropy. On the other hand, motivated by considerations of Axiom A systems and stochastic stability [17], Ruelle has suggested that one should look for measures μ satisfying

$$(*) \qquad h_\mu(f) = \int \chi^+(x)d\mu.$$

For $x \in \Gamma$, let $V(x) = \bigoplus_{\lambda_i(x) \geq 0} E_i(x)$,

and let

$$\phi^u(x) = -\log \operatorname{Jac}(T_x| V(x))$$

Here $\operatorname{Jac}(T_x f|V(x))$ means the volume distortion of $T_x f\colon V(x) \to V(f(x))$ where $V(x)$ and $V(f(x))$ are given the volume elements induced by the standard inner product on $\mathbb{R}^d$. It is known that $\int\phi^u(x)d\mu = -\int\chi^+(x)d\mu$ for every $\mu \in M(f)$. Thus, a measure μ satisfying (*) will satisfy

$$h_\mu(f) + \int\phi^u(x)d\mu = \sup_{\nu \in M(f)} \{h_\nu(f) + \int\phi^u(x)d\nu\}$$

Motivated by relations to thermodynamic formalism [17a], the quantity $\sup_\nu\{h_\nu(f) + \int\phi^u(x)d\nu\} \equiv P(\phi^u)$ is called the pressure of ϕ^u and a measure μ such that $h_\mu(f) + \int\phi^u(x)d\mu = P(\phi^u)$ is called a ϕ^u-equilibrium state for f. The existence of a measure μ satisfying (*) is equivalent to the statement that $P(\phi^u) = 0$ and ϕ^u has an equilibrium state. Since ϕ^u is only Borel measurable, ϕ^u-equilibrium states may not exist.

Recently, F. Ledrappier has proved (for $C^{1+\alpha}f$) that a measure μ satisfying (*) with positive entropy and no zero characteristic exponents must have absolutely continuous measures on unstable disks. This, together with results of Pugh and Shub announced in [14], implies that (1) and (2) hold for sets S of positive Lebesgue measure as soon as one has such a μ.

For C^∞ diffeomorphisms of surfaces, we will give necessary and sufficient conditions for the existence of measures of maximal entropy and ϕ^u-equilibrium states.

Let M^2 be a compact smooth surface, and let $\mathcal{D}^\infty(M^2)$ be the set of C^∞ diffeomorphisms of M^2. For $f \in \mathcal{D}^\infty(M^2)$ and $\chi>0$, let Λ_χ be the set of Lyapunov regular points x with exponents $\lambda_1(x) < \lambda_2(x)$ such that $\lambda_1(x) \leq -\chi$ and $\lambda_2(x) \geq \chi$. For $\ell \in \mathbb{Z}^+$, let $\Lambda_{\chi,\ell}$ be the set of points x such that there is a decomposition $T_xM = E_x^s \oplus E_x^u$ such that for $n \geq 0$,

(1) for $v \in E_x^s$ we have $|T_x f^n(v)| \leq \ell \exp(-\frac{99}{100}\chi n)|v|$

and $|T_x f^{-n}(v)| \geq \ell^{-1} \exp(\frac{99}{100}\chi n)|v|$

(2) for $v \in E^u_x$ we have $|T_x f^n(v)| \geq \ell^{-1} \exp(\frac{99}{100} \chi n)|v|$

and $|T_x f^{-n}(v)| \leq \ell \exp(-\frac{99}{100} \chi n)|v|$

It is not hard to check that $\Lambda_{\chi,\ell}$ is compact and $x \to E^s_x$, E^u_x are continuous on $\Lambda_{\chi,\ell}$. Also, $\Lambda_\chi \subseteq \bigcup_{\ell>0} \Lambda_{\chi,\ell}$.

So, for any $\mu \in M(f)$, such that $\mu(\Lambda_\chi) = 1$, we have $\lim_{\ell\to\infty} \mu(\Lambda_{\chi,\ell}) = 1$.

Let $M_\chi = \{\mu \in M(f): \mu(\Lambda_\chi) = 1\}$. We call M_χ the set of measures with exponents at least χ. Let $\bar{\varepsilon} = (\varepsilon_\ell)_{\ell \geq 1}$ be a non-increasing sequence of non-negative reals such that $\lim_{\ell\to\infty} \varepsilon_\ell = 0$. We call $\bar{\varepsilon}$ a hyperbolicity rate for a measure $\mu \in M_\chi$ if $\mu(M^2\backslash\Lambda_{\chi,\ell}) \leq \varepsilon_\ell$ for all ℓ. Let $M_{\chi,\bar{\varepsilon}} = \{\mu:\mu(M^2\backslash\Lambda_{\chi,\ell}) \leq \varepsilon_\ell\}$ = all measures μ in M_χ with $\bar{\varepsilon}$ as a hyperbolicity rate. Note that $M_{\chi,\bar{\varepsilon}}$ is a closed convex subset of $M(f)$, and for any $\mu \in M_\chi$, there is a sequence $\bar{\varepsilon}$ with $\mu \in M_{\chi,\bar{\varepsilon}}$

Theorem 2. Let $f \in \mathcal{D}^\infty(M^2)$ and let $M_{\chi,\bar{\varepsilon}}$ be the measures with exponents at least χ and hyperbolicity rate $\bar{\varepsilon}$. Then, $\mu \to h_\mu(f)$ is uppersemicontinuous on $M_{\chi,\bar{\varepsilon}}$.

Remark: If $f:M^2 \to M^2$ is only a C^2 diffeomorphism and $\bar{\varepsilon} = (\varepsilon_\ell)_{\ell\geq1}$ has the property that $\lim_{\ell\to\infty} \varepsilon_\ell \log \ell = 0$, then $\mu \to h_\mu(f)$ is uppersemicontinuous on $M_{\chi,\bar{\varepsilon}}$. However, we do not know if each measure $\mu \in M_\chi$ has a hyperbolicity rate $(\varepsilon_\ell)_{\ell \geq 1}$ tending to zero fast enough for $\lim_{\ell\to\infty} \varepsilon_\ell \log \ell = 0$.

Some consequences of theorem 2 follow.

Proposition 3. Suppose $f \in \mathcal{D}^\infty(M^2)$ has a measure μ with $h_\mu(f) > 0$. Then a necessary and sufficient condition for f to have a measure of maximal entropy is that there exist $\chi > 0$ and $\bar{\varepsilon}$ such that $\sup_{\mu\in M_{\chi,\bar{\varepsilon}}} h_\mu(f) = \sup_{\mu\in M(f)} h_\mu(f)$.

Note that the corollary on page 191 of [3] gives a general topological condition for the existence of measures of maximal entropy.

Since $x \to E^u_x$ is continuous on each $\Lambda_{\chi,\ell}$, it follows that $\phi^u(x) = -\log|T_x f|E^u_x|$ is also continuous on each $\Lambda_{\chi,\ell}$. This, together with the fact that ϕ^u is bounded and measurable implies that $\mu \to \int \phi^u d\mu$ is continuous on $M_{\chi,\bar{\varepsilon}}$. Thus, $\mu \to h_\mu(f) + \int\phi^u d\mu$ is also uppersemicontinuous on $M_{\chi,\bar{\varepsilon}}$.

Proposition 4. 1. Let $f \in \mathcal{D}^{\infty}(M^2)$. A necessary and sufficient condition for f to have a ϕ^u-equilibrium state in M_χ with $\chi > 0$ is that there exists an $\bar{\varepsilon}$ such that

$$\sup_{\mu \in M_\chi} \{h_\mu(f) + \int \phi^u d\mu\} = \sup_{\mu \in M_{\chi,\bar{\varepsilon}}} \{h_\mu(f) + \int \phi^u d\mu\}$$

2. A necessary and sufficient condition for f to have a measure μ with $h_\mu(f) > 0$ and $h_\mu(f) = \int \chi^+ d\mu$ is that there exist $\chi > 0$, $\bar{\varepsilon}$, and a sequence $\mu_i \in M_{\chi,\bar{\varepsilon}}$ such that

$$h_{\mu_i}(f) - \int \chi^+ d\mu_i \to 0 \quad \text{as} \quad i \to \infty .$$

Remark: We do not know an example of a C^∞ diffeomorphism f and a $\chi > 0$ for which M_χ is not equal to some $M_{\chi,\bar{\varepsilon}}$.

The condition in Proposition 4.2 can be stated in terms of hyperbolic basic sets for f (see [11] for definition). If Λ is such a set and $\varepsilon > 0$, we can consider the local unstable manifold $W^u_\varepsilon(x)$. The unstable Hausdorff dimension of Λ is defined to be the Hausdorff dimension of $W^u_\varepsilon(x) \cap \Lambda$. This is independent of x and ε. A. Manning has shown that this is the unique number t such that

$$\sup_{\mu \in M(f|\Lambda)} \{h_\mu(f) + t\int \phi^u d\mu\} = 0 .$$

Also, Bowen has proved [1] that if f is C^2, then there is a unique measure $\mu_{\phi^u} \in M(f|\Lambda)$ such that

$$h_{\mu_{\phi^u}}(f) + \int \phi^u d\mu_{\phi^u} = \sup_{\mu \in M(f|\Lambda)} \{h_\mu(f) + \phi^u d\mu\}$$

That is, $f|\Lambda$ has a unique ϕ^u-equilibrium state. Moreover, modifying Katok's methods in [8], one can prove that given $\chi, \bar{\varepsilon}$, there are χ', $\bar{\varepsilon}'$, such that

$$\sup_{\mu \in M_{\chi,\bar{\varepsilon}}} \{h_\mu(f) + \int \phi^u d\mu\} \le \sup_{\substack{\mu \in M_{\chi',\bar{\varepsilon}'} \\ \text{supp } \mu \subset \text{hyperbolic set}}} \{h_\mu(f) + \int \phi^u d\mu\}$$

Putting these things together, the condition in Proposition (4.2) can be restated as follows: there exist $\chi > 0$, $\bar{\varepsilon}$, and a sequence $\{\Lambda_i\} \geq 1$ of compact invariant hyperbolic basic sets such that

(1) the unstable Hausdorff dimension of Λ_i approaches 1 as $i \to \infty$.

(2) the ϕ^u-equilibrium state for $f|\Lambda_i$ is in $M_{\chi,\bar{\varepsilon}}$.

There are explicit formulas for the ϕ^u- equilibrium state of a hyperbolic set Λ, so this last formulation may be of use in constructing diffeomorphisms with strange attractors.

3. Topological entropy and volumes of submanifolds

It is tempting to come to the conclusion that the only results about mappings which are of physical interest concern point sets of positive Lebesque measure. We feel that this is not the case. Sets of zero Lebesgue measure can influence sets of positive Lebesgue measure. For instance, in the map $f_r(x) = rx(1-x)$, $x \in [0,1]$ as r moves from 3.83 to 4 there are values of r for which f_r has invariant measures absolutely continuous with respect to Lebesgue measure. One might think of this by considering the unique attracting invariant set which exists for all r in $[3.83,4]$. (see [5] [19]). At certain values of r this set has Lebesgue measure zero while at other values it fattens up to positive Lebesgue measure. The structure is determined by the orbit of the critical point. Another example is the Poincaré map f_A in the forced oscillation equation in the introduction. The relative positions of the curves $W^s(z_{oA}, f_A)$ and $W^u(z_{oA}, f_A)$ can determine whether or not most points are asymptotically periodic.

In general, the orbit complexity of a mapping can be quantified by a number called the topological entropy. We proceed to discuss a new inequality relating topological entropy to smooth structures, and we briefly consider continuity properties of entropy.

First, we recall the definition and elementary properties of topological entropy. Let $f:M \to M$ be a continuous self-map of the compact metric space M with distance function d. For a positive integer n and positive real ε, one says that a subset $E \subset M$ is (n,ε)-separated if for $x \neq y$ in E there is a $j \in [o,n)$ such that $d(f^j_x, f^j_y) > \varepsilon$. Let $r(n,\varepsilon,f)$ be the maximal cardinality of an (n,ε)-separated set. Define the topological entropy $h(f)$ to be $h(f) = \lim_{\varepsilon \to 0} \limsup_{n \to \infty} \frac{1}{n} \log r(n,\varepsilon,f)$.

One has the following properties.

1. $h(f)$ is a topological invariant; i.e. If $f:M \to M$, $g:N \to N$ are continuous maps and $\phi:M \to N$ is a homeomorphism such that $\phi f \phi^{-1} = g$, then $h(f) = h(g)$.

2. If f is a homeomorphism, then $h(f) = h(f^{-1})$.

3. If K is a compact subset of M and $f(K) \subset K$, then $h(f|K) \leq h(f)$.

4. $h(f) = \sup_{\mu \in M(f)} h_\mu(f)$.

5. For $n > 0$, $h(f^n) = nh(f)$.

Examples:

1. Piecewise monotone maps of an interval. These are continuous maps f for which there are finitely many intervals on which f is monotone. Clearly f^n is also piecewise monotone for $n > 0$. Letting $c(f^n)$ be the number of intervals on which f^n is monotone one has $h(f) = \lim_{n \to \infty} \frac{1}{n} \log c(f^n)$

   Also, $h(f) = \max(0, \lim_{n \to \infty} \frac{1}{n} \log(\text{length } f^n(I)))$ where the length is counted with multiplicities [10] .

2. Horseshoe diffeomorphisms. Let Q be a rectangle in the plane with vertices A,B,C,D, and let $f : Q \to \mathbb{R}^2$ be defined as in figure 4. We take $A'=f(A)$, $B'=f(B)$, $C'=f(C)$, $D'=f(D)$. Let A_1 and A_2 be the two components of $f(Q) \cap Q$ and assume $T_x f = \begin{pmatrix} \alpha & 0 \\ 0 & \alpha^{-1} \end{pmatrix}$ for $x \in f^{-1}A_1$ and $T_x f = \begin{pmatrix} -\alpha & 0 \\ 0 & -\alpha^{-1} \end{pmatrix}$ where $x \in f^{-1}A_2$ and $0 < \alpha < \frac{1}{2}$. If $\Lambda = \bigcap_{n \in \mathbb{Z}} f^n(Q)$ is the largest f-invariant set, then $h(f|\Lambda) = \log 2$.

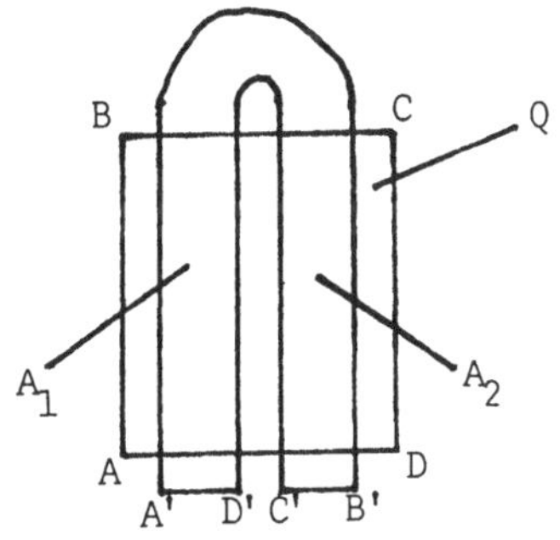

Figure 4

Slightly more complicated mappings have larger entropy as in figures 5a and 5b.

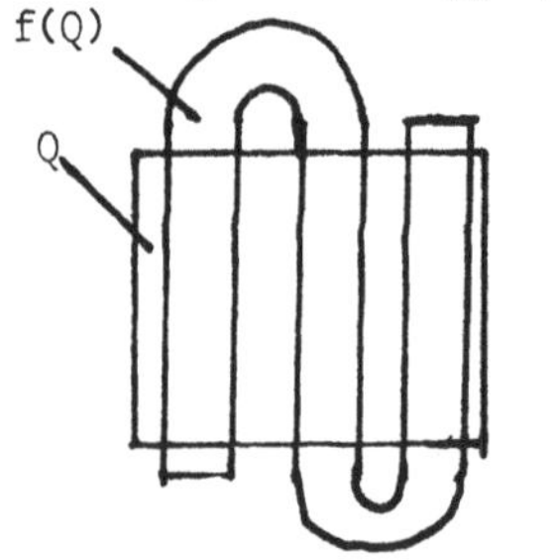

$h(f|\bigcap_{n\in\mathbb{Z}} f^n(Q)) = \log 3$

Figure 5a

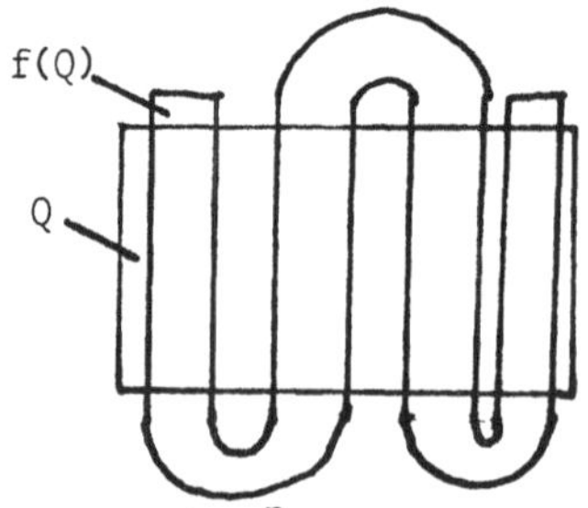

$h(f|\bigcap_{n\in\mathbb{Z}} f^n(Q)) = \log 4$

Figure 5b

3. Generalized horseshoes. Here we have N disjoint rectangles $R_1, \dots, R_N$ whose images by a power f^n of f look as in figure 6.

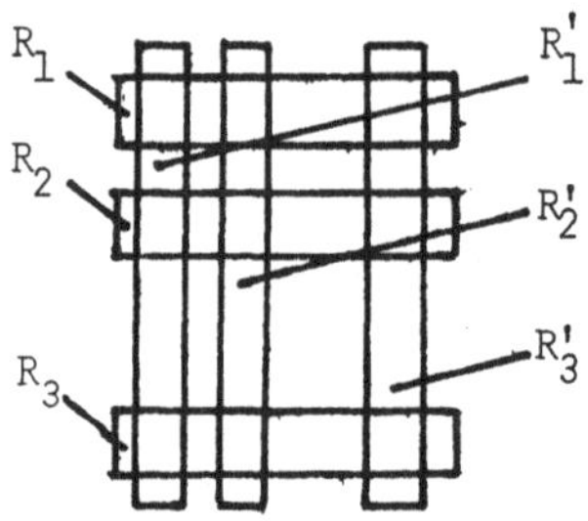

Figure 6

We set $R_i' = f^n(R_i)$. We also assume for x R_i, $T_x f^n = \begin{pmatrix} \alpha_x & \beta_x \\ \delta_x & \gamma_x \end{pmatrix}$ where

$|\delta_x|$ and $|\beta_x|$ are small, $|\alpha_x| < \lambda$, and $|\gamma_x| > \lambda^{-1}$ for some $0 < \lambda < 1$

There is a natural f-invariant set associated to $\bigcup_{i=1}^{N} R_i$, namely,

$$\Lambda = \bigcup_{k=1}^{n} f^k \left(\bigcap_{j\in\mathbb{Z}} f^{nj} \left(\bigcup_{i=1}^{N} R_i\right)\right).$$

It can be shown that $h(f|\Lambda) = \frac{1}{n} \log N$.

It is fair to say that positive topological entropy implies complicated dynamics. One indication of this is the following result due to Katok.

Theorem 5. (Katok). Suppose f is a $C^{1+\alpha}$ diffeomorphism of a compact surface and $h(f) > 0$. Then, given $\varepsilon > 0$ there is a generalized horseshoe Λ for f such that $h(f|\Lambda) > h(f) - \varepsilon$.

Corollary 6: The mapping $f \to h(f)$ is lowersemicontinuous on $\mathcal{D}^{1+\alpha}(M^2)$.

Corollary 7: A $C^{1+\alpha}$ diffeomorphism f on a compact surface has positive topological entropy if and only if f has transverse homoclinic points.

We now state a general result which shows that the topological entropy of a smooth map is dominated by the growth rate of volumes of smooth submanifolds. The results are valid in general Riemannian manifolds, but for simplicity we only consider Euclidean space.

Let $f: \mathbb{R}^p \to \mathbb{R}^p$ be a $C^{1+\alpha}$ map from $\mathbb{R}^p$ to itself. Let D^u be the unit ball in $\mathbb{R}^u$. A C^1 u-disk in $\mathbb{R}^p$ is a C^1 map $\gamma: D^u \to \mathbb{R}^p$. Let Im γ denote the image of γ. For such a γ, let $|\mathrm{Im}\gamma|$ be the u-dimensional volume of Im γ counted with multiplicities. This is also defined for subsets of the image of γ. If $U \subset \mathbb{R}^p$ is an open set, and $n \in \mathbb{Z}^+$, set $W^s(U,n) = \bigcap_{0 \le j < n} f^{-j}(U)$.

Define $G(\gamma,U) = \limsup_{n\to\infty} \frac{1}{n} \log(1+|f^{n-1}(\mathrm{Im}\,\gamma \cap W^s(U,n))|)$. Thus, $G(\gamma,U)$ is the volume growth rate of the $(n-1)^{\underline{st}}$ image of the points in Im γ whose $j^{\underline{th}}$ iterates remain in U for $0 \le j < n$.

Let $\Lambda \subset \mathbb{R}^p$ be a compact f-invariant set. Let L be the set of all non-trivial affine subspaces through points in Λ, and let A be the set of all unit disks in elements in L centered at points in Λ. We call A the family of affine unit disks through Λ. We emphasize that the dimensions of the elements of A run from 1 to p.

Theorem 8. Let $f: \mathbb{R}^p \to \mathbb{R}^p$ be a $C^{1+\alpha}$ map, let Λ be a compact f-invariant set, let U be a neighborhood of Λ, and let A be the family of affine unit disks through Λ. Then, $h(f|\Lambda) \le \sup_{\gamma\in A} G(\gamma,U)$.

This theorem and analogs of it in the complex analytic setting can be used to give simple proofs of the following theorem of M. Gromov.

Theorem 9.(Gromov)1. Suppose $f: \mathbb{R}^p \to \mathbb{R}^p$ is the map $f(x) = (f_1(x),\ldots,f_p(x))$ for $x \in \mathbb{R}^p$ and $f_i(x)$ is a polynomial of degree T. Let Λ be a compact f-invariant set. Then, $h(f|\Lambda) \le p \log T$.

2. Suppose $f: \mathbb{C}^{N+1} \to \mathbb{C}^{N+1}$ is a map from complex (N+1)-space to itself defined by $f(x) = (f_1(x),\ldots,f_{N+1}(x))$ where each f_i is a homogeneous polynomial of degree T and the f_i's have no common zero. Let $\overline{f}: CP^N \to CP^N$ be the induced map on complex projective N-space, Then, $h(\overline{f}) = N \log T$.

Theorem 9 implies, in particular, that the topological entropy of a complex polynomial $f(z) = \sum_{i=0}^{T} a_i z^i$ with $a_T \neq 0$ thought of as a map of the Riemann sphere to itself is precisely log T.

A version of this theorem can be used to prove the following fact.

Proposition 10. Suppose $f : \mathbb{R}^2 \to \mathbb{R}^2$ is a polynomial mapping of degree T. Suppose $\Lambda \subset \mathbb{R}^2$ is a compact invariant set on which f is one-to-one. Then, $h(f|\Lambda) \le \log T$.

This propostion implies that the entropy of any compact invariant set for the Henon map $f(x,y) = (y + 1 - ax^2, bx)$ is less than or equal to log 2.

We now discuss continuity properties of entropy. There is a good theory for piecewise monotone maps $f: I \to I$ where I is a closed interval or the circle.

The following facts hold.

1. $f \mapsto h(f)$ is continuous among piecewise monotone maps f with a fixed number of turning points [10].

2. $\mu \mapsto h_\mu(f)$ is uppersemicontinuous for a fixed piecewise monotone map f [10] The maximal measures have a good structure [6].

3. For dim $M > 0$ and any $0 < r < \infty$, the maps $f \mapsto h(f)$ for $f \in C^r(M,M)$ and $\mu \mapsto h_\mu(f)$ for fixed $f \in C^r(M,M)$ are not uppersemicontinuous.

Here $C^r(M,M)$ is the set of C^r self-maps of M.

The examples in 3, however, are pathological. There is reason to believe that $f \to h(f)$ and $\mu \to h_\mu(f)$ are frequently uppersemicontinuous. In particular, this may hold for C^∞, real analytic, or complex analytic f with the natural topologies.

There are localized versions of theorem 8 giving sufficient conditions for upper - semicontinuity of $\mu \mapsto h_\mu(f)$ and $f \mapsto h(f)$ in terms of volume growth rates of pieces of smooth submanifolds.

Let $f: \mathbb{R}^P \to \mathbb{R}^P$ be a C^2 map and let $U \subset \mathbb{R}^P$ be an open set. For $\varepsilon > 0$, $n \in \mathbb{Z}^+$, and $x \in \bigcap_{j\ge 0} f^{-j}(U)$, set $W^s(x,\varepsilon,n) = \{y \in \mathbb{R}^P: |f^j y - f^j x| < \varepsilon$ for $j \in [o,n)\}$. Here $|f^j y - f^j x|$ means the length of the vector $f^j y - f^j x$. If $\gamma : D^u \to \mathbb{R}^P$ is a C^1 u-disk, let

$$G_\varepsilon(\gamma,U) = \limsup_{n\to\infty} \frac{1}{n} \log \sup_{x \in \cap_{j\ge 0} f^{-j} U} (1+|f^{n-1}(\operatorname{Im} \gamma \cap W^s(x,\varepsilon,n))|)$$

Then, set $\Gamma(\varepsilon,f,U) = \sup_{\gamma \in A} G_\varepsilon(\gamma,U)$ where A is the family of affine unit disks through points of U.

Theorem 11. 1. If $\lim_{\varepsilon \to 0} \Gamma(\varepsilon,f,U) = 0$, then for any compact f-invariant subset Λ of U, the map $\mu \to h_\mu(f)$ on $M(f|\Lambda)$ is uppersemicontinuous.

2. Let Λ be a compact subset of U and let $\Lambda(f,V) = \bigcap_{j \geq 0} f^{-j}V$ be the largest forward invariant subset of V. For g near f set $\Lambda(g,V) = \bigcap_{j \geq 0} g^{-j}(V)$.

If $\lim_{\substack{\varepsilon \to 0 \\ g \to f}} \Gamma(\varepsilon,g,U) = 0$, then $g \to h(g|\Lambda(g,V))$ is uppersemicontinuous at f.

In theorem 11, set $h(g|\Lambda(g,V)) = 0$ if $\Lambda(g,V) = \phi$.

The conditions in theorem 11 are hard to check. However, there is the following new application of the ideas connected with theorem 11.

Theorem 12. 1. $f \mapsto h(f)$ is continuous for C^1 maps of an interval with a uniformly bounded number of turning points (For example, real analytic maps with the C^∞ topology)

2. Let $U \subset \mathbb{C}^1$ be an open set in the complex numbers C^1 and let $f: U \to U$ be holomorphic. Let $V \subset U$ be a compact set, and let $\Lambda(f,V) = \bigcap_{j \geq 0} f^{-j}V$, Then, $f \mapsto h(f|\Lambda(f,V))$ and $\mu \mapsto h_\mu(f|\Lambda(f,V))$ are uppersemicontinuous functions. If $\Lambda(f,V) \subset \text{int } V$, then $f \mapsto h(f)$ is continuous.

The topology on the set of holomorphic functions $f : U \to U$ is the usual one of uniform convergence on compact sets.

REFERENCES

1. R. Bowen, Some systems with unique equilibrium states, Math. Sys. Theory 8 (1974), 193-202.

2. S.N. Chow, J. Hale, and J. Mallet-Paret, An example of bifurcation to homoclinic orbits, J. Diff Eqtns. 37 (1980), 351-373.

3. M. Denker, C. Grillenberger, and K. Sigmund, Ergodic theory on compact spaces, Lecture Notes in Math. 527, Springer-Verlag, NY (1976).

4. A. Fathi, M. Herman, J. Yoccoz, A proof of Pesin's stable manifold theorem, preprint, Bât. 425, Univ. de Paris-Sud, Orsay, France.

5. J. Guckenheimer, One dimensional dynamics, Non-linear Dynamics, NYAcad. of Sci. 357 (1980), 343-348.

6. F. Hofbauer, The structure of piecewise montonic transformations, Jour. Ergodic Theory and Dyn. Sys. 1 (1981), 159-179.

7. P. Holmes, A non-linear oscillator with a strange attractor, Phil. Trans. Roy. Soc. A292 (1979), 419-448.

8. A. Katok, Lyapunov exponents, entropy, and periodic points for diffeomorphisms, Publ. Math. IHES 51 (1980), 137-174.

9. R. Mane, A proof of Pesin's formula, Jour. Ergodic theory and Dyn. Sys. 1 (1981), 95-103.

10. M. Misiurewicz and W. Szlenk, Entropy of piecewise monotone mappings, Asterisque 50 (1977), 299-311.

11. S. Newhouse, Lectures on Dynamical Systems, Progress in Math. 8 (1980), Birkhäuser, Boston, 1-115.

12. V. Oseledec, A multiplicative ergodic theorem, Trans. Mosc. Math. Soc. 1968, 197-231.

13. J. Pesin, Families of Invariant manifolds corresponding to non-zero characteristic exponents, Math. USSR Izvestia 10 (1976), 1261-1305. Lyapunov characteristic exponents and smooth ergodic theory Russ. Math. Surveys 32 (1977), 55-117.

14. C. Pugh and M. Shub, Differentiability and continuity of invariant manifolds, Non-linear Dynamics, Ann. N Y Acad. of Sci. 357 (1980), 322-330.

15. D. Ruelle, An inequality for the entropy of differentiable maps, Bol. Soc. Bras. Mat. 9 (1978), 83-87.

16. D. Ruelle, Ergodic theory of differentiable dynamical systems, Publ. Math. IHES 50 (1980), 275-306.

17. D. Ruelle, Measures describing a turbulent flow, Non-linear Dynamics, Ann. NYAcad. of Sci. 357 (1980), 1-10.

17a. D. Ruelle, Thermodynamic formalism, Encyclopedia of Mathematics and Its Applications 5, Addison-Wesley, 1978.

18. S. Smale, Diffeomorphisms with many periodic points, Differential and Combinatorial Topology (ed. S.S. Cairns), Princeton Univ. Press 196, 63-80.

19. S. Van Strien, On the bifurcations creating horseshoes, Dynamical Sys. and Turbulence, Warwick 1980, Lec. Notes in Math. 898, Springer-Verlag, 316-352.

20. Y. Ueda., Explosion of strange attractors exhibited by Duffing's equation, Non-linear Dynamics, Ann. NY Acad. of Sci. 357, (1980) 422-435.

IMBEDDING OF A ONE-DIMENSIONAL ENDOMORPHISM INTO A TWO-DIMENSIONAL DIFFEOMORPHISM. IMPLICATIONS.

Christian Mira
Equipe "Systèmes Non Linéaires"
I.N.S.A., Av. de Rangueil
31077 Toulouse Cedex (France)

1. INTRODUCTION

This paper considers the properties of the diffeomorphism T_b

(1.1) $$x_{n+1} = y_n + f(x_n, a),\ y_{n+1} = b\, x_n,\ n = 0,1,2,\ldots$$

x, y, being real variables, a, b, real parameters, in relation with those of the endomorphism T_o :

(1.2) $$x_{n+1} = f(x_n, a)$$

T_b has a constant jacobian $J = -b$. For $b = 0$, $n = 1,2,3,\ldots$, the sequence of points, solution of (1.1) is identical to the solution of (1.2). For $f(x, a) \equiv 1 - ax^2$, the first study of T_b was given in [1], with a numerical simulation of the iterative sequence, solution for particular values of (a, b). With the finite number of iterations of the simulation, this solution was interpreted as a "strange attractor". Afterwards many papers were devoted to this problem, and here we only give references concerning the object of this text.

The fact that T_o is imbedded into T_b has implications on the continuity properties of the phase space, and the bifurcations structure, when passing from T_o to T_b and inversely. In particular, it appears that the properties of the conservative case, $b = -1$, are in relation with those of T_o [2, 3]. The continuity properties in the phase space are due to the definition of degenerate invariant curves passing through fixed points, or cycles, for (1.1) with $b = 0$, in the (x, y) plane. These curves are limits of invariant curves of (1.1) when $b \to 0$. The continuity properties related to the bifurcations structure is revealed from the study of the bifurcation curves in the (a, b) plane. Unless the contrary is explicitly stated, we consider here $f(x,a) \equiv 1-ax^2$. T_o has a "box-within-a-box" [4-7] structure for the set of bifurcations on the axis $b = 0$, and at most one attractor at finite distance exists. With $b \neq 0$, $-1 \leq b < 1$, *the "box-within-a-box" bifurcations structure stretches from $b = 0$ to the (a, b) plane* [2,8]*, considered as constituted by a piling up of sheets communicating via cusp type singularities* [10] *of the bifurcation curves* [9].

This particular structure of the bifurcation plane (a, b) explains the possibility to obtain more than one attractor, for a given point (a, b), when $b \neq 0$, $-1<b<1$. All happens as if the (a, b) plane were folded from $b = 0$, giving rise to different sheets, such that a point (a, b) is the superposition of various points, each one corresponding to only one attractor in relation with one sheet.

The purpose of the paper is to describe the transition $T_o \rightleftarrows T_b$. For reasons of length limitation, the symbolism, and the vocabulary, used for the description of the "boxes", and the cycles, are not entirely recalled here, and can be found in [6,7].

2. SHORT DESCRIPTION OF SOME PROPERTIES OF T_o

There exists an interval (box) Ω_1 of a, $-1/4 \leq a \leq 2$, containing all the bifurcation values of T_o. Ω_1 is generated from the birth of two basic fixed points for $a = -1/4$. Ω_1 contains an infinity of intervals (first rank boxes), $\Omega_{k_1}^{j_1}$, $k_1 = 3,4,\ldots$, $j_1 = 1,2,\ldots p_{k_1}, p_{k_1} \to \infty$ if $k_1 \to \infty$. Each $\Omega_{k_1}^{j_1}$ is related to the birth of two basic cycles (k_1, j_1) having the order k_1, and caracterized by a *rotation sequence* [u] depending on the value of j_1, and giving the order of exchange for the k_1 points of the cycle (k_1,j_1) by application of T_o. The k_1 points of this cycle being numbered from 1 to k_1 in the sense of decreasing abscissae, [u] is constituted by the k_1 numbers obtained by successive applications of T_o beginning with 1 [6,7] Relatively to the basic cycles (k_1,j_1) and $T_o^{k_1}$, each $\Omega_{k_1}^{j_1}$ has an organization of bifurcations similar to the structure of Ω_1. $\Omega_{k_1}^{j_1} \subset \Delta_1 \subset \Omega_1$, Δ_1 being the interval outside of the region where takes place the *Myrberg bifurcations chain* of a cycle of order $k.2^i$ into a cycle of order $k.2^{i+1}$, $i = 0,1,2,\ldots$ (here $k = 1$), the multiplier (eigen-value) of the first becoming $S = -1$, when such a bifurcation occurs [11]. A $\Omega_{k_1}^{j_1}$ contains an infinity of $\Omega_{k_1.k_2}^{j_1,j_2}$ intervals (second rank boxes), with the same above organization, in a $\Delta_{k_2}^{j_1}$ interval outside of the region where takes place the Myrberg bifurcations chain with $k = k_1$. $\Omega_{k_1.k_2}^{j_1,k_2}$ contains an infinity of third rank boxes, $\Omega_{k_1.k_2.k_3}^{j_1,j_2,j_3}$ outside of the region of the Myrberg bifurcations chain with $k = k_1.k_2$, ... Considering the power T_o^k, $k = 2^i$, $i = 1,2,3,\ldots$ of T_o, it is possible to define Ω_{2^i} boxes (for more details, cf. [4-7]). This organization of bifurcations is called "box-within-a-box" structure and is of *fractal* type.

Critical points in the sense of Julia-Fatou play an important role in the appearance of the properties of T_o. These points $C(x = 1)$, $C_1(x = 1-a)$, $C_2(x = 1-a(1-a)^2)$, ... are the successive consequences the maximum of f(x, a). When an attractor exists at finite distance,

$\overline{CC_1}$ constitutes an *absorptive segment* containing this attractor which is a cycle, an *invariant (or chaotic) segment*, or a *cyclic invariant segment* [4-7]. So when $a = a_1^* = 2$, one of the two boundaries of Ω_1, $\overline{CC_1}$ is an invariant segment. Let E_C be the set of the critical points, and E'_C the set of the limit points of E_C. Let E be the set of the points of the repulsive cycles, E' the set of the limit points of E. The following proposition [12] gives some essential properties of T_o :

If T_o has an attractive cycle, then a point of E_C or E'_C cannot be a point of $E \cup E'$. Let $\hat{a}$ be the values of $a, -1/4 < a \leq 2$, such that $E \cup E'$ contains points of E_C, or E'_C. Then, for $a = \hat{a}$, T_o has not an attractive cycle at finite distance, and the attractor is either an invariant segment, or a cyclic invariant segment, the segments of which being bounded by critical points. The values $a = \check{a}$, such that E contains one, or several points of E_C, are accumulation points of boxes Ω_k^j, $k \to \infty$, or of boxes within a Ω_k^j. The values $\hat{a}$ different from the $\check{a}$ are accumulation points of the $\check{a}$.

The values $\check{a}$ are either one of the two boundaries of each box Ω having a determined rank (these values are noted a^* in [4-7]), or the accumulation points of a set of Ω boxes, these points not being a boundary of a box Ω [12]. Other accumulation points of Ω boxes are : the values $a^j_{(k)_s}$ corresponding to the limit points of the Myrberg bifurcations chains when $i \to \infty$ (accumulation of $\Omega^j_{k.2^i}$ boxes [4-7]), and the values $a^j_{(k)_o}$ corresponding to the birth of the two basic cycles (k, j), and constituting the first boundary of a Ω_k^j, the second being a_k^{*j}. The rotation sequence, related to the basic cycles of the boxes having an accumulation point of one of the preceding types, obey to noteworthy laws. Some of which are described in [3,12].

3. CONTINUITY PROPERTIES IN THE PHASE PLANE

3.1 Degenerate invariant curves of T_b, b = 0

Consider (1.1), f(x, a) being now a continuous function with respect to its arguments, and b = 0. Let $A(x=\alpha)$ be a fixed point of (1.2). For (1.1), A has the coordinates $x = \alpha$, $y = 0$. Let (C_o) be the curve, $y = -f(x, a)+\alpha$, in the phase plane (x, y). From (1.1), it is easy to verify that the set of the points of (C_o) has the same point A $(\alpha, 0)$ as a first rank consequent. Let $\alpha_{(-i)}$, i = 1,2,..., be a sequence of the antecedents of A, obtained from the inverse mapping T_o^{-1}. When T_o is an endomorphism, generally this sequence is constituted by an increasing number of branches with the rank of the inverse iteration T_o^{-1}, each branch being associated to a given determination of T_o^{-i} (here for simplicity's sake, $\alpha_{(-i)}$ is written with only one index i). Let (C_i) be a curve $y = -f(x, a) + \alpha_{(-i)}$, $i = 1,2,3,\ldots$ The set $\mathcal{C} = \bigcup_{j=0}^{\infty} (C_j)$ is considered as a degenerate ω invariant curve (stable manifold) passing through A, and associated to the multiplier $S_2 = 0$ (the other S_1 being the multiplier of A for (1.2)). If A is a repulsive fixed point for (1.2), it becomes a degenerate saddle for (1.1) with b = 0. Then a degenerate α invariant curve (instable manifold), located on y = 0, passes through A. In the same way, it is possible to define degenerate, ω and α, invariant curves for a saddle cycle of any order.

3.2 Invariant curves of T_b for $b \neq 0$

An invariant curve (C) of T_b, described in the phase plane by $G(x, y) = c$, c = constant, satisfies the functional equation $G(x_{n+1}, y_{n+1}) = G(x_n, y_n)$. The constant c is determined, for example, by means of an initial condition (x_o, y_o). Let $f(x, a)$ be an analytical function with respect to x. Consider a sufficiently small neighbourhood (d) of a saddle fixed point A, such that, inside (d), (C) can be described by a single-valued function $y = \theta(x)$, solution of the functional equation :

$$bx = \theta[f(x, a) + \theta(x)] \tag{3.1}$$

By a change of variables, it is assumed that the coordinates of A are $\bar{x} = \bar{y} = 0$ and that $\bar{f}(\bar{x}, a)$ is the new function f. Let $\bar{y} = \bar{\theta}(\bar{x}) = \sum_{i=1}^{\infty} \theta_i \bar{x}^i$ be the series expansion of the ω invariant curve passing through A. Substituting $\bar{\theta}(x)$ into (1.1) modified by the change of variables, $\bar{\theta}(\bar{x})$ can be written [7] :

$$\bar{\theta}(\bar{x}) = -\sum_{i=1}^{\infty} \bar{x}^i \, |d^{(i)}\bar{f}/d\bar{x}^i| \, / \, i! + bg_1(\bar{x}) + b^2 g_2(\bar{x}) + \ldots \tag{3.2}$$

$g_j(\bar{x})$, $j = 1,2,\ldots$, being functions such that $g_j(0) = 0$. According to Lattes [13], (d) containing A can be choosen sufficiently small in order that $\bar{\theta}(\bar{x})$ may be a convergent series inside (d). From (3.2) $\lim_{b\to 0} \bar{\theta}(\bar{x}) = -\bar{f}(\bar{x}, a)$, and with the variables (x, y) :

$$\lim_{b\to 0} \theta(x) = -f(x, a) + \alpha \tag{3.3}$$

which is the equation of a segment (C_o) of the degenerate ω invariant curve. For $b \neq 0$, (C) is associated to the multiplier S_2, $|S_2| < 1$.

Outside of (d), $y = \theta(x)$ can be continued, by iterating the so obtained segment with T_b^{-1}, in order to have the invariant curve (C). With (C) so defined, the following result appears :

$$\lim_{b\to 0} (C) = \bigcup_{j=0}^{\infty} (C_j) \tag{3.4}$$

When $b = \varepsilon \neq 0$, $|\varepsilon|$ infinitely small value, the segments (C_j) join at the infinite in the direction of the y axis, in order to give the continuous curve (C) (cf. pp. 349, 400 [7]).

Consider now the case $f(x, a) \equiv 1 - ax^2$, and the two fixed points q_1, q_2, $x = [-e \pm (e^2+4a)^{1/2}] / (2a)$, $y = bx$, $e = 1-b$ (- for q_1, + for q_2 in x). The α invariant curve (unstable manifold) passing through q_2 is given in fig.1. α is constituted by two segments: α_1 (continuous line), α_2 (broken line). Let C be the nearest point, from q_2, which has on α_2 a maximum of curvature, C_i, i = 1,2,3,..., the consequents of C. Let C' be the points of the segment $\widehat{C_1C_3}$, having a maximum of curvature, C'_i, i = 1,2,..., its successive consequents. Let $y(C_j)$, $y(C'_j)$ be the ordinates of C_j, C'_j, $C_o \equiv C$, $C'_o \equiv C'$. When $b \to 0$, then $C_j \to C'_j$, $y(C_j) \to 0$, j = 0,1,2,..., and the *points C_j, so defined for the bi-dimensional diffeomorphism (1.1), tends towards the critical points C_j of the one-dimensional endomorphism (1.2)* (cf. Chapter 6 of [7]).

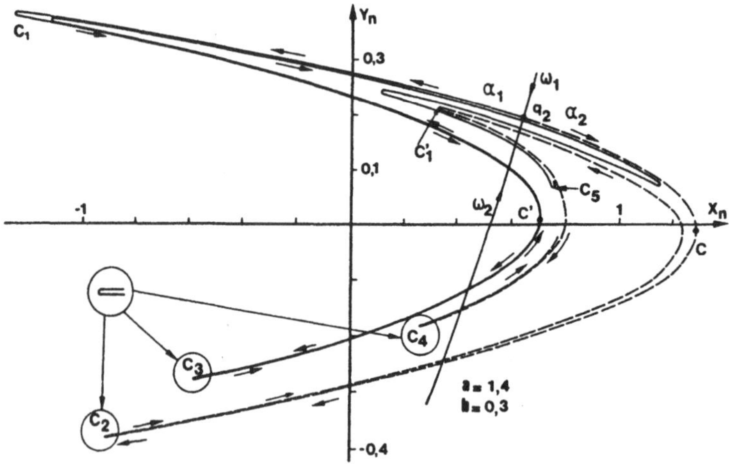

Figure 1

For q_1, a branch α_1 of the α invariant curve follows the α invariant curve of q_2, the other branch α_2 goes to infinite. In particular, the same can be done for the cycles issued from the Myrberg bifurcations chain, with k = 1.

For T_b, b = 0, the absorptive segment $\overline{CC_1}$, on y = 0, is the degenerate α invariant curve of q_2 resulting from the crushing of the α curve of fig. 1, on y = 0. The intersection of $\overline{CC_1}$ with the degenerate ω invariant curve of q_2 give *homoclinic points* of T_o, limits of the corresponding homoclinic points of T_b, when $b \to 0$. So *the consideration of degenerate invariant curves for T_o in the phase plane (x, y) is in connection with the definition of the homoclinic points of a one-dimensional endomorphism first given by Sharkovskij [14] in 1969,* without introduction of the degenerate invariant curves. Considering the cycles of order 2^i, $i = 1,2,\dots$, issued from q_2, defining α, ω degenerate invariant curves for these cycles, and considering the properties of the cycles and of the boxes Ω_{2^i} when $i \to \infty$ (cf. pp. 116-117 of [7]), one has : *the one-dimensional endomorphism T_o has homoclinic points if, and only if, it has cycles of order $\neq 2^i$, $i = 0,1,2,\dots$* [14,15]. The values $\hat{a}$, defined in § 2, correspond to the existence of *structurally unstable homoclinic, or heteroclinic points.* More details about this problem and its implications are given in [7] (pp. 397-400).

4. CONTINUITY PROPERTIES IN THE PARAMETER PLANE (a, b)

The "box-within-a-box" bifurcations structure of T_o stretches from $b = 0$ in the (a, b) plane [2,8,9], by considering the bifurcations curves of T_b. So the box Ω_1 is limited by two boundaries corresponding to $a_{(1)_o} = -1/4$, $a_1^* = 2$ for T_o. The first one, $\Lambda_{(1)_o}$, is the parabola $4a+(1-b)^2 = 0$, which corresponds to the merging of the two fixed points q_1, q_2 of T_b. The second boundary Λ_1^* is defined by a heteroclinic tangency between the ω invariant curve of the saddle fixed point q_1, and the α invariant curve of q_2 when it is a saddle. This heteroclinic tangency is the limit for which the greatest number of the heteroclinic points is attained. Ω_1 contains an infinite number of Ω_k^j boxes, $j = 1, 2,\dots, p_k$, having the form of a ribbon with a variable width, each being associated to a pair of basic cycles (k, j). The intersection of Ω_k^j, with $b = 0$, gives a box which has the same denomination for T_o. The boundaries of the Ω_k^j are defined in the same way as for Ω_1, by considering T_b^k. The symbolism associated to the "box-within-a-box" structure for T_o [7] remains the same for T_b, $b \neq 0$ [2,8,9], the indexes k, j, having the same values.

For $b \neq 0$, a Ω_k^j can overlap a Ω_m^ℓ, with $j \neq \ell$, $k = m$, or $j = \ell$, $k \neq m$, or $j \neq \ell$, $k \neq m$, the region of overlapping corresponding to the existence of two different cyclic attractors respectively of order nk and pm ; $n, p = 1,2,\dots$ So several boxes can overlap making appear the plane (a, b) as constituted by as many sheets. Two boxes, or more, can communicate via cusp type singularities of the boundary $\Lambda_{(k)_o}^j$ (locus of the merging of two basic cycles (k, j)). Two types of communications have been identified [9]. A first one through a region called "*cross-road area*" : an attractive cycle of a Ω_k^j passes into a Ω_k^ℓ, $j \neq \ell$ and can remain attractive when (a, b) follows a given way in the parameter plane. A second type is obtained via a "*spring area*" (form of a cuspidate caustic) : a Ω_k^j overlaps itself, and communicates with a Ω_m^j box, $m = k/n$, $n = 1,2,\dots$ For this type, whatever may be the way used for the communication in the (a, b) plane, a part of this way corresponds always to a repulsive cycle. Fig. 2 gives a partial aspect, $-1/4 \leqslant a \leqslant 3$, of the bifurcation curves*, the Λ_k^j corresponding to a bifurcation of a cycle (k, j) into a cycle of order 2k, when one of the multipliers is -1. More complete information about this problem is given in [2,8,9].

The continuity properties in the plane (a, b) have implications on the relations between the properties of the conservative case b = -1 for T_b, and those of T_o. When b = -1, a change of variables, followed

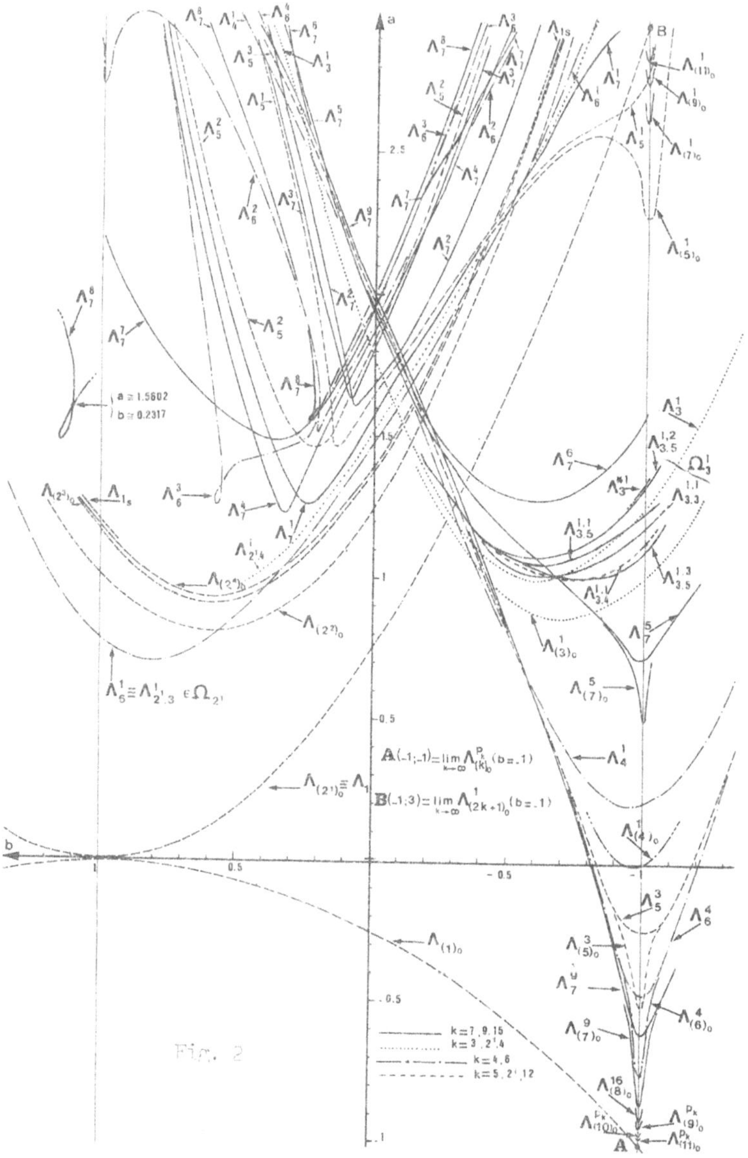

Fig. 2

* For 0<b<1 some of the curves were independtly found by Ushiki (unpublished work).

by a non linear substitution [2], permits to transform (1.1) into a mapping studied before (cf. Chapter 5 of [7]). In particular, this continuity gives the accumulation points of the intersections of the $\Lambda^j_{(k)_o}$ with b = 0 and b = -1, for $k \to \infty$, related to the set of bifurcations of the centre (b = -1) q_2 [2,3]. For $-1 \leq a \leq 3$, this set is such that a pair "saddle cycle k/r - centre cycle k/r", k being the order of the cycle, r the *index of rotation* [7], is released from q_2 (multipliers $S_{1,2} = \exp(j\varphi)$), each time $\varphi = \varphi_o = 2\pi r/k$, r, k, being relatively prime integers [7]. In this way, relations between the rotation numbers r/k of q_2, and the rotation sequences of a set of cycles of T_o, are obtained [2,3]. Such rotation sequences have noteworthy properties [3].

5. CONCLUSION

The above-mentioned properties for $f \equiv 1 - a x^2$, remain the same qualitatively, when f(x, a) is a continuous, and continuously differentiable function, having only one extremum, and satisfying other conditions (cf. p. 121 of [7]). For $f \equiv ax \pm x^3$, from the known bifurcation structure of T_o [7], it is possible to obtain the properties of T_b as for the quadratic case. Consider now the ordinary differential equations of one of the following types : either three-dimensional, autonomous, or two-dimensional with periodical coefficients of the independant variable. Each of these equations has a parameter μ, such that $\mu = o$ gives a one unit decrease of the dimension. The method of sections of Poincaré gives a generalization of T_b, $x_{n+1} = f(x_n, a) + y_n h(x_n, y_n)$, $y_{n+1} = b\, g(x_n, y_n)$ [7], $b = O(\mu^\alpha)$, $\alpha > o$, f, g, h being functions such that this mapping T is a difformorphism [7]. Then T_b can be considered as a first approach to the study of T.

REFERENCES

1. M.Hénon, Commun. Math. Phys., 50, 1976, p. 69-77.
2. C.Mira, 9th International Conference on Non Linear Oscillations (ICNO), Kiev, sept. 1981.
3. C.Mira, C.R. Acad. Sc. Paris, 294, série I, 1982, p. 689-692.
4. I.Gumowski, C.Mira, C.R. Acad. Sc. Paris, 281, série A, 1975, p. 41-44.
5. C.Mira, Proceedings of 7th ICNO, Berlin, sept. 75, Akad. Verlag, Berlin, 1977, p. 81-93.
6. C.Mira, RAIRO Automatique, v. 12, 1978, n° 1, p. 63-94, n° 2, p. 171-190.
7. I.Gumowski, C.Mira, Dynamique Chaotique, Cepadues, Toulouse, 1980.
8. H. El Hamouly, C.Mira, C.R. Acad. Sc. Paris, 293, série I, 1981, p. 525-528.
9. H. El Hamouly, C.Mira, C.R. Acad. Sc. Paris, 294, série I, 1982, p. 387-390.
10. R.Thom, Stabilité structurelle et morphogenèse, W.A. Benjamin, ed., Reading, 1972.
11. P.J.Myrberg, Ann. Acad. Sc. Fenn., série A, 256 (1958), p. 1-10, 268 (1959), p. 1-10, 336 (1963) p. 1-10.
12. C.Mira, C.R. Acad. Sc. Paris, 295, série I, 1982, p. 13-16.
13. S.Lattes, Ann. Di Matematica (3), 13, (1906), p. 1-137.
14. A.N. Sharkovskij, Proceedings of 5th ICNO, Kiev, 1969, p. 541,544.
15. V.V.Fedorenko, A.N. Sharkovskij, Proceedings of 5th Conference on qualitative theory of differential equations (August 1979), Kishinev "Shtinitsa", 1979, p. 174-175.

STRANGE ATTRACTORS FOR DIFFERENTIAL DELAY EQUATIONS

Carles Simó

Facultat de Matemàtiques, Universitat de Barcelona
Gran Via 585, Barcelona 7, Spain.

1. INTRODUCTION

Strange attractors are known for lots of differential equations and maps [3]. Also for DDE some chaotic behavior is observed. For the sunflower equation $\dot{x}(t)+(a/r)x(t)+(b/r)\sin(x(t-r))$, $r>0$ (planar model) it is shown [9,1] that periodic orbits (P.O.) exist for suitable a,b,r. Increasing r produces chaotic behavior not yet analyzed. The model for blood production [3] $\dot{x} = ax_\tau(1+x_\tau^c)^{-1}-bx$, $x_\tau(t) = x(t-\tau)$ has been studied [2] for $a=0.2$, $b=0.1$, $c=10$ and increasing τ. A steady stable solution appears up to $\tau=4.53$. Then and up to $\tau=13.3$ the attractor is a simple P.O. and a cascade of period doubling P.O. accumulates on 16.8. Chaotic behavior appears for larger τ. The existence of a, at least potential, S.A. is related to some transversal homoclinic point [8]. For the equation $\dot{x}(t) = f(x(t-1))$ the existence of P.O. with homoclinic connection is shown for suitable f [10].

2. A BIOLOGICAL POPULATION MODEL. SOME ANALYSIS

May [5] produced some simulation for a model of baleen whales population, namely $\dot{N}=-\mu N+\nu\hat{N}(1-\hat{N}^z)_+$ (2.1) with $\hat{N}(t)=N(t-T)$, $(f(t))_+=f(t)$ if $f(t)>0$, $(f(t))_+=0$ otherwise, $T=2$, $\mu=1$, $\nu=2$, and z a parameter. Our goal is to analyze this example in order to complete and understand May computations. Obviously $N^*=(0.5)^{1/z}$ is a steady solution. If $N=N^*+x$, linearizing (2.1) we get $\dot{x}=-x-(z-1)\hat{x}$ with characteristic equation $(\lambda+1)e^{2\lambda}+(z-1)=0$ (2.2). If λ is a solution of (2.2), $\bar{\lambda}$ also is. Eigenvalues cross the imaginary axis $\lambda=i\eta$ if $\arctan\eta+2\eta=-\pi+2k\pi$ and then $z=1+(1+\eta^2)^{\frac{1}{2}}$. First values are $z_1=2.51980$, $\eta_1=1.14446$, $z_2=5.16977$. For the Hopf P.O. created at z_1 the period is $T_1=5.49006$. We confine ourselves to $z\leq 4.2$ and only one pair of eigenvalues has crossed. If $\lambda=M\exp(i(\frac{\pi}{2}+\varepsilon))-1$ then ε is obtained from

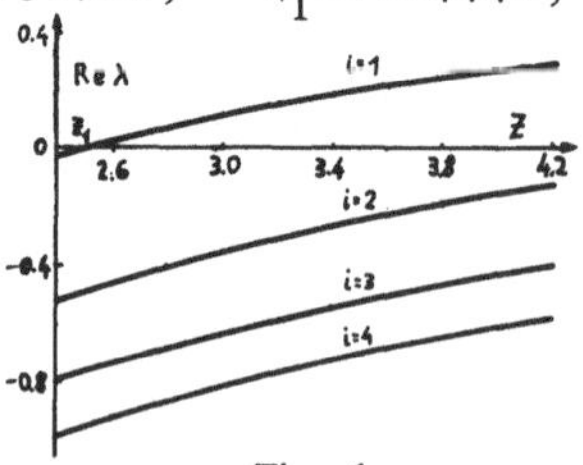

Fig. 1

$$\varepsilon = \arctan\frac{-\ln[e^2(z-1)2\cos\varepsilon/((2k-1.5)\pi-\varepsilon]}{(2k-1.5)\pi-\varepsilon}.$$

Fig. 1 shows the real part of the first four pairs of λ for $z \in$ $\in [2.4,4.2]$. The related eigenvalues of the Poincaré map P (see § 4) are obtained through exp $(T \cdot \mathrm{Re}\, \lambda_i)$ where the "period" $T = 2\pi / \mathrm{Im}\, \lambda_1$ goes from 5.52 at z=2.4 to 5.24 at z=4.2.

3. SUMMARY OF THE RESULT OF NUMERICAL COMPUTATIONS

We integrate (2.1) with different initial conditions: $N(t) = N^* + a + bt + ct^2$, $t \in [-2,0]$, a,b,c small or the i.c. given from the previous attractor (for a near z). The results agree. The plot $\hat{N}(t)$ versus $N(t)$ was studied and also the values of $(t, N(t))$ when $N(t) = N^*$. Using successive intersections and P several types of behavior were found (see Table 1). There k P.O. means a k P.O. of P and k S.A. a S.A. which has k pieces. We must note that for high values of z small loops appear in the orbit (see fig. 2,c,d). The intersections produced by these loops must not to be taken into account.

z	type of attractor	z	type of attractor
2.520-3.299	1.P.O.	3.68-3.69	1 S.A.
3.300-3.596	2 P.O.	3.70	5 P.O.
3.597-3.635	4 P.O.	3.71-3.74	1 S.A.
3.636-3.6415	8 P.O.	3.75	3 P.O.
3.642-3.6428	16 P.O.	3.76	3 S.A.
3.643-3.6432	32 P.O.	3.77-3.80	1 S.A.
3.64375	8 S.A.	3.81	7 P.O.
3.65	4 S.A.	3.82-3.93	1 S.A.
3.66	6 P.O.	3.94-4.20	1 P.O.
3.67	2 S.A.		

Table 1

4. A POINCARE MAP. A RELATED LORENZ MAP

Let $\mathscr{C}$ be the space of bounded continuous functions on $[-2,0]$ with sup norm and $\bar{\mathscr{C}} = \{\phi \in \mathscr{C} \mid f(-2) = N^*, f'(-2) < 0\}$. Starting with $\phi \in \bar{\mathscr{C}}$ we compute the Poincaré map P w.r.t. $\bar{\mathscr{C}}$. Let $(\phi_k, t_k), (\phi_{k+1}, t_{k+1})$ be successive points and times. To visualize it we draw $(\phi_{k+1}(0), t_{k+1} - t_k)$. It is a good display if $\Phi_t \phi$ has at most one unstable dimension and is

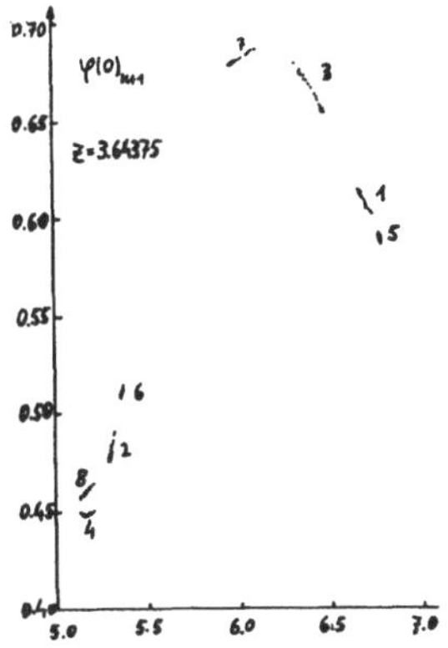

Fig. 2

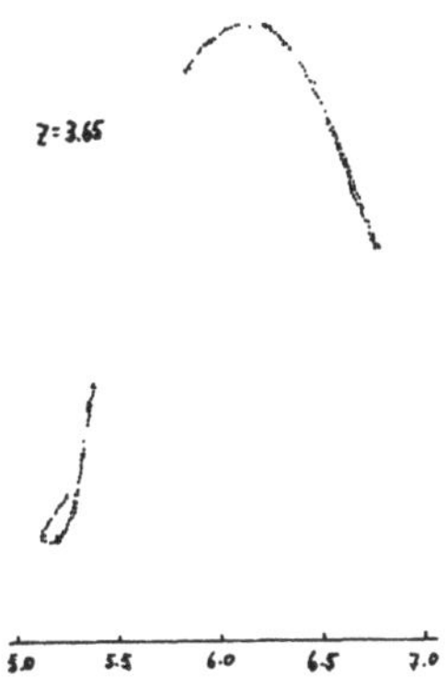

Fig. 3

Fig. 4

Fig. 5

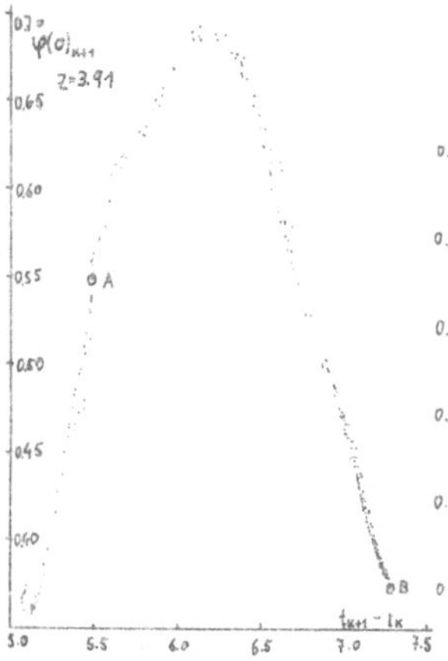

Fig. 6

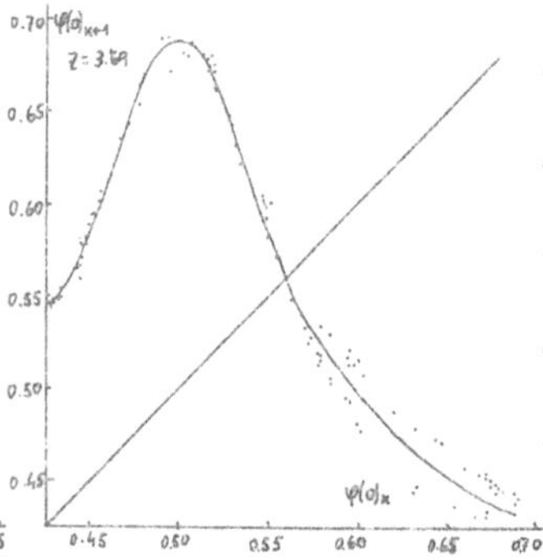

Fig. 7

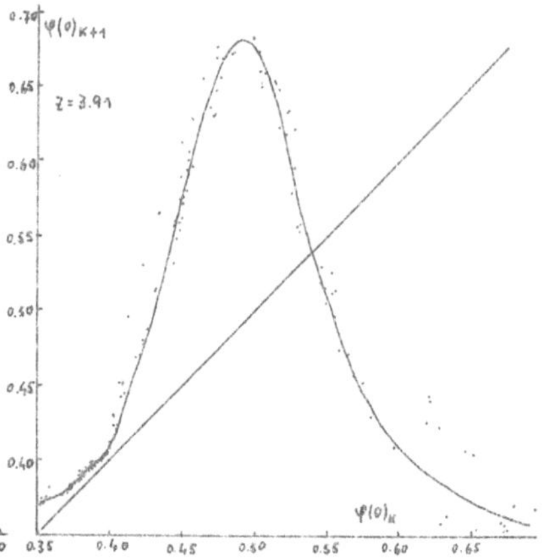

Fig. 8

strongly stable in all but two dimensions. Compare with results for N^*: for $z\in[3.6,4.2]$, $\exp(T\cdot \mathrm{Re}\,\lambda_3)\in[0.069,0.122]$. The skipped dimensions produce some dispersion in the 2-D representation.

Figures 2 to 6 show 2-D plots of the S.A. (after 10^3 iterates of P). In fig. 2,5 the numbers show consecutive pieces. In fig. 4 numbers 0 to 11 are consecutive points. A saddle, arising from the P.O. created by Hopf at z_1 is near point 1. In fig. 6 A is again this P.O. and B another saddle which destroys the attractor when it becomes a sink.

We can plot $\phi(0)_{k+1}$ versus $\phi(0)_k$ to get 1-D maps [4] (see fig.7,8). It is clear that this is a bad representation. However it is apparent that an unstable fixed point exist and that for z=3.91 intermittency [7] is present (compare with fig. 6 where many points are found near B).

5. ANALYSIS OF THE BEHAVIOR

First a cascade of period doublings is found. Afterwards, an inverse cascade of period halving S.A. Each halving is due to homoclinic tangencies of bifurcates from A [8]. This process is broken by intervals where a stable k P.O. (and after the direct and inverse related cascades) appears. The creation of a potential S.A. (whichs appears as $\overline{W}^u_A$ for some fixed or periodic A) is done through homoclinic tangency. The destruction by heteroclinic one when $\overline{W}^u_A \cap W^s_C \neq \emptyset$ for some W^s_C in the boundary of a sink.

Near z=3.94 a fixed point B goes from saddle to sink. The destruction of the S.A. is clear looking for a 2-D model of intermittency.

Questions: High order P.O. sinks can destroy the S.A. (Newhouse phenomenon [6]). Estimate the size of their bassin. How does the dimension reduction affect the qualitative description of the attractor? What happens when dim $W^u > 1$?

ACKNOWLEDGEMENTS

The author wishes to express his gratitude to Guillermo Gomez from the UNAM for interesting him in this problem during a stay in Mexico. The computing facilities of UNAM and U. Barcelona were used.

REFERENCES

1. A. Casal and A. Somolinos, Rev. Real Acad. Ciencias Ex., Fis. y Nat. LXXIII, 503 (1979).
2. D. Farmer, Ph.D. Thesis, Santa Cruz (Cal.)(1982).
3. L. Glass and M.C. Mackey, in Bifurcation Theory and Applications in Scientific Disciplines, Ann. N.Y. Acad. Sci. 316, 214 (1979).
4. E.N. Lorenz, J. Atmos. Sci. 20, 130 (1963).
5. R.M. May, in Nonlinear Dynamics, Ann. N.Y. Acad. Sci. 357, 267 (1980).
6. S.E. Newhouse, in Nonlinear Dynamics, Ann. N.Y. Acad. Sci. 357, 292 (1980).
7. Y. Pomeau and D. Manneville, Commun. Math. Phys. 74, 189 (1980).
8. C. Simó, J. Statist. Phys. 21, 465 (1979).
9. A. Somolinos, Quarterly of Appl. Math. 35, 465 (1978).
10. H.O. Walter, Nonlinear Analysis. Theory, Methods and Appl. 5, 775 (1981).

STOCHASTIC PERTURBATIONS OF SOME STRANGE ATTRACTORS

Gerhard Keller

Institut für Angewandte Mathematik
Universität Heidelberg
Im Neuenheimer Feld 294
D-6900 Heidelberg (F.R.G.)

1. INTRODUCTION

For a class of 2-dimensional, time-discrete dynamical systems including generalized baker's transformations [1] and similar mappings [2] we show that the asymptotic measure has an exponential mixing rate and is stable under small stochastic perturbations. The approximation of arbitrary systems by Markov-systems [3,4] is a special case.

2. THE ASYMPTOTIC MEASURE

Let $I=[0,1]$ and $J\subseteq\mathbb{R}$ be compact intervals. We consider the system $S:I\times J\to I\times J$, $(x,y)\mapsto(Tx,V_xy)$, where $T:I\to I$ is a piecewise C^2-monotonic transformation [5] with $\inf|T'|=:\alpha>1$ which we assume to be mixing [6], and where the $V_x:J\to J$ are λ-contractive, $0<\lambda<1$ ($|V_xy-V_xz|\leq\lambda\cdot|y-z|$). An example is [2]

$$Tx=2x\ (0\leq x\leq 1/2),\quad Tx=T(1-x)\ (1/2\leq x\leq 1),\quad V_xy=\sin\pi x+\lambda y. \tag{2.1}$$

Denote by M the space of all probabilities on $I\times J$ with absolutely continuous projection on I; i.e., those measures m which can be desintegrated by $m(A)=\int_I m(A|x)\cdot g_m(x)dx$, where the $m(.|x)$ are conditional probabilities on J and g_m is a probability density on I. For $m,\hat{m}\in M$, $x\in J$, and $f\in C(J)$ let $\Delta(m,\hat{m},f|x):=|g_m(x)\cdot\int f(y)dm(y|x)-g_{\hat{m}}(x)\cdot\int f(y)d\hat{m}(y|x)|$ and define

$$d(m,\hat{m}):=\int\sup\{\Delta(m,\hat{m},f|x):\ f \text{ Lipschitz conts., } Lip(f)\leq 1,\ \|f\|_\infty\leq L\}dx, \tag{2.2}$$

where L is the length of J. (M,d) is a complete metric space. The adjoint of S, defined by $\int h(x,y)d(S^*m)(x,y)=\int h(S(x,y))dm(x,y)$ $(h\in C(I\times J))$, acts continuously on M and has good contraction properties.

<u>Theorem 1</u> (on asymptotic measures)

a) There is an S-invariant measure $\bar{m}\in M$ with $\lim_{n\to\infty} d(S^{*n}m,\bar{m})=0$ for all $m\in M$.

b) If $h_1,h_2:I\times J\to\mathbb{R}$ are Hölder-continuous and if $\int h_1d\bar{m}=\int h_2d\bar{m}=0$, then $\int h_1(x,y)h_2(S^n(x,y))d\bar{m}(x,y)\longrightarrow 0$ exponentially fast for $n\to\infty$.

c) $\lim_{n\to\infty}\frac{1}{n}\Sigma_{k=0}^{n-1}h(S^k(x,y))=\int hd\bar{m}$ Lebesgue-a.e. for each $h\in C(I\times J)$.

The proof of a) and b) uses corresponding results for the base-transformation T [7,8] and the contraction properties of S^*. c) is an easy corollary to a). Because of a) the system $(S,\bar{m})$ is mixing.

3. PERTURBED SYSTEMS

A stochastically perturbed system S_K assigns to each point (u,v) a probability density on $I\times J$, $S_K:(u,v)\mapsto S(u,v)=(Tu,V_u v)\mapsto K(Tu,x)\cdot\tilde{K}(V_u v,y)$ where K and $\tilde{K}$ are stochastic kernels on I and J, i.e. $K,\tilde{K}\geq 0$ and $\int_I K(u,x)dx=\int_J \tilde{K}(v,y)dy=1$. In order to simplify the results we only consider two cases (more general kernels have been investigated in [9]). First,

$$K(u,v)=\tilde{K}(u,v)=k(v-u)+k(-v-u)+k(2-v-u) \quad \text{with} \qquad (3.1)$$
$$k(t)=0\ (|t|\leq 1),\ \int k(t)dt=1,\ k(t)=k(-t),\ k(s)\geq k(t) \text{ for } 0\leq s\leq t,$$

which is a convolution-kernel with "elastic reflections" at the endpoints of the interval. The functions $k_n(t)=n\cdot k(nt)$ generate a sequence of kernels $K_n,\tilde{K}_n$ converging weakly to the identity-kernel. A second class of kernels produces Markov-approximations [3,4]:

$$K_n(u,v)=n \text{ if } \frac{i-1}{n}\leq x,y\leq\frac{i}{n} \text{ for some } i,\quad K_n(u,v)=0 \text{ otherwise}, \qquad (3.2)$$

$\tilde{K}_n$ is defined analogously on $J\times J$.

Our main result on perturbed systems is:

<u>Theorem 2</u>: Suppose $|T'|\geq\alpha>2$ or $T(I_i)=[0,1]$ for each monotonicity interval of T. For the kernels (3.1) and (3.2) (and for many others!) one has:

a) There is an absolutely continuous probability $\bar{m}_K$ on $I\times J$, invariant under S_K, with $\lim_{n\to\infty} d(S_K^{*n}m,\bar{m}_K)=0$ for all $m\in M$.

b) For the above described sequences of kernels, $d(\bar{m}_{K_n},\bar{m})=0\left(\frac{\ln n}{n}\right)$, if the Banach-space-valued map $V:I\to C(J)$, $u\mapsto V_u$ is of bounded variation (e.g. $V_u=V_0$ for $x\leq\frac{1}{2}$, $V_u=V_1$ for $x>\frac{1}{2}$ or (2.1)).

It is not clear to which extent these results are valid in the case $1<\alpha\leq 2$ [9]. We conjecture that at least for kernels of the type (3.2) the restriction $\alpha>2$ of theorem 2 is not necessary. A partial result is

<u>Theorem 3</u>: For a transformation $T(x)=a+2(b-a)x$ if $x\leq\frac{1}{2}$, $T(x)=T(1-x)$ if $x\geq\frac{1}{2}$, with $2(b-a)>\sqrt{2}$, the results of theorem 2 remain valid.

The proof of theorem 2 takes advantage of analogous results for the base transformation T [9], while in the proof of theorem 3 the symmetry of the transformation T has been used in a crucial way. Similar results can be obtained for small deterministic perturbations of the system S.

REFERENCES

1. J.C. Alexander and J.A. Yorke, The fat baker's transformation, Preprint, Univ. of Maryland (1982).
2. D. Mayer and G. Roepstorff, Strange attractors and asymptotic measures of discrete-time dissipative systems, Preprint RWTH Aachen (1982).
3. H.J. Scholz, Dissertation Universität Essen (1980).
4. T.Y. Li, J. Approx. Th. 17, 177 (1976).
5. A. Lasota and J.A. Yorke, Trans. Amer. Math. Soc. 186, 481 (1973).
6. R. Bowen, Isr. J. Math. 28, 161 (1977).
7. G. Keller, C. R. Acad. Sci. Paris, Série A, 291, 155 (1980).
8. F. Hofbauer and G. Keller, Math. Z. 180, 119 (1982).
9. G. Keller, Monatsh. Math. (To appear).

SOLUTIONS OF STOCHASTIC DIFFERENTIAL EQUATIONS AND FRACTAL TRAJECTORIES

Bruce J. West

Center for Studies of Nonlinear Dynamics
La Jolla Institute
P. O. Box 1434
La Jolla, CA 92038

1. MOTIVATION AND BACKGROUND

The motivation for studying the behavior of simple dynamical systems and discrete mappings which exhibit chaotic behavior, aside from intrinsic interest, is the hope that this chaos property is generic and can provide insight into the statistical properties of actual physical systems. The chaotic solutions of these equations, called strange attractors, have been proposed as models for a number of physical phenomena in which the underlying physical process is stochastic and intermittent, i.e. clustered in space and/or time, e.g. homogeneous isotropic turbulence[1] the Belousov-Zhabotinskii chemical reaction[2], and many others.[3] Thus one of the basic problems that emerges from these studies is to determine the "statistical" properties of strange attractors. One measure of these properties that has become popular since Mandelbrot introduced it into the lexicon of scientific terms is that of a fractal dimension[4], i.e. the Hausdorff-Besicovich dimension D which strictly exceeds the topological dimension. For a bounded set of points in R^d the fractal dimensionality is defined by the number N of d-balls of size ε needed to cover the set. If this number increases as $N(\varepsilon) = \text{constant}\ \varepsilon^{-D}$ as $\varepsilon \to 0$ then D is the fractal dimensionality of the set.

In this lecture I will not discuss the use of fractal dimensions in characterizing strange attractors. My intent is to demonstrate how the concept has emerged in the solution to certain stochastic differential equations (s.d.e.'s) and to suggest possible relations between the statistical properties of the solutions to these s.d.e.'s and those of strange attractors. This strategy has only recently been adopted and so what I will discuss is primarily the properties of a class of *dynamic* Markoff processes whose trajectories have a fractal dimensionality. One of the most interesting features of the example we study is that the fractal dimensionality can be expressed in terms of the control parameters characterizing the underlying physical process. In addition many of the asymptotic properties of the solution to the s.d.e. can be related to those of Lévy processes (define below).

The distribution functions for the class of Markoff processes considered here satisfy the Bachelier-Smoluchowski-Chapman-Kolmogorov (BSCK) chain condition.[5] For translationally invariant Markoff processes with statistically independent increments in continuous space this condition is

$$P(x_2-x_1,t) = \int P(x_2-x,t-\tau)\; P(x-x_1,\tau)\; dx \tag{1.1}$$

where $P(x_2-x_1,t)$ is the probability density that the value of the process $x(t)$ changes from x_1 to x_2 in time t. The general solution to (1.1) on the interval $-\infty \le x \le \infty$ was obtained by Lévy[6] in terms of the characteristic function, i.e. the Fourier transform of the probability density, and is

$$\phi(k,t) = e^{-\gamma t|\underline{k}|^{\mu}} \quad ; \quad 0 < \mu \le 2 \quad . \tag{1.2}$$

We will refer to statistical processes characterized by (1.2) as symmetric Lévy processes in one, two and three dimensions. The parameter γt is real, positive and linear in the time t for translationally invariant processes[7]. More general time dependencies of the parameter have been investigated for non-translationally invariant Markoff processes[8].

There are a number of interesting properties of the Lévy distribution (1.3) that concern us here. Among these are: (1) It is a member of the class of infinity divisible distributions[9] and therefore is a fixed point of a renormalization group. Stated somewhat differently it is a fixed point of the convolution integral (1.1)[10] in d = 1,2,3. (2) It has the general scaling property, that for a real parameter λ,[11] $P(\lambda^{1/\mu}r,\lambda t) = \lambda^{-d/\mu}\, P(r,t)$; d = 1,2,3. (3) Its asymptotic form is a power law[4,7], i.e. at fixed t, $\lim_{r\to\infty} P(r) \sim \text{constant}/r^{d+\mu}$; d = 1,2,3, and $r = |x|$ for d=1. (4) The probability that r is outside some hypersphere of large radius R at fixed t is $\text{Prob}\,(r>R) \sim \text{constant}/R^{\mu}$.

Property (2) implies that if r(t) is a random variable with density P(r,t) then the two quantities $r(\lambda t)$ and $\lambda^{1/\mu}r(t)$ have the same distribution, in particular $\lambda = 1/t$. Note that $\mu = 2$ corresponds to Brownian motion and P(r,t) is then a zero-centered Gaussian distribution. Property (3) implies that the moment $\langle r^{\alpha}\rangle$ is finite if $\alpha < \mu$ and is infinite if $\alpha \ge \mu$ for $\mu < 2$.† Property (4) indicates that the r-process is that of a hyperbolic random variable. Such hyperbolic distributions preserve self-similarity and have trajectories with fractal

---

†No such expansion exists for the Gaussian distribution which has all moments finite.

dimensionality D. Thus, Lévy process with exponent μ have trajectories with Hausdorff-Besicovitch (fractal) dimensionality $D = \mu$[4,7]. This has also been shown for the non-symmetric but translationally invariant Lévy processes.[11]

We now present a s.d.e. to describe the evolution of a particular hyperbolic random variable. The statistical properties of the solution to this s.d.e. are described by an inverse power law probability distribution. This distribution enables us to associate a fractal dimensionality with the process described by the s.d.e. in a straightforward way. The simplest mechanical system we have found for which a hyperbolic random variable arises in an uncontrived manner is a linearly damped harmonic oscillator with additive fluctuations and a stochastic frequency.

2. LINEAR STOCHASTIC OSCILLATOR

The linear stochastic oscillator we study in this section satisfies the equations

$$\dot{x} = p \; ; \quad \dot{p} + 2\lambda p + \omega^2(t)x = f_1(t) \tag{2.1}$$

where $x(t)$ is the oscillator displacement, $p(t)$ is the oscillator momemtum (the mass has been set to unity), λ is a dissipation parameter, and $f_1(t)$ is a zero-centered delta-correlated Gaussian process with spectral strength D_1 independent of $x(t)$ and $p(t)$. The time dependent frequency $\omega^2(t)$ $\left(\equiv\Omega_0{}^2 + f_2(t)\right)$ is also a stochastic function with mean value $\Omega_0{}^2$ and zero-centered delta-correlated Gaussian fluctuations of spectral strength D_2 represented by $f_2(t)$. In order that (2.1) describe an oscillator for all time the fluctuations $f_2(t)$ cannot be Gaussian, i.e., $\omega^2(t) > 0$ always. The present choice of Gaussian statistics therefore implies that $\omega^2(t)$ can have negative values at some times in some realizations so that the evolution is not always oscillatory. This system is known to have an energetic instability in some region of the system parameter space[12]. Such a system has been used as an approximation to the Lorenz system in which the fluctuations model a degree of freedom that has been eliminated from the description.[13]

A description of the stochastic oscillator equivalent to (2.1) is given by the phase space equation of evolution. With the statistical assumptions made above this is a Fokker-Planck equation and is[12]

$$\frac{\partial}{\partial t} P(x,p,t) = \left\{-p\frac{\partial}{\partial x} + \frac{\partial}{\partial p}\left(2\lambda x + \Omega_0^2 x\right) + \left(D_1 + D_2 x^2\right)\frac{\partial^2}{\partial p^2}\right\} P(x,p,t) \tag{2.2}$$

where $P(x,p,t)dx\,dp$ is the probability that $x(t)$ and $p(t)$ have the values x and p at time t, respectively. An explicit solution to (2.2) has not been obtained even in the $t\to\infty$ limit. We therefore consider a transformation of variables that will enable us to find a solvable phase space equation.

We use the procedure of Stratonovich[14] to construct an equation of evolution for the energy envelope function $E = \frac{1}{2}p^2 + u(x)$, where $u(x)$ is the potential energy. The procedure is based on the notion that the average energy varies more slowly than the average displacement of the oscillator. This separation of time scales implies weak damping and in the x,E (rather than x,p) representation we can eliminate the rapid variable x from the phase space description. This technique is fairly standard and leads to the Fokker-Planck equation for the singlet distribution function $W(E,t)$[15], in terms of the scaled energy $\varepsilon \equiv 2\lambda E/D_1$ and the scaled time $\tau = 2\lambda t$

$$\frac{\partial}{\partial\tau}W(\varepsilon,\tau) = \frac{\partial}{\partial\varepsilon}\left\{\left[(1-2\beta)\varepsilon-1\right] + \frac{\partial}{\partial\varepsilon}\left[\varepsilon\ (1+\beta\varepsilon)\right]\right\} W(\varepsilon,\tau) \tag{2.3}$$

where $\beta \equiv D_2/4\lambda\Omega_0{}^2$. The s.d.e. corresponding to (2.3) is given by

$$\frac{d\varepsilon}{d\tau} = \frac{1}{2} - (1-\beta)\ \varepsilon + \left[\varepsilon(1+\beta\varepsilon)\right]^{1/2} f(\tau) \quad . \tag{2.3'}$$

The steady state solution to (2.3) is obtained by setting the left hand side equal to zero and integrating to obtain

$$W_{ss}(\varepsilon) \equiv W(\varepsilon,\infty) = \frac{(1-\beta)}{\beta(1+\beta\varepsilon)^{1/\beta}} \quad ; \ \beta < 1 \quad . \tag{2.4}$$

For $\beta > 1$ the solution is not normalizable and $W_{ss}(\varepsilon)$ is then not a probability density.

The probability that the energy envelope exceeds a value E for large E is

$$\text{Prob}(\varepsilon > E) = \int_E^\infty W_{ss}(\varepsilon)d\varepsilon \sim \text{constant } E^{1-1/\beta} \tag{2.5}$$

So that the ε-process is that of a hyperbolic random variable. Thus the trajectory of the energy envelope of a linear harmonic oscillator with frequency Ω_0, dissipation rate λ and linear multiplicative fluctuations of spectral strength D_2 has a "fractal" dimensionality $D = 4\lambda\Omega_0^2/D_2 - 1$. Note that the steady state distribution is only defined for $\beta < 1$ so that $D > 0$ always, but for $\beta < 1/3$ it is possible for D to exceed 2.

We can also interpret this result in the context of a random walk by recalling the model developed by Hughes et al.[16] They showed that if the transition probabilities in a random walk are given by a probability a^{-n} of taking a step $\pm\, b^n\Delta$ where Δ is the step length and (a,b) are real, then in the continuum limit the process is Lévy with exponent $\mu=\ln a/\ln b$. They argue that in one dimension the walk is persistent if $2>\mu\geq 1$ and is transient if $0<\mu<1$. If $D>1$ the set of sites visited will fill the one dimensional space. Thus $\mu>1$ implies persistence. If $D<1$, the set of sites visited is not dense and the walk is transient. That is to say the sites visited are clustered. Thus is the present case we interpret the energy trajectories described by (2.5) as clustered for $1>\beta>\frac{1}{2}$ for which $D<1$. Physicists tend to think of fluctuations as small perturbations about some average behavior. However in the case of fractals the fluctuations are of equal importance on all scales because there is no intrinsic to the fluctuations. Such behavior is observed in second order phase transitions and radar wave scattering in the ionosphere, for example.

The s.d.e. resulting from a Hamiltonian formulation of this problem in which the oscillator is put in contact with a heat bath of linear oscillators gives quite different results. If the coupling between the oscillator and heat bath is linear, then in the Markoff limit one obtains the usual Wang-Uhlenbeck oscillator. This is to say, a linearly damped harmonic oscillator driven by internally generated Gaussian white noise with the spectral level of the noise and dissipation rate satisfying a fluctuation-dissipation relation. Nothing new is learned in this case. If, however, the coupling between the oscillator and heat bath is *not* linear, then in the Markoff limit of the resulting generalized Langevin equation one obtains an oscillator with internally generated multiplicative fluctuations.[17] The dissipation corresponding to these fluctuations is determined to be nonlinear and in the Markoff limit there is again a fluctuation-dissipation relation of the first kind. For a quadratic coupling between the oscillator and heat bath the dissipation is found to have the form $\lambda_1 x^2 p$ and with this dissipation the steady state solution of the Fokker-Planck equation is the canonical distribution for thermal equilibrium, i.e., exp $[-\varepsilon]$[18]. Thus, the self-consistent Hamiltonian description of a multiplicative oscillator does *not* exhibit a fractal dimensionality. We conclude from this that the parametric fluctuations leading to a fractal trajectory of the energy envelope is a consequence of externally generated fluctuations.

It is one thing to establish that a given process is that of a hyperbolic random variable and quite another to measure the fractal dimensionality associated with the trajectory of the process. In the next section we present analytic asymptotic estimates for two measures of the fractal dimensionality of a hyperbotic random variable.

3. EXTREMA PROPERTIES

In this section we present analytic results for the mean first passage time and mean maxima of certain Fokker-Planck process corresponding to the s.d.e. of interest. We also discuss rigorous results for some related statistical properties of Lévy processes. Extreme value statistics provide information about the behavior of the wings of the steady state distribution function for the process in contrast to the usual second order statistics such as the variance and the auto-correlation functions which probe the region near the maximum of this function. For Lévy processes and Fokker-Planck processes for hyperbolic random variables these latter measures are infinite so that it is *only* through the extrema statistics that the statistical properties can be determined. We stress that the analytic results for the Fokker-Planck processes are based on asymptotic analyses of *dependent* processes not the more familiar independent process.

We begin our discussion by recalling some of what is known about the process generated by the intersection of a Lévy trajectory $\varepsilon(t)$ with a given level $\varepsilon = E$. This process would seem to bear some relation to the distribution of points on a Poincaré surface of section. We refer to the set of level crossing as the zero-set and it has been known for a long time that the statistics of the zero-set are identical to those of the first passage time for a Lévy process to achieve a specified level.[19] If $Q(y,\tau)$ is the first passage time distribution for the transition $0 \to y$ in time τ then[5]

$$P(\varepsilon,t) = \int_0^t P(0,t-\tau)\, Q(\varepsilon,\tau)\, d\tau \quad . \tag{3.1}$$

The convolution for Laplace transforms ($\mathscr{L}$) enable us to write directly from (3.1)

$$\text{Prob}\left(\varepsilon(\tau) \text{ touches } \varepsilon = E\ ,\ 0 < \tau < t\right) = \mathscr{L}^{-1}\,\frac{1}{u}\left\{\frac{\tilde{P}(E,u)}{\tilde{P}(0,u)}\right\} \tag{3.2}$$

where $\tilde{P}(E,u) = \mathscr{L}P(E,t)$ and $\mathscr{L}^{-1}$ is the universe Laplace transform operator. If $P(E,t)$ is a Lévy distribution with exponent $1 \le \mu < 2$ Kac[20] has shown (using a tauberian theorem) that

$$\text{Prob}\left(\varepsilon(\tau) \text{ does not touch } \varepsilon = E,\ 0<\tau<t\right) \sim f(\mu)\,|E|^{\mu-1}\, t^{\frac{1}{\mu}-1} \tag{3.3}$$

where $f(\mu)$ is a known function. Equation (3.3) is sufficient to establish that the first passage time has a fractal dimension $D = 1 - 1/\mu$, but in addition a more detailed analysis[19] shows that $Q(E,t)$ is a Lévy process with exponent $\beta = 1 - 1/\mu$. Thus the fractal dimensionality of the level crossings is also $D = 1 - 1/\mu$ and the zero-set distribution is Lévy.

An example of the dissimilarity of the original process $E(t)$ and that of the level crossings is provided by Brownian motion. In this case $\mu=2$ for $E(t)$ and the distribution is a zero-centered Gaussian with variance γt. The Lévy exponent for the zero-set is $\beta = 1-1/\mu = 1/2$ yielding the distribution

$$P(z) = \begin{cases} \dfrac{1}{\sqrt{2\pi z^3}} \exp\left[-\dfrac{1}{2z}\right] & z \ge 0 \\ 0 & z < 0 \end{cases} \tag{3.4}$$

with $z = E^2/\gamma t$. Thus even though Brownian motion is space filling in $d = 1$ and 2 with $D = 2$, the zero-set with $D = \frac{1}{2}$ is clustered.

The distribution (3.4) can be used to calculate the mean first passage time

$$T_1(E) = \int_0^\infty t\, P(E,t)\, dt = E^2/\gamma \tag{3.5}$$

where $P(E,t)\, dt = P(z)dz$. The result (3.5) is well known for the mean first passage time for a Brownian trajectory to intercept a level $\varepsilon=E$ in $d = 1,2,3$. In general the mean first passage time for a Lévy trajectory to achieve a level ε scales as $T_1(\varepsilon) \sim \varepsilon^\mu$ where μ is the Lévy exponent so that $D=\mu$ as noted earlier. Thus we can write the fractal dimensionality as $D \sim \ln T_1(\varepsilon)/\ln \varepsilon$ independent of d. Below we show that the same relation can be used for the s.d.e. discussed in the preceeding section.

The mean first passage time for a dynamic process can be calculated solely from a knowledge of the steady state solution to the Fokker-Planck equation. This is due to the fact that it is possible to write an exact differential equation (the Pontryagin equation) for $T_1(\varepsilon)$ starting from the backward form of the Fokker-Planck equation with the appropriate boundary conditions. This equation can be integrated exactly so that the mean time for the oscillator trajectory to intersect a circle of radius E in phase space for the first time is given by[21]

$$T_1(E) = \int_{\varepsilon_0}^{E} d\varepsilon \left[\int_0^\varepsilon W_{ss}(\varepsilon')d\varepsilon'\right] \Big/ \left[\tfrac{1}{2}M(\varepsilon)W_{ss}(\varepsilon)\right] \tag{3.6}$$

where $M_2(\varepsilon)$ is the state-dependent diffusion coefficient in (2.3), i.e., $M_2(\varepsilon) = \varepsilon(1 + \beta\varepsilon)$, ε_0 is the initial energy of the oscillator and $E \gg \varepsilon_0$. Inserting the steady-state solution (2.4) into (3.6) and integrating yields the asymptotic result

$$T_1(E) \sim \text{constant } E^{\frac{1-\beta}{\beta}} + O(\ln E/E^{\frac{1-\beta}{\beta}}) \quad . \tag{3.7}$$

Equation (3.7) supports the qualitative arguments of Mandelbrot relating the fractal dimensionality of a trajectory to the time spent in a given volume of space, i.e. $T_1(\varepsilon) \sim \varepsilon^D$, so that the fractal dimensionality is $D = (1-\beta)/\beta$.

The mean first passage time, however, is not a sharply defined quantity because its distribution is very broad [cf. (3.4)]. Maxima moments, on the other hand, are a sensitive measure of the fractal dimensionality.[15]

We define the random variable $\hat{\varepsilon}(t)$ to be the greatest excursion of the energy envelope $\varepsilon(t)$ from zero up to time t, i.e.,

$\hat{\varepsilon}(t) \equiv \max \left\{ \varepsilon(\tau),\ 0 \le \tau \le t \right\}$. Let $\varepsilon_\nu(t)$ be the average of $\hat{\varepsilon}^\nu(t)$ over an ensemble of response histories of the oscillator. It has been shown that the long time behavior of the mean maximum $\varepsilon_\nu(t)$ is given by the asymptotic relation[21]

$$\varepsilon_\nu(t) \sim \int_0^\infty dE\ E^{\nu-1} \left\{ 1 - \exp\left[-t/T_1(E) \right] \right\} \tag{3.8}$$

where $T_1(E)$ is defined by (3.6). An asymptotic analysis of (3.9) yields the scaling result $\varepsilon_\nu(t) \sim$ constant $t^{\nu\beta/(1-\beta)}$ that is, $\varepsilon_\nu(t) \sim t^{\nu/D}$. This asymptotic result has been shown elsewhere[7] to be exact for the class of translationally invariant Lévy processes in d = 1, 2, and 3.

4. DISCUSSION

It should be emphasized that although the asymptotic properties of the hyperbolic random variable described by the s.d.e. and that described by the Lévy distribution are the same, the evolution of the two systems are distinct. The Lévy process does not satisfy a diffusive equation of the Fokker-Planck type, but rather a more general Master equation, i.e.

$$\frac{\partial}{\partial t} P(\varepsilon,t) = \frac{\gamma}{\pi} \sin(\pi\mu/2)\ \Gamma(1+\mu) \int_0^\infty \frac{d\varepsilon'}{|\varepsilon-\varepsilon'|^{1+\mu}} P(\varepsilon',t);\ d=1, \tag{4.1}$$

with similar expressions in d=2 and d=3. Of course the solution to (4.1) given by (1.2) does *not* have steady state, i.e. $\lim_{t\to\infty} \phi(k,t) = 0$. However we have examined the response of a dissipative linear system to Lévy fluctuations described by (1.2) and found the response to be stable. Consider the s.d.e.

$$\frac{d\varepsilon(t)}{dt} + \lambda\varepsilon(t) = L(t) \tag{4.2}$$

where $\lambda > 0$ and $L(t)$ is a differential Markoff process specified by (1.2). The characteristic function for the statistics of the solution to (4.2) with the initial distribution $P(\varepsilon,t_0|\varepsilon_0,t_0) = \delta(\varepsilon)$ is

$$\phi(k,t) = e^{-\sigma_\mu^2(t-t_0)|k|^\mu} \tag{4.3}$$

where $\sigma_\mu^2(t-t_0) = \gamma/\mu\lambda\ \left(1-\exp[-\mu\lambda(t-t_0)]\right)$, $t>t_0$. This Lévy process does have a steady state with a "variance" $\gamma/\mu\lambda$.

Thus for large ε both s.d.e.'s (2.3') and (4.2) have power law probability distributions as $t\to\infty$. The exponents and coefficients are equal when $\mu = 1/\beta-1$ and $\gamma/\lambda = \pi\mu^2\ [\sin\mu\pi/2\ \Gamma(1+\mu)]^{-1}$ yielding equivalent statistical properties. The approach to this steady state behavior can however, be quite different in the two cases. Inasmuch as we are unable to solve the Fokker-Planck equation (2.3) for arbitary t, questions regarding the transient behavior of the s.d.e. (2.3') cannot be answered. Whereas (4.3) completely specifies the transient behavior of the Lévy process.

We conclude by summarizing the content of the lecture. Stochastic processes described by fractal trajectories are characterized by hyperbolic random variables so that the distribution function has a long tail. The existence of this tail indicates that much of the information about the process is contained in the wings of the distribution. In particular low order moments diverge. Thus in order to determine the dimensionality of the process measures must be introduced which emphasize this tail region. The mean time for a process to first emerge from a closed region of radius ε is one such measure. For a translationally invariant Markoff process, i.e. a Lévy process, the mean first passage time is given exactly by $T_1(\varepsilon) = c_\mu \varepsilon^\mu$ where μ is the Lévy exponent and the constant C_μ can be determined exactly. This same scaling law has been shown to hold more generally for certain s.d.e. having nontranslationally invariant solutions.[8] A second measure, which has a distribution narrower than that of the mean first passage time are the maxima moments $\varepsilon_\nu(t)$. Again one obtains an exact scaling law for a Lévy process $\varepsilon_\nu(t) = D_\mu t^{\nu/\mu}$ where D_μ is a constant which can be determined exactly.

These two measures of the fractal dimensionality are useful in general because $T_1(\varepsilon)$ and $\varepsilon_\nu(t)$ can be determined solely from a knowledge of the steady state solution of the Fokker-Planck equation corresponding to the s.d.e. Thus if the s.d.e. is that of a hyperbolic random variable then the fractal dimensionality can be determined immediately.

Finally the statistics of the zero-set for a Lévy process are the same as for the first passage time to a given level. There is also an apparent similarity between the zero-set process and the points on a Poincaré surface of section generated for a Hamiltonian system. Thus the above association may enable one to relate the distribution of times between successive interceptions of a surface of section by a trajectory in a Hamiltonian system to the distribution of points on the surface of section. Such an assumption may be tested using the extrema measures discussed above.

ACKNOWLEDGEMENT

We thank the Air Force Office of Scientific Research for support of this research and V. Seshadri for many stimulating discussions.

REFERENCES

1. H.A. Rose and P.L. Sulem, Le Journal de Physique, 441 (1978).
2. R.H. Simoyi, A. Wolf and H.L. Swinney, Phys. Rev. Lett. 49, 245 (1982)
3. See e.g. the papers in Nonlinear Dynamics, ed. R.H.G. Helleman, Ann. New York Acad. Sci. 357, (1980).
4. B.B. Mandelbrot, Fractals, Form, Chance and Dimension, Freeman San Francisco, (1977).
5. E.W. Montroll and B.J. West in Fluctuation Phenomena, eds. E.W. Montroll and J.L. Lebowitz, North-Holland, Amsterdam (1979).
6. P. Lévy, Procès Stochastiques et Mouvement Brownien, Paris (1948).
7. V. Seshadri and B.J. West, Proc. Natl. Acad. Sci. 79, 4501, (1982).
8. B.J. West and V. Seshadri, Physica 113, 203 (1982).
9. B.V. Gnedenko and A.M. Kolmogorov, Limit Distributions for Sums of Independent Random Variables, Addison-Wesley, Cambridge (1954).
10. G. Jona-Lasinio, Nuovo Cimento 26B, 99 (1975).
11. S.J. Taylor, J. Math. and Mech. 16, 1229 (1967).
12. B.J. West, K. Lindenberg and V. Seshadri, Physica 102A, 470 (1980); ibid 105A, 445 (1981); Phys. Rev. A 22, 2171 (1980).
13. E. Knobloch, J. Stat. Phys. 20, 695 (1979).
14. R.L. Stratonovich, Topics in the Theory of Random Noise, vol. 1, Gordon and Breach, New York (1967).
15. V. Seshadri, B.J. West and K. Lindenberg, J. Sound and Vibration 68, 553 (1980).
16. B.D. Hughes, M.F. Shlesinger and E.W. Montroll, Proc. Nat. Acad. Sci. USA, 78, 3287 (1981); J. Stat. Phys. 28, 111 (1981).
17. R. Zwanzig, J. Stat. Phys. 9, 215 (1973).
18. K. Lindenberg and V. Seshadri, Physica 109A, 483 (1981).
19. R.M. Blumenthal and R.K. Getoor; Illinois J. Math. 6, 308 (1962).
20. M. Kac, Publ. Inst. Statist. Univ. Paris, 6, 303 (1957).
21. K. Lindenberg, K.E. Shuler, J. Freeman and T.J. Lie, J. Stat. Phys. 12, 217 (1975).

CONTINUOUS BIFURCATION AND DISSIPATIVE STRUCTURES ASSOCIATED WITH A SOFT MODE RECOMBINATION INSTABILITY IN SEMICONDUCTORS

Eckehard Schöll

Institut für Theoretische Physik
Rheinisch-Westfälische Technische Hochschule
D-5100 Aachen
W. Germany

1. INTRODUCTION

There exist numerous physical, chemical and biological examples of nonlinear dynamical systems which exhibit bifurcation and dissipative structures far from equilibrium[1]. Here we consider a semiconductor, subject to generation, recombination, and diffusion of carriers, with an externally applied electric field. This field heats the electrons up, and drives the system far from thermal equilibrium. The steady state is determined by the balance of various nonlinear generation and recombination processes, among which impact ionization of free or bound carriers represents the essential autocatalytic step[2]. The dissipative dynamical system described by the semi-conductor transport equations may give rise to soft-mode instabilities, saddle-node and Hopf bifurcations, limit cycle oscillations, bifurcation of spatially inhomogeneous dissipative structures, and nonequilibrium phase transitions between low and high current states[3-6].

The semiconductor generation-recombination instabilities considered here represent particularly instructive models for self-organization and dissipative structures in systems far from equilibrium, since they are easily accessible to both analytical treatment and experimental observation. The various bifurcating solutions show up as different branches of the current-voltage characteristics[7], and the dissipative spatial structures can be observed directly from the recombination radiation emitted by the enhanced electron concentration of high current filaments[8].

2. GENERATION-RECOMBINATION MODEL

As a specific example we consider an n-type semiconductor with donor concentration N_D and compensating acceptor concentration $N_A < N_D$. The acceptors are all assumed to be occupied by electrons. The generation and recombination processes between the conduction band and the donor ground and excited level are shown in Fig. 1. The symbols denote the various rate constants. The impact ionization coefficients X_1^*, X_1 are increasing functions of the applied electric field E; either X_1^*, X_1 or E may be taken as control parameters.

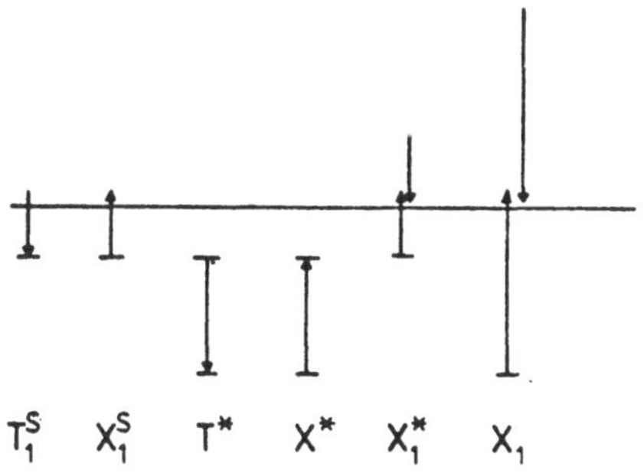

Fig. 1: Generation-recombination processes

We normalize the concentrations of free electrons (n), electrons bound in the donor ground state (n_D), and the excited state (n_{D*}) by the effective donor concentration $P_A := N_D - N_A$:

$$\nu := n / P_A \quad , \quad \nu_D := n_D / P_A \quad , \quad \nu_{D*} := n_{D*} / P_A \tag{2.1}$$

and introduce dimensionless time and space variables

$$\tau := \Gamma_0 t \quad , \quad \underline{\xi} := \underline{x} / L_D \tag{2.2}$$

with the dielectric relaxation frequency $\Gamma_0 := 4\pi e \mu P_A / \varepsilon_M$ and the Debye length

$$L_D := [\varepsilon_M D / (4\pi e \mu P_A)]^{1/2}.$$

D, μ and ε_M are the diffusion constant, mobility, and static dielectric constant, respectively. The dimensionless electric field is

$$\underline{\varepsilon} := (\mu L_D / D) \underline{E} . \tag{2.3}$$

The concentrations and fields are coupled by the following continuity equations

$$\frac{\partial}{\partial \tau} \nu - \underline{\nabla} \cdot (\underline{\varepsilon} \nu + \underline{\nabla} \nu) = \varphi_1 (\nu, \nu_D, \nu_{D*}) \tag{2.4a}$$

$$\frac{\partial}{\partial \tau} \nu_D = \varphi_2(\nu, \nu_D, \nu_{D*}) \tag{2.4b}$$

$$\frac{\partial}{\partial \tau} \nu_{D*} = \varphi_3(\nu, \nu_D, \nu_{D*}) \tag{2.4c}$$

and Maxwell's electrostatic equation:

$$\underline{\nabla} \cdot \underline{\varepsilon} = 1 - \nu - \nu_D - \nu_{D*} . \tag{2.4d}$$

The generation-recombination rates are defined by

$$\varphi_1 := \left[X_1^S \nu_{D*} - T_1^S P_A \left(\frac{N_D}{P_A} - \nu_D - \nu_{D*}\right)\nu + X_1^* P_A \nu_{D*} \nu + X_1 P_A \nu_D \nu\right] / \Gamma_0 \tag{2.5a}$$

$$\varphi_2 := \left[-X^* \nu_D + T^* \nu_{D*} - X_1 P_A \nu_D \nu\right] / \Gamma_0 \tag{2.5b}$$

$$\varphi_3 := -\varphi_1 - \varphi_2 \tag{2.5c}$$

Of course φ_1, φ_2 and φ_3 depend implicitly upon ε via the impact ionization coefficients X_1^*, X_1.

The spatially homogeneous steady states of (2.4) are given by

$$\begin{aligned} \varphi_1(\nu, \nu_D, \nu_{D*}) &= \varphi_2(\nu, \nu_D, \nu_{D*}) = 0 \\ \nu + \nu_D + \nu_{D*} &= 1 \end{aligned} \tag{2.6}$$

For suitable values of the other rate constants there exists a domain in the (X_1, X_1^*) control parameter space where (2.6) gives three homogeneous states, corresponding to two stable nodes and a saddle in phase space[5]. At the boundaries of this domain saddle-node and pitchfork bifurcations occur. Employing a particular field dependence of X_1, X_1^*, one obtains a curve in the (X_1, X_1^*) plane, along which $\nu(\varepsilon_0)$ and hence the current $j_0 = \nu(\varepsilon_0)\,\varepsilon_0$ is given as a function of the externally applied electric field ε_0. Thus an S-shaped current voltage characteristic may be produced.

3. BIFURCATION OF INHOMOGENEOUS STEADY STATES

We linearize the transport equations (2.4) around the homogeneous steady state $\underline{\Phi} := (\underline{\varepsilon}_0, \nu, \nu_D, \nu_{D*})$ for small perturbations

$\delta \underline{\phi}(\underline{\xi},\tau)$ and Fourier transform the resulting linear system:

$$\delta \underline{\phi}(\underline{\xi},\tau) = \delta \underline{\tilde{\phi}}(\underline{k},\lambda)\, e^{i\underline{k}\cdot\underline{\xi}}\, e^{\lambda\tau} \tag{3.1}$$

For transverse perturbations $(\underline{k} \parallel \delta\underline{\varepsilon} \perp \varepsilon_0)$ one obtains the dispersion relation

$$0 = (\lambda-\lambda_1)(\lambda-\lambda_2)(\lambda+\nu) + k_\perp^2(\Delta + \lambda\Theta + \lambda^2) \tag{3.2}$$

where

$$\Delta := \left[(X_1^S + T_1^S P_A\nu + X_1^* P_A\nu)(X^* + X_1 P_A\nu) + T^*(T_1^S + X_1)P_A\nu\right] / \Gamma_0^2 \tag{3.3}$$

$$\Theta := \left[X^* + T^* + X_1^S + (T_1^S + X_1 + X_1^*)P_A\nu\right] / \Gamma_0 \tag{3.4}$$

and λ_1, λ_2 are the non-zero(always real!) eigenvalues of the Jacobian matrix

$$A_{ij} := \frac{\partial \varphi_i}{\partial \nu_j} \qquad (\nu_1 := \nu,\ \ \nu_2 := \nu_D,\ \ \nu_3 := \nu_{D*}) \tag{3.5}$$

The stability of $\underline{\phi}$ against spatially homogeneous, charge conserving perturbations is determined by λ_1, λ_2. For stable steady states $\lambda_1, \lambda_2 < 0$, whereas for unstable steady states $\lambda_1 > 0 > \lambda_2$.

At the boundaries of the bistability domain in the (X_1, X_1^*) control parameter plane the eigenvalue λ_1 tends to zero, resulting in a soft mode instability. This corresponds to values ε_h and ε_{th} of the external field ε_0 where the current-voltage characteristic has points of inflexion with infinite differential conductivity $dj_0/d\varepsilon_0$, similar to a "fold" in catastrophe theory. It reflects in a direct and experimentally observable way the saddle-node bifurcation.

The spectrum of normal modes given by (3.2) is shown in Fig. 2 for an unstable homogeneous steady state. It consists of real modes $(\mathrm{Im}\,\lambda = 0)$ modulated in space transversally to the applied electric field ε_0. The branch $\lambda^{(1)}$ becomes undamped for $k_\perp < k_\perp^b$. With increasing ε_0 the wavevector $k_\perp^b$ changes as represented in Fig. 3.

This leads to the successive bifurcation of a family of spatially inhomogeneous stationary solutions from the homogeneous unstable steady state, with increasing ε_0.

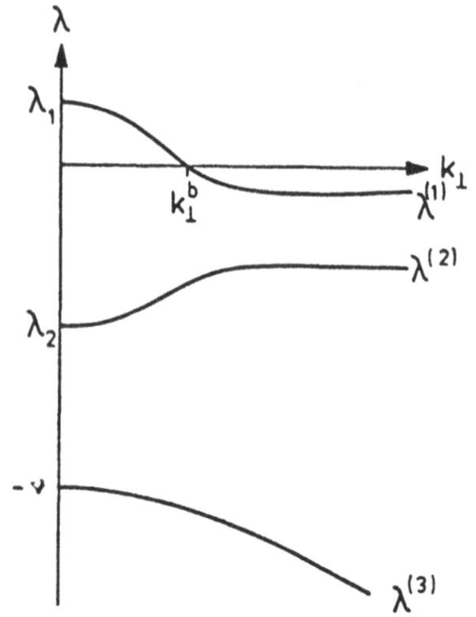

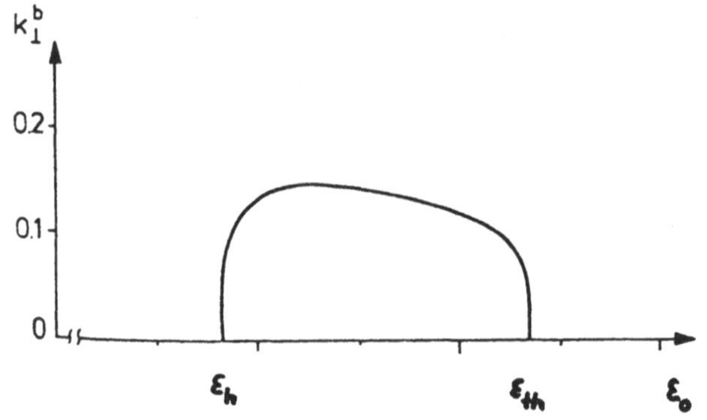

Fig. 2: Modes of the unstable homogeneous steady state

Fig. 3: Bifurcation line

The solutions are modulated with period

$$\Lambda_\perp(\varepsilon_0) = \frac{2\pi}{k_\perp^b(\varepsilon_0)} = 2\pi\sqrt{\frac{\Delta}{-\nu\lambda_1\lambda_2}} \tag{3.6}$$

In an infinitely extended sample this family is continuous, while for periodic boundary conditions of period L there exists a discrete set of bifurcation solutions with $\Lambda_n = L/n \quad (n \in \mathbb{N})$.

Similarly, for longitudinal perturbations $(\underline{k} \parallel \delta\underline{\varepsilon} \parallel \underline{\varepsilon}_0)$ of the homogeneous steady state, a family of travelling waves, propagating parallel to the applied field with period

$$\Lambda_\parallel(\varepsilon_0) = \frac{2\pi}{k_\parallel^b(\varepsilon_0)} \approx 2\pi\sqrt{\frac{\hat{\mu}}{\lambda_1}\left[1+\varepsilon_0^2\,\frac{(2\lambda_1+\Theta)-\hat{\mu}(2\lambda_1-\lambda_2+\nu)}{(\lambda_1-\lambda_2)(\lambda_1+\nu)}\right]} \tag{3.7}$$

and phase velocity

$$v := \hat{\mu}\varepsilon_0 \approx \frac{\lambda_1^2+\lambda_1\theta+\Delta}{(\lambda_1-\lambda_2)(\lambda_1+\nu)}\,\varepsilon_0 \tag{3.8}$$

bifurcates from the steady state[6].

4. DISSIPATIVE STRUCTURES

The transversal stationary spatial patterns which emerge from the bifurcation of inhomogeneous solutions can be computed directly from the full nonlinear transport equations (2.4). Elimination of ν_3, ν_{3*} by (2.4b,c) yields, for a plane geometry,

$$\frac{d}{d\xi}\nu = -\nu\mathcal{E}_\perp \quad , \quad \frac{d}{d\xi}\mathcal{E}_\perp = \varrho(\nu,\mathcal{E}_0) \tag{4.1}$$

Here ξ and $\mathcal{E}_\perp$ are the spatial coordinate and the internal electrical field perpendicular to the external field $\mathcal{E}_0$. The dimensionless charge density

$$\varrho(\nu,\mathcal{E}_0) := (a\nu^3 + b\nu^2 + c\nu + d) \,/\, (\Gamma_0^2\Delta(\nu)) \tag{4.2}$$

with

$$a := -(T_1^S + X_1^*)X_1P_A^2$$
$$b := [X_1^*X_1P_A - (X_1^S + T_1^SN_A + T^*)X_1 - X^*X_1^* - T_1^S(T^* + X^*)]\,P_A$$
$$c := [(X_1^S + T^*)X_1P_A + X^*X_1^*P_A - X_1^SX^* - T_1^SN_A(T^* + X^*)]$$
$$d := X_1^SX^*$$

is a nonlinear function of ν and depends implicitly upon the control parameter via X_1 and X_1^*. The dynamical system (4.1) is "Hamiltonian" in terms of the variables $\ln\nu$, $\mathcal{E}_\perp$ and ξ, and a first integral is given by

$$\frac{\mathcal{E}_\perp^2}{2} + \int \varrho(\nu,\mathcal{E}_0)\, d(\ln\nu) = \text{const.} \tag{4.3}$$

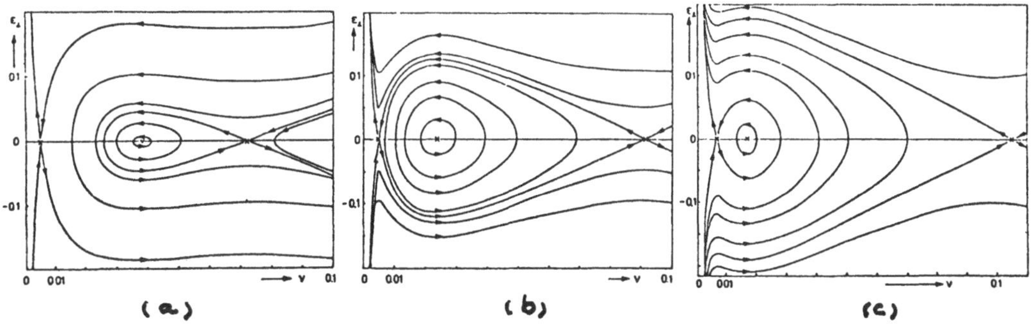

Fig. 4: Phase portraits of (4.1) for three different values of the control parameter $\mathcal{E}_0$ increasing from (a) to (c).

For $\mathcal{E}_h < \mathcal{E}_0 < \mathcal{E}_{th}$ (4.1) has three fixed points: two saddles corresponding to the stable homogeneous steady states, and a center corresponding to the unstable homogeneous steady state. The trajectories in phase space include oscillatory, solitary and unbounded inhomogeneous steady states (Fig. 4). The solitary solutions are homoclinic orbits (saddle-to-saddle loops) corresponding to hump-shaped electron concentration profiles, and - at a special value $\mathcal{E}_{co}$ of the control parameter $\mathcal{E}_0$ - heteroclinic orbits, i.e. saddle-to-saddle separatrices connecting the two stable homogeneous steady states in a monotonic, kink-shaped concentration profile. The required condition upon $\mathcal{E}_0$

$$\int_{\nu_1}^{\nu_3} g(\nu, \mathcal{E}_0)\, d(\ln \nu) = 0 \tag{4.4}$$

can be visualized as a spatial coexistence condition for the two homogeneous phases of the system: the high and the low conductivity states. Similar coexistence conditions were derived by Schlögl for a chemical reaction model with nonequilibrium phase transitions[9].

The stability of the inhomogeneous solutions $\underline{\phi}(\xi) = (\mathcal{E}_0 + \mathcal{E}_\perp(\xi), \nu(\xi), \nu_D(\xi), \nu_{D^*}(\xi))$ against transversal perturbations $\delta\underline{\phi}(\xi,\tau) = (\delta\mathcal{E}_\perp(\xi), \delta\nu(\xi), \delta\nu_D(\xi), \delta\nu_{D^*}(\xi))\, e^{\lambda\tau}$ follows from the linearization of (2.4):

$$\begin{aligned}
(\lambda + \nu)\,\delta\mathcal{E}_\perp + \left(\mathcal{E}_\perp + \frac{d}{d\xi}\right)\delta\nu &= 0 \\
A_{21}\,\delta\nu + (A_{22} - \lambda)\,\delta\nu_D + A_{23}\,\delta\nu_{D^*} &= 0 \\
A_{31}\,\delta\nu + A_{32}\,\delta\nu_D + (A_{33} - \lambda)\,\delta\nu_{D^*} &= 0 \\
\frac{d}{d\xi}\delta\mathcal{E}_\perp + \delta\nu + \delta\nu_D + \delta\nu_{D^*} &= 0
\end{aligned} \tag{4.5}$$

We transform (4.5) into the self-adjoint form of a generalized eigenvalue problem

$$\frac{d}{d\xi}\left[\frac{1}{\nu(\nu+\lambda)}\frac{d}{d\xi}\delta\nu\right] + \frac{1}{\nu}\left[\frac{d}{d\xi}\left(\frac{\mathcal{E}_\perp}{\nu+\lambda}\right) - \frac{H(\lambda)}{G(\lambda)}\right]\delta\nu = 0 \tag{4.6}$$

with

$$\begin{aligned}
G(\lambda) &:= \lambda^2 + \Theta\lambda + \Delta \\
H(\lambda) &:= \lambda^2 + \left[\Theta + \frac{X_1^S T_1^S N_D (X^* + X_1 P_A \nu)}{\Gamma_0^3 \Delta}\right]\lambda - \Delta \frac{dg}{d\nu}
\end{aligned} \tag{4.7}$$

Since the inhomogeneous steady states $\underline{\phi}(\xi)$ are states of broken translation symmetry, there exists a Goldstone mode $\delta\underline{\phi}(\xi) := \frac{d}{d\xi}\underline{\phi}$ associated with the eigenvalue $\lambda = 0$. Applying an Oscillation Theorem of Sturmian Theory[10], it can be shown that for the kink-shaped solutions $\lambda = 0$ is the highest eigenvalue, securing stability, whereas for hump-shaped solitary or oscillatory solutions there exist larger eigenvalues

$\lambda > 0$, which yields instability [11].

5. CONCLUDING REMARK

The dissipative structures found in this semiconductor model are associated with nonequilibrium phase transitions between various branches of the current voltage characteristics (Fig. 5).

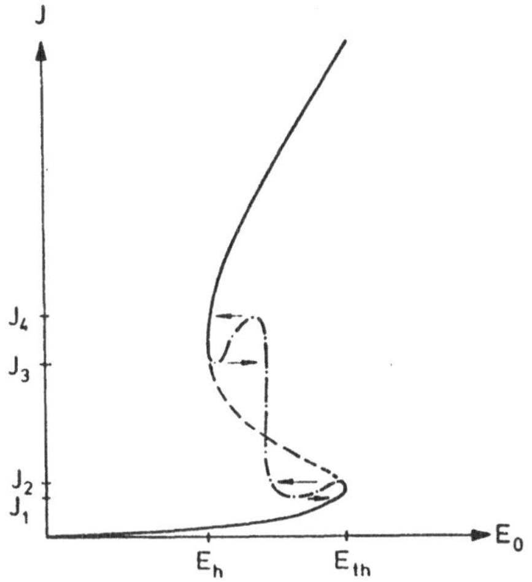

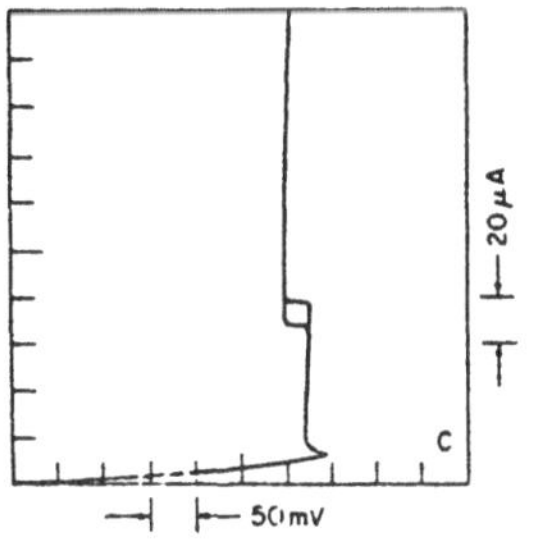

Fig. 6: Typical measured current-voltage characteristic in GaAs [7].

Fig. 5: Extended current-voltage characteristics. Full lines: Stable homogeneous steady states. Dash-dotted: solitary steady states. Dashed: unstable homogeneous steady states.

I am indebted to Prof. F. Schlögl and W. Renz for valuable discussions.

REFERENCES

1. Haken, H.: Synergetics. Berlin, Heidelberg, New York: Springer 1977
   Nicolis, G., Prigogine, I.: Self-Organization in Nonequilibrium Systems. New York: Wiley 1977
2. Schöll, E.: Proc. R. Soc. A 365, 511 (1979)
3. Robbins, D.J., Landsberg, P.T., Schöll, E.: Phys. Status Solidi (a) 65, 353 (1981)
4. Pimpale, A., Landsberg, P.T., Bonilla, L.L., Velarde, M.G.: J. Phys. Chem. Solids 42, 873 (1981)
5. Schöll, E.: Z. Phys. B - Condensed Matter 46, 23 (1982)
6. Schöll, E.: Z. Phys. B 47 (In press, 1982)
7. Stillman, G.E., Wolfe, C.M., Dimmock, J.O.: Semicond. Semimet. 12, 169 (1977)
8. Barnett, A.M.: Semicond. Semimet. 6, 141 (1970)
9. Schlögl, F.: Z. Phys. 253, 147 (1972)
10. Ince, E.L.: Ordinary Differential Equations. New York: Dover 1956
11. Schöll, E.: To be published

ON THE CHARACTERIZATION OF CHAOTIC MOTIONS

Itamar Procaccia, Peter Grassberger* and H.G.E. Hentschel

Department of Chemical Physics
The Weizmann Institute of Science
Rehovot 76100, Israel

1. INTRODUCTION

In this paper we demonstrate that scaling ideas and fractal statistics can be very useful in dealing with chaotic motions. In particular we shall concentrate on two topics:

i. The characterization of strange attractors[1,2]. We shall offer a method of analyzing and characterizing chaotic experimental signals which stems from motions on strange attractors. The method can distinguish between these and random "noisy" signals, and can be applied also to very high (or infinite) dimensional systems which possess lower dimensional attractors.

ii. The assessment of the fractal nature of fully developed turbulence[3,4]. Here we shall show how ideas similar to those treated in part 1 can be applied to the important process of turbulent diffusion, and how to infer from it the fractal dimension of fluid turbulence.

2. THE CHARACTERIZATION OF STRANGE ATTRACTORS

The main idea advanced here is that one can characterize a strange attractor on the basis of a time series of either one variable, a few variables, or as many variables as there are degrees of freedom. Consider a system with F degrees of freedom and a long time series of N points $\vec{X}_i \equiv \{X_i^1, X_i^2 \ldots, X_i^F\}$; $i=1,\ldots,N$. The N points can be for example the value of $\vec{X}$ recorded at given times $t+i\tau$, with τ arbitrary. Having a record of the time series we consider the correlation integral $C(\ell)$ defined as

$$C(\ell)=\lim_{N\to\infty} \frac{1}{N^2} * \left\{ \begin{matrix} \text{number of points } (i,j) \text{ whose distance} \\ |X_i - X_j| \text{ is less than } \ell \end{matrix} \right\} \quad (2.1)$$

The correlation integral is related to the correlation function

$$c(\vec{r}) = \lim_{N\to\infty} \frac{1}{N^2} \sum_{i,j=1}^{N} \delta^F(\vec{X}_i - \vec{X}_j - \vec{r}) \tag{2.2}$$

by

$$C(\ell) = \int_0^{\ell} d^F r\, c(\vec{r}) \tag{2.3}$$

The point is that the correlation integral scales with ℓ. In other words, for small ℓ

$$C(\ell) \sim \ell^{\nu} \tag{2.4}$$

and the "correlation exponent" ν can be taken as a most useful measure of the "strangeness" of strange attractors.

In Fig. 1 we show two examples of the quality of the scaling behavior in low dimensional systems. Other examples are collected in table 1, including the values of two previously used characteristics of strange attractors, i.e. their fractal and information dimension. These measures and their relation to ν warrant a discussion.

Table 1.

	ν		σ
Hénon map a=1.4, b=0.3	1.25±.01	1.26 (ref. 5)	-
Kaplan-Yorke map α=0.2	1.42±.02	1.431 (ref. 5)	-
Logistic Eq., b=3.5699456···	0.500±.005	.538 (ref. 6)	.5170976 (ref. 2)
Lorenz Eq.[a] parameters as in ref.7	2.05±.01	2.06±.01 (ref. 5)	-
Rabinovich[b] Fabrikant Eq. (see 3 of ref. 8)	2.19±.01	-	-

The fractal dimension is obtained by covering phase space with F dimensional cubes of edge size ℓ and counting the number of boxes $M(\ell)$ which contain a piece of the attractor. The limit

$$D = -\lim_{\ell\to 0} \{\ell n[M(\ell)]/\ell n\, \ell\} \tag{2.5}$$

is the fractal dimension[9]. The information dimension is obtained from the information entropy $S(\ell)$,

$$S(\ell) = -\sum_{i=1}^{M(\ell)} p_i\, \ell n\, p_i \tag{2.6}$$

where p_i is the probability for $\vec{X}(t)$ to fall into the i'th cube. For all attractors studied so far, $S(\ell)$ increases logarithmically with $1/\ell$ as $\ell\to 0$, and we shall accordingly make the ansatz

$$S(\ell) \simeq S_0 - \sigma\, \ell n\, \ell\ . \tag{2.7}$$

The constant σ is referred to as the information dimension[10]. Whenever the attractor is covered uniformly with a long time series, then one can show that

$$\nu = \sigma = D \; . \qquad (2.8)$$

However in many cases there are some parts of the attractor which are visited more frequently then others. We refer to such neighborhoods as having higher "seniority". In such cases one can only show that

$$\nu \leqslant \sigma \lesssim D. \qquad (2.9)$$

Details of the arguments leading to Eq. (2.9) can be found in refs. 1,2 . The main difference between these various measures lies, however, in the feasibility of their calculation. It is well documented[7,10,11] that the evaluation of D and σ, and in fact any measure which is obtained from a box counting algorithm (like the capacity) is exceedingly difficult and requires prohibitively long time series and large storage. For higher dimensional systems (D>2) no convergence with 800000 (!) points has been reported[7]. In contrast the ν measure is easily computed and a few thousand points are usually sufficient to obtain a good estimate of its numerical value.

In cases where $\nu \neq D$ we believe that ν is in fact dynamically more relevant than D. The reason is that D is ignorant of seniority and weighs all parts of the attractor equally. The ν measure, on the other hand, is sensitive to the frequency with which different neighborhoods are visited.

The greatest advantage of the ν measure lies however in its usefulness in characterizing chaotic signals which come from very high dimensional systems, such as experimental systems. We shall demonstrate this usefulness by discussing infinite dimensional delay differential equations. The particular example studied here is due to Mackey and Glass, and reads

$$\dot{X}(t) = \frac{a\, X(t-\tau)}{1+[X(t-\tau)]^{10}} - b\, X(t) \qquad (2.10)$$

where a=0.2, b=0.1 and τ is a given time delay. The fact that this is an infinite dimensional equation is most easily seen[10] from the necessity in specifying initial conditions over a whole interval of length τ.

For the numerical investigation Eq. (2.10) is turned into an F-dimensional set of difference equations with F=600-1200. Details are described in ref. 2. The time series was chosen as $\{X(t), X(t+\tau), X(t+2\tau)\dots\}$. The situation is thus analogous to monitoring <u>one variable</u> out of F variables in an experimental set up.

The method now goes as follows[2]: A new d-dimensional space is constructed by using vectors $\vec{\xi}_i = (X(t_i), X(t_i+\tau)\dots X(t_i+(d-1)\tau))$ which are then inserted in Eq.(1.1) to calculate the correlation integral. One starts with, say d=2 and sees whether a plot of $\ell n\, C(\ell)$ vs. $\ell n\, \ell$ yields a straight line. If it does not, one tries d=3 and repeats the process. Once a straight line is obtained, <u>one goes anyhow to a higher d</u> to verify that the slope is not changing. This is <u>important</u> in order to verify that there are no large unnoticed corrections to the scaling behavior (Eq. (1.4)). In many cases they are present (e.g. even the map $X_{n+1}=4X_n(1-X_n)$ shows logarithmic corrections), but their importance decreases with increasing d (see ref. 2). In table 2 we show the results of this procedure for various values of τ. The higher is τ the more "chaotic" is the system in the sense that the dimension of the attractor is larger. In the table we also show the number of positive Lyapunov exponents, the fractal dimension (where its computation converges) and the value of the Kaplan-Yorke conjecture for the dimension of the attractor[13].

Table 2. Estimates of the correlation exponent ν for the Mackey-Glass equation with a=.2, b=.1 . Values for other entries are from ref. 13.

τ	number of positive λ_i	ν	D	D_{KY}
17.0	2	1.95±.03(d=3)	2.13±.03	2.10±.02
		" (d=4)		
		" (d=5)		
23.0	2	2.38±.15(d=3)	2.76±.06	2.82±.03
		2.43±.05(d=4)		
		2.44±.05(d=5)		
		2.42±.1 (d=6)		
30.0	3	2.9 ±.3 (d=4)	>2.94	3.58±.04
		3.0 ±.2 (d=5)		
		3.0 ±.2 (d=6)		
		2.8 ±.3 (d=7)		
10.0	6	5.8 ±.3 (d=10)	-	∿10.0
		6.6 ±.2 (d=12)		
		7.2 ±.2 (d=14)		
		7.5 ±.15(d=16)		

The relation of the conjecture by Kaplan and Yorke to the ideas presented here deserves elaboration. This will demonstrate the usefulness of the ν measure in gaining some theoretical insight into the nature of strange attractors. The conjecture of Kaplan and Yorke relates the Lyapunov exponents to the dimension of the attractors.

As is well known the Lyapunov exponents are determined by the rate of deformation of an infinitesimal F dimensional ball to an ellipsoid with principal axes ε_i by the action of the flow:

$$\lambda_i = \lim_{t\to\infty} \lim_{\varepsilon_i(0)\to 0} \frac{1}{t} \ell n[\varepsilon_i(t)/\varepsilon_i(0)] . \tag{2.11}$$

In a dissipative system one has $\sum_{i=1}^{F} \lambda_i<0$. Let us order the Lyapunov exponents in descending order

$$\lambda_1 \geq \lambda_2 \geq \lambda_3 \geq \ldots\ldots \geq \lambda_F \quad , \tag{2.12}$$

and let us denote by j the largest integer for which $\sum_{i=1}^{j} \lambda_i>0$. Kaplan and Yorke conjectured that the quantity D_{KY},

$$D_{KY} = j + \sum_{i=1}^{j} \lambda_i / |\lambda_{j+1}| \tag{2.13}$$

is equal to D. Later it has been suggested that D_{KY} equals σ.

That this conjecture cannot be always true can be seen by considering the following counter example: Take a Horseshoe embedded in 3-dimensional space (Fig. 2). Assume that each step of evolution consists of stretching in the X direction in a unit time by a factor of 2 (thus $\lambda_x = \ell n\ 2$) and that the y and z directions are both squeezed. We consider two cases:

i. The y direction is squeezed by 1/4 ($\lambda_y = -2\ell n\ 2$) and the z direction by 1/8 ($\lambda_z = -3\ \ell n\ 2$).

ii. The inverse situation. $\lambda_z = -2\ \ell n\ 2$ and $\lambda_y = -3\ \ell n\ 2$.

In both these cases however the folding is back onto y. It is easy to see that in both these cases Eq. (2.13) would predict $D_{KY} = 3/2$. However this is the correct result for $D = \sigma = \nu$ in case (i) but incorrect for case (ii) where one finds easily $D = 4/3$.

The point is that the Lyapunov exponents contain information about stretching and contracting under the action of the flow, but they say very little about folding.

Although case (ii) in the counter-example discussed above is not generic (while case (i) is), it should be taken as a warning to be extremely careful.

Therefore we consider it worthwhile to present arguments which show that in general

$$\nu \leqslant D_{KY} \tag{2.14}$$

connecting thus the Kaplan-Yorke formula to the correlation exponent. To see (2.14) consider the correlation function $(\vec{\Delta})$ where $\vec{\Delta}$ is the vector distance between pairs of points in Eq. (2.2). By picking $\vec{\Delta}(0)$ to be small enough and a time t large enough we have for the coordinates along the principal axes

$$\Delta_i(t) = \Delta_i(0) e^{\lambda_i t} \tag{2.15}$$

and by the conservation of number of trajectories we obtain an equation of motion for $c(\vec{\Delta}(t))$

$$c(\vec{\Delta}(t)) = \frac{\partial[\vec{\Delta}(0)]}{\partial[\vec{\Delta}(t)]} c(\vec{\Delta}(0)) = e^{-t \sum_{i=1}^{F} \lambda_i} c(\vec{\Delta}(0)) \tag{2.16}$$

To proceed further, we need a scaling ansatz which generalizes the scaling law

$$c(|\vec{\Delta}|) \sim |\vec{\Delta}|^{\nu - F} \quad . \tag{2.17}$$

Observing that the attractor is locally a topological product of an R^n with Cantor sets, and that the relevant axes are the principal axes, we associate with each axis an exponent ν_i, $0 \leqslant \nu_i \leqslant 1$, and make the ansatz

$$c(\vec{\Delta}) = \prod_{i=1}^{F} c_i(\Delta_i) \tag{2.18}$$

with

$$c_i(x) \sim \begin{cases} x^{\nu_i - 1} & \text{with} \quad 0<\nu_i<1 \\ \delta(x) & \text{if } \nu_i=0 \end{cases} \tag{2.19}$$

If $\nu_i=0$, this means that the motion along this axis dies asymptotically (example: directions normal to a limit cycle). Directions with $\nu_i=1$ are the unstable directions, with a continuous density. Directions with $0<\nu_i<1$, finally, are either Cantorian or, in exceptional cases, directions along which the distribution is continuous but singular at $\Delta=0$. Notice that $\nu_i>1$ is impossible.

Substituting Eq. (2.18) into (2.16), we find

$$\sum_{i=1}^{F} \lambda_i \nu_i = 0 \; . \tag{2.20}$$

In addition we have, from Eqs. (2.18) and (2.17),

$$\sum_{i=1}^{F} \nu_i = \nu , \tag{2.21}$$

and

$$0 \leqslant \nu_i \leqslant 1 \; . \tag{2.22}$$

It is now easy to find the maximum of ν subject to the constraints (2.21)-(2.22). It is obtained when

$$\nu_i = \begin{cases} 1 & \text{for} \quad i \leqslant j \\ 0 & \text{for} \quad i \geqslant j+2 , \end{cases} \tag{2.23}$$

and

$$\nu_{j+1} = \frac{1}{|\lambda_{j+1}|} \sum_{i=1}^{j} \lambda_i \quad . \tag{2.24}$$

Inserting the solution into Eq. (2.21), we obtain indeed

$$\nu \lesssim j + \frac{\sum_{i \leqslant j} \lambda_j}{|\lambda_{j+1}|} \equiv D_{KY} \quad , \tag{2.25}$$

From the derivation it is clear that the Kaplan-Yorke conjectures $\sigma=D_{KY}$ or $D=D_{KY}$ cannot be expected to hold when either the attractor is Cantorian in more than one dimension, or if the folding was in a direction which is not the minimally contracting one.

To summarize this part, we have introduced the ν measure to characterize strange attractors. We gave a few examples for its usefulness in high dimensional systems and in gaining new theoretical insight. More details can be found in ref. 2.

3. TURBULENT DIFFUSION AND THE FRACTAL NATURE OF FULLY DEVELOPED TURBULENCE

In this part we show that the scaling concepts discussed before find their place in other domains of chaotic behavior, and in particular in fully developed fluid turbulence. We shall concentrate on the fractal nature of turbulence and the way it gets manifested by turbulent diffusion.

It is well known that turbulence is intermittent at smaller scales[14]. As energy is cascaded from large to small scales the turbulent activity gets concentrated in smaller and smaller fractions of space. Regions where the activity exceeds a given threshold form a fractal[9,15,16].

The most commonly quoted manifestation of the intermittent nature of turbulence is the long tail in the correlation function of the viscous dissipation, $\varepsilon(\vec{r})$[17]. Experimentally one finds[14,18]

$$\langle\varepsilon(\vec{r})\varepsilon(\vec{r}+\vec{\ell})\rangle = \bar{\varepsilon}^2(\ell_0/\ell)^{\mu} , \tag{3.1}$$

where ℓ is in the inertial range, $\ell_d \ll \ell \ll \ell_0$ and ℓ_0, ℓ_d are the stirring and dissipation length scales respectively. $\bar{\varepsilon}$ is the mean energy input per unit mass per unit time. Experimentally one finds[17,18] $0.25 \lesssim \mu \lesssim 0.50$. Notice the close analogy of (3.1) with the previous scaling law $c(\ell) \sim (1/\ell)^{F-\nu}$. Theoretically one can adopt a model of "fractally homogeneous turbulence"[3,4,15]. In the notation of the previous discussion this would mean $\nu \equiv d-\mu=\sigma=D$, where d is the spatial dimension, and D the fractal dimension of the active region. In the case of fractally homogeneous turbulence we have recently estimated theoretically $2.50 \lesssim D \lesssim 2.75$ in agreement with experiment[3]. A very important question is the effect of this fractal nature on the observable properties of turbulent transport processes. Concentrating on the relative diffusion of test particles, we find the intermittency corrections to Richardson's "$\frac{4}{3}$ law"[19] and to other laws which stem from Kolmogorov's similarity theory[5,14]. A re-examination of the available experimental data shows agreement with our analysis. It is thus possible to suggest turbulent diffusion as an interesting probe of the fractal nature of turbulence.

Consider then the relative motion of two particles immersed in a turbulent medium without affecting its properties. Denoting the separation between the particles by $\vec{R} \equiv \vec{r}_1 - \vec{r}_2$, the three quantities of major interest would be $d\langle R^2\rangle/dt$, $\langle(d\vec{R}/dt)^2\rangle$ and $\langle R^2(t)\rangle$, where pointed brackets denote averaging over many realizations of the experiment. All theories based on dimensional analysis in the inertial range would predict[14]

$$dR^2/dt \sim \bar{\varepsilon}^{1/3}R^{4/3};\ \langle(d\vec{R}/dt)^2\rangle \sim \bar{\varepsilon}^{2/3}R^{2/3};\ R^2(t) \sim \bar{\varepsilon}t^3 \tag{3.2}$$

where here and below $R \equiv \langle R^2\rangle^{1/2}$. The corrections due to intermittency are dimensionless. We shall write

$$dR^2/dt \sim \bar{\varepsilon}^{1/3}R^{4/3}(R/\ell_0)^{\alpha};\ \langle(d\vec{R}(dt)^2\rangle \sim \bar{\varepsilon}^{2/3}R^{2/3}(R/\ell_0)^{\beta};\ R^2(t) \sim \bar{\varepsilon}t^3(t/t_0)^{\gamma} \quad (3.3)$$

where $t_0 \equiv [\ell_0^2/\bar{\varepsilon}]^{1/3}$.

The sign of α, β and γ can be determined by physical considerations alone. If turbulence were space filling, the laws of (3.2) would hold. Clearly if deviation from this behavior exists, it would become more pronounced at smaller length scales, where turbulence becomes very spotty. If the test particles are caught in an inactive region, their relative diffusion would become molecular and thus negligible[19,20]. Therefore, the smaller is the separation between particles, the more susceptible they are to intermittency and the reduction in their relative diffusion is thus greater. Since $R \ll \ell_0$, α and β must then be positive. The exponent γ is completely determined by α, and by integrating dR^2/dt we find $\gamma = 9\alpha/(2-3\alpha)$. Evidently α must be smaller then 2/3.

Theoretical estimates for the numerical value of the exponents are obtained by rewriting the above quantities in terms of velocity correlation functions. Using $\vec{R}(t) = \vec{R}(0) + \int_0^t \vec{V}(\tau)d\tau$ where $\vec{V}$ is the relative velocity we find

$$dR^2/dt = 2\int_0^t \langle\vec{V}(t)\cdot\vec{V}(\tau)\rangle d\tau\ ; \qquad \langle(d\vec{R}/dt)^2\rangle = \langle\vec{V}(t)\cdot\vec{V}(t)\rangle \quad (3.4)$$

The second of these quantities is simpler to estimate, being a one time correlation function. In fact, this correlation function is precisely the square of the velocity difference across a distance R at time t. For fractally homogeneous turbulence it is simply

$$\langle[\vec{v}(\vec{r}) - \vec{v}(\vec{r}+\vec{R}(t))]^2\rangle \sim v_R^2(R/\ell_0)^{\mu} \quad (3.5)$$

where v_R is the velocity difference across distance R in <u>an active region</u>. The reason for Eq. (3.5) is that $\vec{v}(\vec{r})$ and $\vec{v}(\vec{r}+\vec{R})$ are correlated only if they belong to the same active region. The weight of such an occurrence on a fractal whose dimension is D is[9] $(R/\ell_0)^{d-D} = (R/\ell_0)^{\mu}$. The velocity difference across a length R can be found by equating $\bar{\varepsilon}$ to the rate of transfer on length scales R, which <u>in an active region</u> is[15] v_R^3/R. Thus $\bar{\varepsilon} \sim (R/\ell_0)^{\mu}v_R^3/R$ and $V_R \sim \bar{\varepsilon}^{1/3}R^{1/3}(R/\ell_0)^{-\mu/3}$. Consequently

$$\langle(d\vec{R}/dt)^2\rangle \sim \bar{\varepsilon}^{2/3}R^{2/3}(R/\ell_0)^{\mu/3}\ . \quad (3.6)$$

We comment that since $0.25<\mu<0.5$ the correction to dimensional analysis is quite sizeable here.

The quantity dR^2/dt is slightly more difficult to obtain. We rewrite the integral of the time correlation function as

$$\int_0^t d\tau\langle\vec{V}(t)\cdot\vec{V}(\tau)\rangle d\tau = \langle\vec{V}(t)\cdot\vec{V}(t)\rangle \int_0^t d\tau g\left(\frac{t-\tau}{t_R}\right) \quad (3.7)$$

where t_R is the correlation time between velocity differences across a scale R in an active region,

$$t_R \sim R/V_R \sim \bar{\varepsilon}^{-1/3} R^{2/3} (R/\ell_o)^{\mu/3} \tag{3.8}$$

and t=0 is the time origin for the inertial subrange-dominated phase of relative diffusion. In writing Eq. (3.7) we have assumed that the dominant contribution to the integral comes from $\tau \sim t$, and thus $R(\tau) \sim R(t)$. A change of variables leads to

$$\langle \vec{V}(t)\cdot\vec{V}(t)\rangle t_R \int_o^{t/t_R} ds\, g(s) \sim \begin{cases} \langle \vec{V}(t)\cdot\vec{V}(t)\rangle t_R & t \gg t_R \\ \langle \vec{V}(t)\cdot\vec{V}(t)\rangle t & t \ll t_R \end{cases} \tag{3.9}$$

Using Eqs. (3.4), (3.7) and (3.9) we then find

$$d\langle R^2\rangle/dt \sim \begin{cases} \bar{\varepsilon}^{1/3} R^{4/3} (R/\ell_o)^{\mu/6} & t \ll t_R \\ \bar{\varepsilon}^{1/3} R^{4/3} (R/\ell_o)^{2\mu/3} & t \gg t_R \end{cases} \tag{3.10}$$

these results can be straightforwardly integrated to yield

$$R^2(t) \sim \begin{cases} \bar{\varepsilon} t^3 (t/t_o)^{3\mu/(4-\mu)} & t \ll t_R \\ \bar{\varepsilon} t^3 (t/t_o)^{3\mu/(1-\mu)} & t \gg t_R \end{cases} \tag{3.11}$$

One should stress that the two regimes of t compared to t_R are not short and long time regimes. In fact when intermittency does not exist t scales like t_R. With intermittency included, $t_R = Ct(t/t_o)^{\delta}$ where δ is shown to be positive by using Eq. (3.8), and C is a dimensionless constant. There are experiments in which C can be estimated. When we have no such possibility we shall assume $C \sim O(1)$ and thus for $t \ll t_o$, $\delta > 0$ leads to $t \gg t_R$.

In this paper we shall present only the comparison of this theory with the original results of Richardson. Richardson has plotted the diffusivity dR^2/dt over 5 orders of magnitude of R. In Fig. 2 we replot the data, excluding the lowest point which pertains to molecular diffusivity. The line with a slope of 4/3, in addition to a line of least squares fit are shown. The latter yields a slope of 1.57 . If one excludes the highest point one finds a slope of 1.48 . Assuming C to be O(1) we use the second of Eqs. (3.10) to compare with these results (i.e. the regime $t \gg t_R$). Thus $.15 \lesssim \frac{2\mu}{3} \lesssim .24$, or $.22 \lesssim \mu \lesssim .36$, in agreement with other estimates of μ.

Turbulent diffusion is by no means the only process that can be understood on the basis of the above approach. Additional considerations and many more examples such as intermittency corrections to turbulent sound scattering and the scintillation of light sources can be found in ref. 21.

ACKNOWLEDGEMENTS

This work has been supported in part by the Israel Commission for Basic Research. PG thanks the Minerva Foundation for financial support.

*Permanent address: Physics Department, University of Wuppertal, W. Germany.

REFERENCES

1. P. Grassberger and I. Procaccia, submitted to Phys.Rev.Lett.
2. P. Grassberger and I. Procaccia, submitted to Physica D.
3. H.G.E. Hentschel and I. Procaccia, Phys.Rev.Lett. in press.
4. H.G.E. Hentschel and I. Procaccia, Phys.Rev.A, in press.
5. D.A. Russel, J.D. Hanson and E. Ott, Phys.Rev.Lett. 45, 1175 (1980).
6. P. Grassberger, J.Stat.Phys. 26, 173 (1981).
7. H.S. Greenside, A. Wolf, J. Swift and T. Pignataro, Phys.Rev.A 25, 3453 (1982).
8. M.I. Rabinovich and A.L. Fabrikant, Sov.Phys.JETP 50, 311 (1979).
9. B.B. Mandelbrot in Fractals-Form Chance and Dimension, Freeman, San Francisco (1977).
10. J.D. Farmer, Physica 4D, 366 (1982).
11. H. Froehling, J.P. Crutchfield, D. Farmer, N.H. Packard and R. Shaw, Physica 3D, 605 (1981).
12. M.C. Mackey and L. Glass, Science 197, 287 (1977).
13. J.L. Kaplan and J.A. Yorke in Functional Differential Equations and Approximations of Fixed Points, p. 204; ed. by H.O. Peitgen and H.O. Walther, Lecture Notes in Math. 730, Springer Verlag (1979).
14. A.S. Monin and A.M. Yaglom in Statistical Fluid Mechanics, M.I.T. Press, Cambridge (1975).
15. U. Frisch, P.L. Sulem and M. Nelkin, J.Fluid.Mech. 87, 719 (1978).
16. H. Mori and H. Fujisaka in Systems Far From Equilibrium, ed. L. Garrido, Springer Verlag, Berlin (1980).
17. M. Nelkin, Phys.Fluids 24, 556 (1981).
18. R.A. Antonia, N. Phan Thien and B.R. Satyaparakash, Phys.Fluids 24, 554 (1981).
19. L.F. Richardson, Proc.Roy.Soc.A 110, 709 (1926).
20. G.T. Csanady in Turbulent Diffusion in the Environment, Reidel, Dodrecht, Holland (1973).
21. H.G.E. Hentschel, I. Procaccia and N. Rozenberg, preprint.

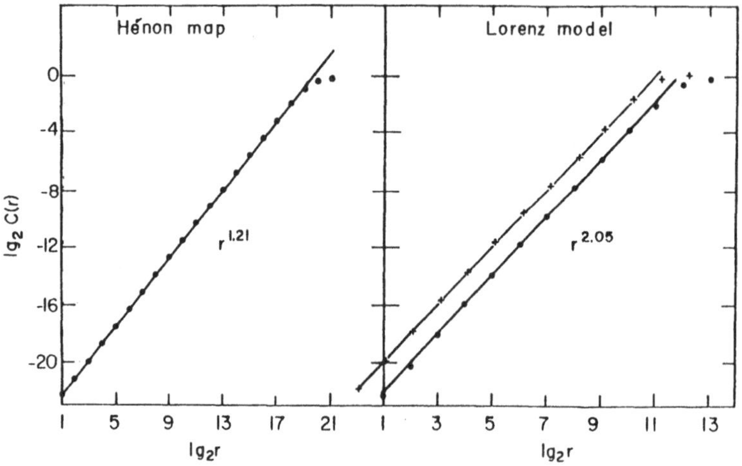

Figure 1. Correlation integrals for Hénon map (a) and Lorenz model (b) on doubly logarithmic scales. In (b) the upper line was computed from a single variable time series. In both panels the scale r is arbitrary. In panel a the slope $\nu=1.21$ is influenced by corrections to scaling. The true slope is $\nu=1.25$, see ref. 2.

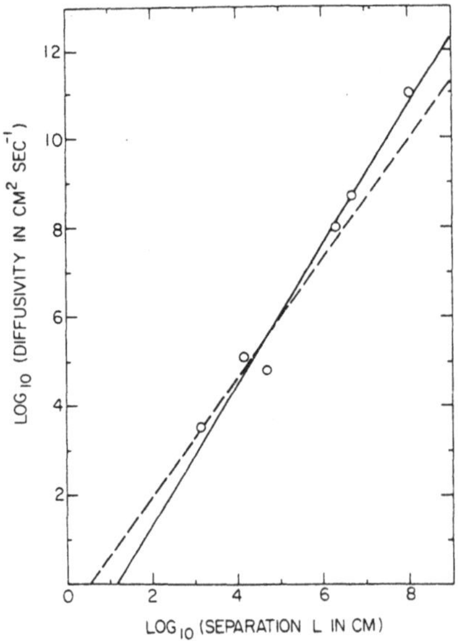

Figure 2. Turbulent diffusivity in the atmosphere as a function of the length scale. The dashed line is the original line suggested by Richardson with a slope of 4/3. The continuous line is a least square fit with a slope of 1.57 . The difference is attributed to an intermittency exponent $\mu\sim.36$.

COMPLEX BIFURCATIONS IN A PERIODICALLY FORCED NORMAL FORM.

Claude Baesens

Université Libre de Bruxelles
Campus Plaine CP 231
Boulevard du Triomphe, B-1050 Bruxelles

1. INTRODUCTION

The purpose of this communication is to show that in certain classes of systems controlled by two parameters, chaotic motion can be viewed as a local problem both in parameter and in phase space.

2. PERIODIC PERTURBATION OF A NORMAL FORM

The system studied here is the normal form describing the vicinity of the codimension-2 bifurcation where both roots of the caracteristic equation vanish simultaneously[1], perturbed by an additive periodic forcing

$$\frac{dx_1}{dt} = x_2$$

$$\frac{dx_2}{dt} = \varepsilon_1 + \varepsilon_2 x_1 + x_1^2 - x_1 x_2 + A\cos\omega t \tag{2.1}$$

We consider the vicinity of the bifurcation point $(\varepsilon_1, \varepsilon_2) = (0,0)$ and suppose a long period small amplitude forcing such that, by an appropriate scaling, the system (2.1) can be viewed as the perturbation of a reference, exactly soluble, Hamiltonian system possessing a separatrix loop for negative values of ε_1 and ε_2.

We then apply a traditional perturbative analysis to see how the solutions of the Hamiltonian system are effected by the perturbation. At the first order, we obtain a solvability condition :

$$\Delta_T(t_0) = \int_{-T/2}^{T/2} dt\, x_{20}(t-t_0) \cdot (x_{10}(t-t_0) x_{20}(t-t_0) + A\sin\omega t) = 0. \tag{2.2}$$

If $2\pi/\omega$ is not rationally related to the period of the reference solution $(x_1(t-t_0), x_2(t-t_0))$, the limit $T \to \infty$ should be taken. In particular this is so when the reference solution is on the homoclinic trajectory. In this latter case, $\Delta_{T\to\infty}(t_0)$ is nothing but the Melnikov integral which expresses the distance between the stable and the unstable manifolds of the saddle[2]. From (2.2), we then obtain in the parameter space spanned by $\varepsilon_1, \varepsilon_2, A, \omega$ the locus where a homoclinic bifurcation occurs and the domain of existence of homoclinic points implying complex non-periodic behavior[3]. However, if $2\pi/\omega$ is rationally related to the period of the reference solution, T should be taken equal to the largest common period. In this case, following the analysis of Chow and Hale[4], we obtain in the parameter space, the locus of subharmonic bifurcations to attracting period-$2\pi N/\omega$ solutions (for each N) and their domain of existence. In fig.1 we have represented the threshold for subharmonic bifurcations of the amplitude of the forcing $A_c(N)$ as a function of N for some fixed values of ε_1, ε_2 and ω. We observe that A_c decreases sharply until it reaches a minimum value for N = 8, well <u>before</u> the homoclinic bifurcation, and subsequently, it increases to tend to the homoclinic threshold A_c^h when $N\rightarrow\infty$, as predicted by Chow and Hale.

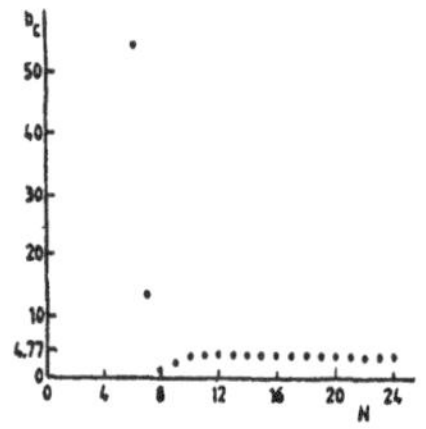

Fig.1 : Subharmonic bifurcation threshold $b_c (=A_c \times 10^{-5})$ as a function of N. For comparison, the homoclinic threshold is then $b_c^h = 4.77$

Numerical simulations corroborate the theoretical predictions[5]. Moreover, they show that the attracting subharmonic solutions evolve to chaotic ones via period-doubling bifurcations.

3. DISCUSSION

We have shown that in a general class of dynamical systems (those amenable to the normal form), the introduction of a small amplitude long period forcing induces complex nonperiodic behavior. In the space spanned by the system's parameters ε_1, ε_2, A, ω, both the onset and the domain of nonperiodic and subharmonic behavior can be localized. The most important aspect of this localization is the fact that complex behavior can occur infinitely close to the singular points of the flow, as it subsists even in the limit of small values of the system's intrinsic parameters $_1$, $_2$.

We constructed some examples of chemical systems locally reducible to the normal form above, both for isothermal kinetics, and for continuously stirred tank exothermic reactions[5].

REFERENCES

1. J. Guckenheimer , in the present Proceedings.
2. V.K. Melnikov, Trans. Moscow Math. Soc. 12(1),1 (1963)
3. S. Smale, Bull. A.M.S. 73,747(1967)
4. S.N. Chow and J.K. Hale, "Methods of bifurcation theory", Springer-Verlag (1982)
5. C. Baesens and G. Nicolis, preprint submitted to Physica D (1982)

TOPOLOGICAL ENTROPY AND SCALING BEHAVIOUR

J. Dias de Deus, R. Dilão and J. Taborda Duarte

Centro de Física da Matéria Condensada

Av. Prof. Gama Pinto 2

1699 Lisboa Codex (Portugal)

In one dimensional maps $f_\mu : [-1,1] \circlearrowleft$ with fixed end points and depending monotonically on $\mu \in [0,1]$, the growth number (G.N.) A takes the constant value 1 in the first Feigenbaum region and values between 1 and 2 in the chaotic region, $\mu \in [\mu_\infty, 1]$. We analize the behaviour of the G.N. and of the topological entropy (T.E.) h in $[\mu_\infty, 1]$. A method for computing the G.N. was developed [1] and the scaling behaviour in the parameter μ can be transposed for the T.E. .

Let $\mu_k = \sup\{\mu \in [0,1] : f_\mu^k(0)=0\}$, $k \geq 2$. It can be shown [2] that $\mu_k \to 1$ as $k \to +\infty$ and $(\mu_k - \mu_{k-1})/(\mu_{k+1} - \mu_k) \to \Delta = f_1'(1)$. In the intervals (μ_k, μ_{k+1}), $k \geq 2$, there exists one and only one point $\mu = \eta_k$ so that $f_\mu^{k+1}(0) = f_\mu^{k+2}(0)$. The map f_{η_k} has the kneading sequence $CI_0\underbrace{I_1 \dots I_1}_{k-1}(I_0)^\infty$ and the G.N. is given by the largest root of a polynomial obeying the recurrence relation

$$\bar{P}_{\eta_k}(A) = A\,\bar{P}_{\eta_{k-1}}(A) - 2\ , \quad k \geq 2$$

where $\bar{P}_{\eta_2}(A) = A^2 - 2$. Developing $\bar{P}_{\eta_k}(A)$ in Taylor series in the neighbourhood of $A = 2$ ($\mu=1$) we obtain, for $A(\eta_k)$, the assymptotic expression

$$A(\eta_k) = 2 - \frac{2}{\frac{3}{2}\,2^k - 2}$$

With $1-\eta_k = c\,\Delta^{-k}$, after substitution in the last expression and in the continuous approximation we have for T.E.

$$h_1(\mu) = \log_2\left(2 - \frac{2}{\bar{c}(1-\mu)^{-\frac{\ln 2}{\ln \Delta}} - 2}\right) \qquad (1)$$

The same kind of analysis can be carried for the ergodic points that converge to μ_∞, η_2^j, $j=1,2,..$. In this case we have exactly

$$h(\eta_2^j) = 1/2^j \quad , j \geqslant 1.$$

With $\eta_2^j - \mu_\infty = c' \delta^{-j}$ we obtain

$$h_{\mu_\infty}(\mu) = \bar{c'} (\mu - \mu_\infty)^t \qquad (2)$$

where $t = \ln 2/\ln\delta$. An analogous expression was obtained by Huberman and Rudnick [3] for the Lyapunov characteristic exponent λ. In Fig.1 we compare (1) and (2) with the characteristic exponent of the map $2\mu(x^2 -1)+1$.

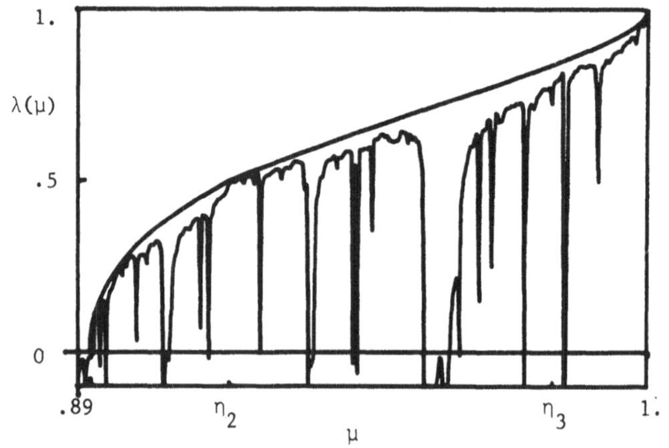

Fig.1

REFERENCES

1. J. Dias de Deus, R. Dilão and J. Taborda Duarte,Topological entropy and dynamics of the interval, preprint CFMC-E3/82; Phys. Lett. 90A,1(1982); Topological entropy, characteristic exponents and scaling behaviour in dynamics of the interval, preprint CFMC-E14/82.
2. J. Dias de Deus and J. Taborda Duarte, Commun. Math. Phys. 84,251(1982).
3. B. Huberman and J. Rudnick, Phys. Rev. Lett. 45,154(1980).

ON THE ANALYTIC STRUCTURE OF CHAOS IN DYNAMICAL SYSTEMS

Tassos Bountis*

Department of Mathematics and Computer Science
Clarkson College of Technology
Potsdam, N. Y., 13676 (USA)

ABSTRACT

A number of new and exciting results on the chaotic properties of dynamical systems have been recently obtained by studying their movable singularities in the complex time plane. New, integrable systems were identified by requiring that their solutions admit only poles. Allowing for logarithmic singularities, it has been possible to distinguish between "strongly" and "weakly" chaotic Hamiltonian systems, while in some cases natural boundaries with self-similar structure have been found. The analysis is direct, widely applicable and is illustrated here on some simple examples.

1. INTRODUCTION

The subject of self generated chaotic behavior in simple dynamical systems is by now widely recognized as one of the most interesting and intensively studied areas of mathematical physics.[1-5] The fact that most deterministic systems of Nonlinear Mechanics are *non-integrable,* (i.e., non-solvable by standard methods) and possess large classes of solutions with truly *random* properties, is generally appreciated. What is perhaps less well known is that to date important questions such as how to *identify* integrable dynamical systems, or how to determine the *size* of "regions" of chaotic behavior in phase space, remain largely unresolved.

In this paper, we review the results of a recently proposed, direct method aiming to provide an answer to these questions. This method rests on a detailed analysis of the movable singularities of the (unknown) solutions of a given dynamical system in the *complex*

---

*1982-83: On leave at the Institute for Theoretical Physics, Valckenierstraat 65, 1018 XE, Amsterdam, The Netherlands.

time plane (a singularity is called movable if its location depends on the initial conditions[6]). It has already been applied to a great number of dynamical systems, conservative as well as dissipative, and has yielded several new and interesting results.

The singularity analysis described in this paper was originally applied by Kovalevskaya[7] to the equations of motion of a rotating rigid body. By demanding that the solutions have *only movable poles* in the complex t-plane she discovered one more set of parameter values for which the equations could be integrated and solved exactly in terms of known functions. In the early 1900's, the French mathematician P. Painlevé and his co-workers[8,6] found *all second* order o.d.e's having the so-called Painlevé property: i.e., their *general* solution admits no movable singularities other than *poles*. Of a total of 50 such equations 44 were solved exactly in terms of elementary functions, while for the remaining 6, classes of solutions as well as many properties of their general solution have since been discovered.[6,8,9]

Interest in these equations was revived in 1978 by Ablowitz, Ramani and Segur,[10,11] who established a deep connection between *second* order o.d.e's with the Painlevé property and nonlinear evolution equations solvable by Inverse Scattering Transforms.[12] Segur[13] was subsequently able to integrate a *third* order system of o.d.e's-the Lorenz equations[14]-in some special cases, in which the solutions have only movable poles in the complex t-plane. Bountis, Segur and Vivaldi[15] demonstrated that the Painlevé property can be used to identify integrable Hamiltonian systems of two or more degrees of freedom (see Section 2).

At the same time, the singularity analysis of *non-integrable* dynamical systems was being studied. Bountis and Segur[16,17] attempted to relate poles at *lowest* order (plus lnt terms at higher orders) with the presence of *"small scale"* chaos; and lnt terms at *lowest* order (plus ln lnt terms at higher orders) with the presence of *"large scale"* chaotic regions. Their results are summarized in section 3. On the other hand, Tabor, Weiss and Chang, using algorithms developed by Chang and Corliss,[18] investigated the *location* of singularities of the Lorenz equations[19] and the Hénon-Heiles system[20,21] in connection with the chaotic properties of the solutions in *real* time. In the Hénon-Heiles case they discovered that the singularities formed *natural boundaries* in the complex plane with a remarkable self-similar structure. These results are briefly described in section 4.

In section 4, we also discuss some recent, interesting results of Ramani, Grammatikos, Dorizzi et al.[22,23]. They discovered a new class of 2-degree of freedom Hamiltonian systems, with a second integral *quadratic* in the momenta, and established a connection between these Hamiltonians and a "weaker" form of the Painlevé property, whereby *rational* powers are allowed in the series expansions[22,23] Finally, some concluding remarks are offered in section 5.

2. INTEGRABILITY AND THE PAINLEVÉ PROPERTY

The main idea is simple: chaotic motion, i.e., classes of solutions, which depend extremely sensitively on the choice of initial conditions, is expected to have a rather complicated representation in terms of functions highly singular in the (complex) time variable. On the other hand, regular motion is usually described by functions which are analytic everywhere, except perhaps for some poles in the complex t-plane. Thus, given a set of equations of motion, like those of the famous Hénon-Heiles problem[24,25]

$$\ddot{x} = - Ax + 2xy \tag{2.1a}$$

$$\ddot{y} = - By + \varepsilon y^2 + x^2 \tag{2.1b}$$

it might be interesting to investigate for which values of ε, A, B the *general* solutions of (2.1) behave like

$$x \sim a\tau^p \quad , \quad y \sim b\tau^q \quad , \quad \tau \equiv t - t_0 \tag{2.2}$$

near a *movable* singularity $t = t_0$, where t_0 is a *free* constant (the location of the singularity) and p,q are *integers*, not both positive.[15,20]

Inserting (2.2) in (2.1) one finds that lowest order terms can be "balanced" in two ways:

$$\begin{aligned} &(i) \quad p = q = -2, \\ &(ii) \quad p > -2 \quad , \quad q = -2 \end{aligned} \tag{2.3}$$

For case (i) we write

$$x = 3(2-\varepsilon)^{\frac{1}{2}}\tau^{-2} +..+ \alpha\tau^{-2+r} +... \; ; \; y = 3\tau^{-2} +...+ \beta\tau^{-2+r}+... \tag{2.4}$$

substitute in (2.1), obtain linear equations for α, β and set the determinant of their coefficient matrix equal to zero, so that arbitrary constants may enter in (2.4). This yields[15,20]

$$(r^2 - 5r + 12 - 6\varepsilon)(r+1)(r-6) = 0 \; . \tag{2.5}$$

If r were *not* integer, fractional powers of τ would enter in (2.4) and the solutions of (2.1) would be multivalued with branch point singularities. For the same reason we demand that the condition obtained by inserting (2.2) in (2.1) for case (ii), cf. (2.3), i.e.

$$\varepsilon p(p-1) = 12 \quad , \tag{2.6}$$

be satisfied by *integer* $p > -2$. There are only three values of ε for which *both* r and p in (2.5) and (2.6) are integers: $\varepsilon = 1$, 2, and 6.

These three values of ε must now be checked in detail. This means that the Laurent series expansions, like (2.4), must be carried out in order to verify that all arbitrary constants expected at order r, cf. (2.5), do enter with all the proper compatibility conditions satisfied. Should this *not* be the case at one value of r, *logarithmic* terms will be needed to capture the arbitrary constant expected at that order[16,17,21]. This is what happens for $\varepsilon = 2$, at $r = 0$, cf. (2.5). The $r = 0$ free constant enters there if instead of (2.2) we write[15,16]

$$y \sim 3\tau^{-2} \quad , \quad x \sim (-15/2)^{1/2}\, \tau^{-2}\, (\ln \tau)^{-1/2} \quad , \tag{2.7}$$

as $\tau \to 0$. Thus the $\varepsilon = 2$ case is *not* Painlevé and is, in fact, non-integrable; worse, it has ln τ singularities *at lowest order* and ln lnτ terms are required at higher order, to capture all 4 arbitrary constants. We conjecture in section 3, that such singularities are connected with "widespread" chaotic behavior, which the $\varepsilon = 2$ Hénon-Heiles does exhibit already at low energies.[16-18]

The other two values $\varepsilon = 1$, 6 indeed possess the Painlevé property. There is also a third case, $\varepsilon = 16$, which was shown to have the Painlevé property,[20] in the new variables $X(\equiv x^2)$ and y. These three cases are the only ones proved to be integrable to date and are summarized again, below:

(a) $\varepsilon = 1$, $A = B$: Here, eqns. (2.1) easily "uncouple" in (x+y), (x-y) variables and integrability has long been known.[26] The other two cases are much less obvious.

(b) $\varepsilon = 6$, any A, B: The second integral was first written down by J. M. Greene[27] and has been independently derived by Hall[28] and Grammatikos et al.[29].

(c) $\varepsilon = 16$, $B = 16A$: Here also the second integral has been recently derived as a polynomial quartic in the momenta.[28,29]

Several other integrable Hamiltonians have been identified by the Painlevé property.[15,30] We, notably, mention here the Toda lattice, whose integrability has been, in every case, verified independently by other methods.[31-33]

3. LOGARITHMIC SINGULARITIES AND CHAOTIC BEHAVIOR

As was mentioned already in section 2, it is possible to find that, in order to capture a free constant entering at a given order, one must include logarithmic terms in the series expansions of the solutions. This is what happens in the $\varepsilon = 2$ Hénon-Heiles case, and also in the $\varepsilon = 1$ and $\varepsilon = 16$ cases, if $A \neq B$ and $B \neq 16A$, respectively. The important point, however, seems to be whether these $\ln \tau$ terms enter at *lowest* order or not. If they do (and $\ln \ln \tau$ terms are needed at higher orders to capture all free constants), the chaotic properties of the system appear to be much "stronger" than if the lowest order behavior of the solutions is pole-like and only $\ln \tau$ terms enter at higher orders.[16,17]

The above statement is suggested by a number of examples of 2 degree of freedom Hamiltonian systems which have been analyzed in some detail. Such systems have the advantage that their chaotic properties (e.g., "size" of chaotic regions at a given value of the total energy) can be numerically studied and compared by the surface of section method.[1-4]

Specifically, pole-like behavior at *lowest* order (with $\ln \tau$ terms at higher orders) has been found in the analysis of the following Hamiltonians:

$$H_1 = \frac{1}{2}(\dot{x}^2+\dot{y}^2) + \frac{1}{4}[x^4+y^4+\frac{1}{4}(x-y)^4] \tag{3.1}$$

$$H_2 = \frac{1}{2}(p_1{}^2+p_2{}^2+p_3{}^2) + e^{q_1-q_2} + e^{q_2-q_3} - q_1+q_3 \tag{3.2}$$

$$H_3 = \frac{1}{2}(\dot{x}^2+\dot{y}^2 + Ax^2+By^2) + \frac{1}{4}(x^2+y^2)^2 + y \tag{3.3}$$

The explicit series expansions have been given for H_1 and H_2 in ref. 16 and for H_3 in ref. 17. On the other hand $\ln\tau$ behavior at lowest order (with $\ln \ln \tau$ terms at higher orders) was found in the $\varepsilon = 2$ Hénon-Heiles case

$$H_4 = \frac{1}{2}(\dot{x}^2+\dot{y}^2+x^2+y^2) - x^2y - \frac{2}{3}y^3 \quad , \tag{3.4}$$

and in

$$H_5 = \frac{1}{2}(\dot{x}^2+\dot{y}^2) + \ln(x^2+y^2+b^2) + y \quad , \tag{3.5}$$

for which the explicit series expansion have been written down in refs. 16, 17, respectively.

Preliminary computer experiments indicated that Hamiltonians H_1, H_2 and H_3 were integrable.[16,17] Prompted, however, by the singularity analysis, more careful numerical investigation revealed the presence of "chains of islands" with "thin" chaotic regions in between these islands indicating non-integrability. On the other hand, similar surface of section experiments with Hamiltonians H_4 and H_5 immediately showed "large scale" regions of chaotic behavior already at low energies.[15,16]

In an effort to make these statements more precise, we compared several Hamiltonians, like H_3 and H_5 above, describing the motion in potentials whose cylindrical symmetry had been "perturbed" by the same linear term in y. Already at *one* energy unit above their equilibrium energy the chaotic regions in the surface of section of H_5 were significantly larger than they were in that of H_3.[17]

It is important to note, however, that the presence of logarithms or any other type of "bad" singularities, does not necessarily preclude integrability.[17] It is still, at present, only a suggestive criterion which must be complemented by other numerical or analytical tests. A simple example of this is provided by the Lotka-Volterra equations of competing species[8]

$$\dot{x} = ax - xy \tag{3.6a}$$

$$\dot{y} = -by + xy \tag{3.6b}$$

where a, b are positive constants. A straightforward singularity analysis of (3.6), along the lines of section 2, gives (as $\tau \to 0$)

$$x = -\frac{1}{\tau} + a_0 - \frac{a+b}{2}\ln\tau - \frac{(a+b)^2}{12}\tau(\ln\tau)^2 + O(\tau) \quad , \tag{3.7a}$$

$$y = \frac{1}{\tau} + a_0 + \frac{a-b}{2} - \frac{a+b}{2}\ln\tau + \frac{(a+b)^2}{12}\tau(\ln\tau)^2 + O(\tau) \quad . \tag{3.7b}$$

where a_0 is the *second* free constant. As is well known[8] (3.6) possesses one integral for *all* a,b:

$$x^b y^a e^{-(x+y)} = \text{const.} \quad , \tag{3.8}$$

and hence the presence of logarithms in (3.7) does not imply the non-existence of integrals. Had we insisted, however, that (3.6) satisfy the Painlevé property and *no* $\ln\tau$ terms enter in (3.7), this would have required

$$a + b = 0 \quad .$$

In that case, (3.6) actually possesses a *second* integral: $(x+y)e^{-at}$ = const., and x,y can be obtained in closed *form* in terms of elementary functions.

Finally, it is interesting to note that the divergent terms in (3.7) can be used to *derive* the integral (3.8): Adding (3.7a,b) we find

$$x + y \sim 2a_0 + \frac{a-b}{2} - (a+b)\ln\tau \quad , \tau \to 0$$

whence

$$e^{x+y} \sim C\,\tau^{-(a+b)} \tag{3.9}$$

where C is arbitrary (since a_0 is). Combining now (3.9) with $x^a \sim \tau^{-a}$ and $y^b \sim \tau^{-b}$, cf. (3.7), the integral (3.8) immediately follows.

4. THE RESULTS OF OTHER RESEARCH GROUPS

Besides the work of this author and his co-workers, interesting results using the singularity analysis have also been obtained, recently, by a number of other researchers. Here we summarize the main results of two groups, which are referred to as "the La Jolla group" and "the Paris group". It is hoped that this summary with the accompanying references will help the reader assess both the progress made so far, as well as the future possibilities of this direction of research in Nonlinear Dynamics.

a. Main Results of the La Jolla Group

Among the first to undertake an investigation of dynamical systems, using the singularity approach, were M. Tabor and J. Weiss who started by analyzing the equations[19]

$$\dot{x} = \sigma(y - x) \quad , \quad \dot{y} = -xy + Rx - y \quad , \quad \dot{z} = xy - Bz \qquad (4.1)$$

introduced by Lorenz in his famous paper on aperiodic (or "turbulent") motion in deterministic systems.[14] They rediscovered all the integrable cases with the Painlevé property found earlier by Segur[13] and made a thorough study of the more general case, where the solutions of (4.1) admit logarithmic singularities of the form $\tau^j(\tau^2 \ln\tau)^k$.

Using an algorithm developed by Chang and Corliss[18], Tabor and Weiss numerically located the singularities of the Lorenz system and pointed out several interesting correlations between their "arrangement" in the complex t-plane and the behavior of the solutions in real time. For example, "lines" of equally spaced singularities corresponding to periodic motion "break up" into a random-looking pattern in turbulent regimes, where the *position* of the singularities and their mutual *spacing* are correlated with the local *maxima* and the *amplitude* of the oscillations, respectively.[19]

The La Jolla group then turned to Hamiltonian systems and made a thorough investigation of the analytic structure of the Hénon-Heiles equations[20,21] for all values of a nonlinearity parameter $\lambda[= -1/\varepsilon$, cf. (2.1)]. One of their most striking results was the discovery of a *natural boundary* of singularities with a self-similar structure. In particular, they found for $1/48 < \lambda < \infty$ an infinite number of singularities near the real time axis, and about each one of them a double spiral of other singularities, "closing in" to form a dense "wall" (or natural boundary) in the complex time plane.

Natural boundaries (through which no analytic continuation is possible), have also been found recently in non-integrable area-preserving mappings on the plane[35], and are expected to play an important role in the study of non-integrable Hamiltonian systems. Their precise connection, however, with the chaotic properties of such systems is only now beginning to be explored.

b. Main Results of the Paris Group

In their first paper on this subject, Grammatikos et al.[29] rediscovered, using the singularity analysis, the three integrable cases of the Hénon-Heiles listed in Section 2. They also derived the second integral in the $\varepsilon = 16$ case by *assuming* the form

$$f_1 \dot{x}^4 + f_2 \dot{x}^3 \dot{y} + g_1 \dot{x}^2 + g_2 \dot{x}\dot{y} + h = C \quad , \tag{4.2}$$

and solving for f_1, f_2, g_1, g_2, h (as functions of x,y) from a set of p.d.e's resulting from $dC/dt = 0$. This integral had also been obtained independently by Hall, using a similar approach.[28] In another paper, Dorizzi et al.[33] considered the 3-particle free-end Toda Hamiltonian

$$H = \frac{p_1^2}{2m_1} + \frac{p_2^2}{2m_2} + \frac{p_3^2}{2m_3} + e^{\varepsilon(q_1-q_2)} + e^{q_2-q_3} \quad , \tag{4.3}$$

which had been shown earlier to possess the Painlevé property in three cases[15]. Here also they derived the missing integral in every case by writing it in the form[36]

$$C = \sum_{n=0}^{N} \sum_{k=0}^{n} f_k^n \dot{x}^{n-k} \dot{y}^k \quad , \tag{4.3}$$

$[x \equiv \varepsilon(q_1-q_2)$, $y \equiv q_2-q_3]$ and solving for $f_k^n(x,y)$ from (linear) p.d.e.'s obtained by setting $dC/dt \equiv 0$.

Now, Whittaker[35] had found a class of 2 degrees of freedom Hamiltonians with a second integral *quadratic* in the momenta $\dot{x},\dot{y}$, by solving a linear, second order p.d.e. for the potential $V(x,y)$, when some parameter $\alpha \neq 0$. Letting $\alpha = 0$, the Paris group derived and *solved* a different p.d.e. for V:

$$2y\, V_{xy} + 3V_x + x(V_{xx} - V_{yy}) = 0 \quad , \tag{4.4}$$

thus obtaining a *new* class of integrable Hamiltonians with a second integral quadratic in the momenta.[22,23]

Not all of these Hamiltonians, however, satisfy the Painlevé property of Section 2. An important discovery of the Paris group was that the family of polynomial potentials

$$V = \sum_{k=0}^{[n/2]} 2^{n-2k}\, c_{n-k}^{k}\, x^{2k}\, y^{n-2k} \tag{4.5}$$

which are *solutions* of eq. (4.4), actually satisfy a "weaker" form of the Painlevé property[22,23]: The series expansions of the solutions x,y near a singularity contain *only* powers of $(t-t_0)^{1/p}$, where

$p = n-2$, n being the degree of the polynomial (4·5). No "worse" singularities - like irrational powers of $(t-t_0)$, or $\ln(t-t_0)$ etc. - are allowed. However, these powers of $(t-t_0)^{1/p}$ cannot be simply transformed away by raising x and y to the power p, and this constitutes indeed a "weakening" of the original Painlevé property. Ramani et al[22] have already used this "weak" form of the Painlevé criterion to identify new, integrable Hamiltonians with potential $V(x,y)$ *quartic* in x and y.

5. CONCLUDING REMARKS

In this paper, we have attempted to review the main results of several research groups studying the integrability and chaotic properties of dynamical systems from the point of view of their movable singularities in the complex t-plane. We have seen that the Painlevé property (in both its original and "weaker" form) has provided us with a much needed *direct* method for identifying integrable dynamical systems. On the other hand, the singularity analysis has shown that natural boundaries can be present in non-integrable Hamiltonian systems, while poles at lowest order [with $\ln(t-t_0)$ at higher orders], or $\ln(t-t_0)$ at lowest order [with $\ln\ln(t-t_0)$ at higher orders] can distinguish between weakly and strongly chaotic Hamiltonians, respectively.

Encouraging as all this may seem, it is still, at present, a collection of interesting results for which the proper mathematical foundation is lacking. That certain connections do exist between the real time motion of a dynamical system and its singularities in the complex t=plane is clearly suggested by the evidence presented here. The precise nature, however, of these connections poses some very challenging questions, for which no answers are, as yet, available.

6. REFERENCES

1. R. H. G. Helleman in Fundamental Problems in Statistical Mechanics, vol. 5, North Holland (1981).
2. M. V. Berry in Topics in Nonlinear Dynamics, A.I.P. Conf. Proc., vol. 46, A.I.P. (1978).
3. A. J. Lichtenberg and M. A. Lieberman, Regular and Stochastic Motion, to appear (1982).
4. J. Ford in Fundamental Problems in Statistical Mechanics, vol. 3, North Holland (1975).
5. B. V. Chirikov, Physics Reports 52, 265 (1979).
6. E. L. Ince, Ordinary Differential Equations, Dover (1956).
7. S. Kovalevskaya, Acta Mathematica 14, 81 (1890).
8. H. T. Davis, Introduction to Nonlinear Diff. and Integral Equations, Dover (1962).

9. A. S. Fokas and M. J. Ablowitz, J. Math. Phys., to appear (1982).
10. M. J. Ablowitz, A. Ramani and H. Segur, Lett. al Nuovo Cimento, 23 (9), 333 (1978).
11. M. J. Ablowitz, A. Ramani and H. Segur, J. Math. Phys. 21(4), 715 and J. Math. Phys. 21(5), 1006 (1980).
12. M. J. Ablowitz and H. Segur, Solitons and the Inverse Scattering Transform, SIAM series in Appl. Math. (1982).
13. H. Segur, Lectures at International School of Physics "Enrico Fermi", Varenna, Italy (1980).
14. E. Lorenz, J. Atmos. Sci. 20, 130 (1963).
15. T. Bountis, H. Segur and F. Vivaldi, Phys. Rev. A, 25, 1257 (1982).
16. T. Bountis, H. Segur in Mathematical Methods in Hydrodynamics and Related Dynamical Systems, A.I.P. Conf. Proc. 88, A.I.P. (1982).
17. T. Bountis, AIAA/AAS Astrodynamics Conference Reprint AIAA-82-1443 (1982).
18. Y. F. Chang and G. Corliss, J. Inst. Maths. Applics. 25, 349 (1980).
19. M. Tabor and J. Weiss, Phys. Rev. A, 24, 2157 (1981).
20. Y. F. Chang, M. Tabor, J. Weiss, J. Math. Phys. 23(4), 531 (1982).
21. J. Weiss, in same volume as reference 16.
22. A. Ramani, B. Dorizzi, B. Grammatikos, Ecole Polytech., Palaiseau, preprint A497.0482 (1982).
23. B. Dorizzi, B. Grammatikos, A. Ramani, CNET, Issy les Moulineaux, preprint A500.0582 (1982).
24. M. Hénon, C. Heiles, Astron. J., 69(1), 73 (1964).
25. G. Contopoulos, Astrophys. J. 138, 1297 (1963); see also Astron. J. 68, 1 (1963).
26. Y. Aizawa, N. Saitô, J. Phys. Soc. Jpn. 32, 1636 (1972).
27. J. M. Greene, private communication.
28. L. S. Hall, Lawrence Livermore Lab., preprint UCID-18980 (1981).
29. B. Grammatikos et.al., Phys. Lett. A, to appear (1982).
30. H. Yoshida, Astronomy Dept., Tokyo Univ. preprint (1981).
31. H. Flaschka, Phys. Rev. B 9, 1924 (1974); Progr. Theor. Phys. 51, 703 (1974); see also J. Moser, Adv. Math. 16, 197 (1975).
32. O. I. Bogoyavlenski, Commun. Math. Phys. 51, 201 (1976); see also B. Kostant, M.I.T. Math. Dept., preprint (1981).
33. B. Dorizzi et al., "New Integrals of Motion for the Toda System", preprint (1982).
34. I. C. Percival, J. M. Greene, Princeton Plasma Physics Lab. preprint PPPL-1744 (1981).
35. E. T. Whittaker, Analytical Dynamics, Chpt. III, Dover (1937).
36. The highest power N, in (4.3), is not known à priori. To help overcome this problem Hall (see ref. 28) numerically studies projections of orbits from which it is often possible to infer the value of N from the number of times $\dot{x}$ (or $\dot{y}$) changes sign over one "period" of the motion.

TYPE-III-INTERMITTENCY IN A SMOOTH PERTURBATION OF THE LOGISTIC SYSTEM

G.Mayer-Kress , H.Haken

1.Institut für theoretische Physik
Universität Stuttgart
Pfaffenwaldring 57/4, D-7000 Stuttgart 80

1. INTRODUCTION

Maps on the interval have been used to model universal features in the transitions to chaotic behavior in physical [1], chemical [2], and biological [3] systems. A detailed theory exists for the period-doubling-route [4], which is associated with "flip"-bifurcations [5] and for the type-I-intermittency [6,7,8], which occurs at saddle-node bifurcations. For maps with negative Schwarzian derivative [9] these are the only possible scenarios [4].

2. PERTURBATION OF THE LOGISTIC FUNCTION

We introduced a two-parameter family of convex maps [10], which acts as a smooth perturbation of the logistic parabola and for which the Schwarzian derivative can become positive. It is given by:

$$f_{r,b}(x) := \frac{32r}{b+8}\, x(1-x)(1+bx(1-3x+2(2-x)x^2)) \qquad (2.1)$$

where r is the value of $f_{r,b}$ at its critical value x_c and b measures the perturbation of the logistic parabola $f_{r,0}$.

3. NUMERICAL RESULTS

For system (2.1) we additionally observe type-III-intermittency, where a flip-bifurcation of a fixed point is not followed by a doubling of the period, but by a transition to a chaotic attractor, which contains the unstable fixed-point [10]. The statistical properties of the system are quite different from the case of type-I-intermittency. There is no upper bound for the length of the laminar phase. Very long chaotic bursts can appear at the same parameters where we observe long laminar phases. This behavior is similar to the situation of a bistable system, where external fluctuations induce transitions between a fixed point and a chaotic attractor. The ratio between laminar and chaotic motion is controlled by the global properties of the function, which determine the probability with which the system is reinjected into the laminar region close to x^* [7]. In our case it depends both on r and on b. It becomes maximal if the critical point x_c is mapped on the fixed point x^*. Thus we obtain a critical parameter configuration (r^*,b^*) by the conditions:

$$f^3_{r^*,b^*}(x_c) = x^* \qquad f'_{r^*,b^*}(x^*) = -1. \tag{3.1}$$

We numerically determined the critical behavior of the system by evaluating its Lyapunov exponents $\lambda(r)$ [11] at fixed values of b (fig.1). For $\Delta b = b-b^* < 3.5\cdot 10^{-4}$ there exists a power law behavior of λ:

$$\lambda(r) \sim (r-r^*)^{\kappa} \tag{3.2}$$

with the critical exponent $\kappa \cong 0.79 \pm 0.01$. For $\Delta b \cong 3.5\cdot 10^{-4}$ and Δr small, we get a transition to larger values of λ because of the change in the reinjection density (see broken line in fig.1)

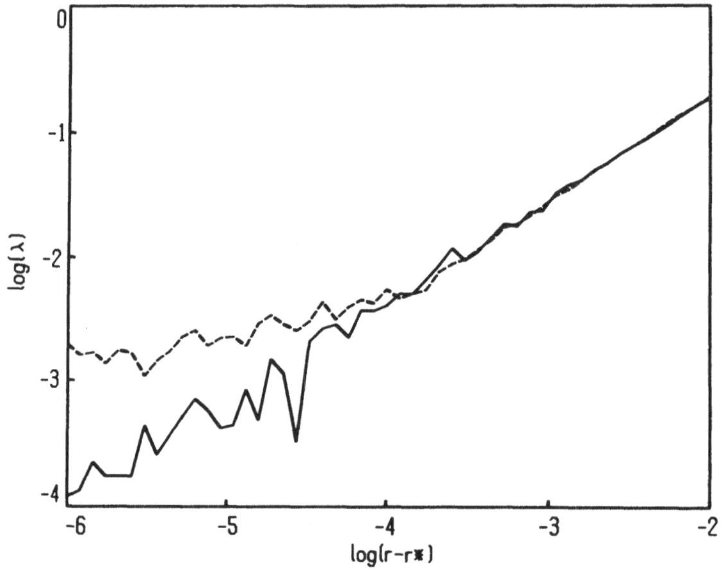

Figure 1: Logarithm of Lyapunov exponents versus bifurcation parameter for $\Delta b = 10^{-6}$ (solid line) and $\Delta b = 3.5\cdot 10^{-4}$ (broken line).

REFERENCES

1. S.Fauve,A.Libchaber,in "Chaos and Order in Nature", H.Haken ed. Springer Series in Synergetics Vol.11,(Springer 1981)
2. J.C.Roux,J.S.Turner,W.D.McCormick,H.L.Swinney in "Nonlinear Problems :Present and Future", A.R.Bishop ed.,(Amsterdam 1981)
3. R.M.May,Nature,261,459,(1976)
4. P.Collet,J.-P.Eckmann,"Iterated maps on the interval as dynamical systems",(Birkhäuser 1980)
5. J.-P.Eckmann,Rev.Mod.Phys.,53,643,(1981)
6. Y.Pomeau,P.Manneville,Commun.Math.Phys.,77,189,(1980)
7. J.-P.Eckmann,L.Thomas,P.Wittwer,J.Phys.A,14,3153,(1981)
8. J.E.Hirsch,B.A.Huberman ,D.J.Scalapino,Phys.Rev.A,25,519,(1982)
9. M.Misiurewicz, in "Chaos from deterministic systems",Les Houches Summerschool 1981,R.H.G.Hellemann,G.Joos (eds.) (to appear)
10. G.Mayer-Kress,H.Haken,in:Int.Symp. on Synergetics,ed.by H.Haken, Springer Series in Synergetics, (Springer 1982)
11. R.Shaw,Z.Naturforsch.,36A,80,(1981)

IRREVERSIBLE EVOLUTION OF DYNAMICAL SYSTEMS

Servet Martinez

Departamento de Matemáticas

and

Enrique Tirapegui

Departamento de Física
Facultad de Ciencias Físicas y Matemáticas
Universidad de Chile (Santiago)

Following some basic ideas of the theory of irreversibility of Prigogine et al [1] the group of Brussels has recently developped a canonical method to associate a Markov process "equivalent" to the reversible evolution of certain dynamical systems [2,3] exhibiting intrinsic randomness. We study here this process working for simplicity with the Baker transformation. The dynamical system is (Ω,μ,B) where $\Omega=[0,1[\times[0,1[\bmod 1$, μ is the Lebesgue measure on the sets of Ω (we write $\bar{\mu}$ for the Lebesgue measure on lines) and $\omega\in\Omega\to\omega'=B\omega$, where B is the Baker transformation [4]. We note $\phi_\Delta(\omega)$ the characteristic function of the set Δ. In $L^2(\Omega,\mu)$ we define the unitary evolution operator $Uf(\omega)\equiv f(B^{-1}\omega)$. Let $\alpha=\{\Delta_0,\Delta_1\}$, $\Delta_0=[0,\frac{1}{2}[\times[0,1[$, $\Delta_1=[\frac{1}{2},1[\times[0,1[$ be the independent and generating partition. The partition $\bigvee_{-\infty}^{n} B^j\alpha=\xi_n$ generates the σ-algebra $\bar{\xi}_n$, we call F_n the measurable functions with respect to $\bar{\xi}_n$ [3] and $M_n=F_n-F_{n-1}$. The M_n are orthogonal subspaces and $\bigoplus_{-\infty}^{\infty} M_n=\{1\}^\perp$, the subspace orthogonal to the constant functions. Let E_n be the orthogonal projector on M_n, $E_n=R_n-R_{n-1}$, $R_nf=E[f|\bar{\xi}_n]$ the conditional expectation of $f(\omega)$ with respect to the σ-algebra $\bar{\xi}_n$. One has $(n\geq 0)$

$$(1)\quad R_nf(x,y)=2^n\sum_{k=0}^{2^n-1}\phi_{[\frac{k}{2^n},\frac{k+1}{2^n}[}(y)\int_{\frac{k}{2^n}}^{\frac{k+1}{2^n}}dy'\,f(x,y')\ ,$$

$$(2)\quad R_{-n}f(x,y)=2^{-n}\sum_{k=0}^{2^n-1}\int_0^1 dy\,f(x+\frac{k}{2^n},y)\ .$$

The Markov process is constructed using a bounded self-adjoint operator $\Lambda = \sum_{n\in Z} \lambda_n E_n + P_o$, $\lambda_n = (1+a^n)^{-1}$, $P_o f = \int d\mu(\omega) f(\omega)$. The diagram ($W^\star = \Lambda \cup \Lambda^{-1}$) **(a>1)**

$$\begin{array}{ccc} L^2(\mu) & \overset{U}{\rightarrow} & L^2(\mu) \\ \Lambda\downarrow & & \downarrow\Lambda \\ L^2(\mu) & \overset{W^\star}{\rightarrow} & L^2(\mu) \end{array}$$

is commutative and $W^\star$ defines the Markov process [2,3] (note that Λ^{-1} is self-adjoint but unbounded).

One has $W^{\star m} = U^m \sum_{n\in Z} \nu_n^{(m)} E_n + P_o$, $\nu_n^{(m)} = \frac{\lambda_{n+m}}{\lambda_n} < 1$, and its adjoint $W^m = U^{-m} \sum_{n Z} \nu_{n-m}^{(m)} E_n + P_o$. One easily checks that $\Lambda, W, W^\star$, map positive Borel functions on positive Borel functions, $\|\Lambda\| \leq 1$, $\|W^m\| \leq 1$, $\|W^{\star m}\| \leq 1$, i.e. $\{W^m\}$, $\{W^{\star m}\}$, $m \geq 0$, are uniformly bounded.

The transition probability $\omega^o \to \Delta$ at time m is $p_m(\omega^o,\Delta) \equiv W^m \phi_\Delta(\omega^o)$. The probability distribution at time m, $\rho_m(\rho_o;\omega)$, which at time zero is $\rho_o(\omega)$, $P_o\rho_o = 1$, is $\rho_m(\rho_o;\omega) = W^{\star m}\rho_o(\omega)$. For $\omega = (x,y)$ we consider the sets $\Delta_{p,q}(\omega) = \Delta_p(x) \times \Delta_q(y)$, $x = \sum_{j\geq 1} x(j)2^{-j}$, $y = \sum_{j\geq 1} y(j)2^{-j}$, $x_n = \sum_{j=1}^{n} x(j)2^{-j}$, $y_n = \sum_{j=1}^{n} y(j)2^{-j}$, $x_o = y_o = 0$, $\Delta_p(x) = [x_p, x_p + 2^{-p}[$, $\Delta_q(y) = [y_q, y_q + 2^{-q}[$, $p,q \geq 0$. We define the operator $S = \sum_{\bar{N}} \sigma_n E_n$, $n \in Z$, $\sigma_n \in \mathbb{R}$, $|\sigma_n| \leq 1$, as $\lim S_N$, $N \to \infty$, $S_N = S_N^{(1)} + S_N^{(2)}$, $S_N^{(1)} = \sum_{n=0} \sigma_n E_n = \sum_{n=0}^{N} \bar{\sigma}_n R_n + \sigma_{N+1} R_n$, $S_N^{(2)} = \sum_{n=1}^{N} \sigma_{-n} E_{-n} = \sum_{n=1}^{N} \bar{\sigma}_{-n} R_{-n} - \sigma_{-N} R_{-(N+1)}$, $\bar{\sigma}_n = \sigma_n - \sigma_{n+1}$, $\bar{\sigma}_{-n} = \sigma_{-n} - \sigma_{-n+1}$, $n \geq 0$. Then ($S^{(i)} = S_N^{(i)}$, $N \to \infty$, $\omega = (x,y)$, $\omega^o = (x^o,y^o)$)

$$(3) \quad S^{(1)}\bar{\phi} = 2^{-q}\phi_{\Delta_p(x)}(x^o) \sum_{n=o}^{q-1} 2^n \bar{\sigma}_n \phi_{\Delta_n(y)}(y^o) + \sigma_q \bar{\phi},$$

$$(4) \quad S^{(2)}\bar{\phi} = 2^{-q} \sum_{n=1}^{p-1} 2^{-n} \bar{\sigma}_{-n} \phi_{\tilde{D}_n^p(x)}(x^o) - \sigma_{-p+1}\mu(\Delta_{p,q}(\omega)),$$

where $\bar{\phi} \equiv \phi_{\Delta_{p,q}(\omega)}(\omega^o)$, $\tilde{D}_n^p(x) = \bigcup_{k=o}^{2^n-1} (\Delta_p(x) + 2^{-n}k)$.

We call $\underline{\omega} \in \{0,1\}^Z$ the representation of $\omega = (x,y)$ as $\{\ldots\underline{\omega}(-1), \underline{\omega}(o), \underline{\omega}(1), \ldots\}$, $\underline{\omega}(-k) = y(k)$, $\underline{\omega}(k-1) = x(k)$, $k \geq 1$. The sets $D_n^p(\omega) = \cap B^{-j}\Delta_{\underline{\omega}(j)}$, $j \in [-(p-1), n]$, are $D_n^p(\omega) = \Delta_p(x) \times \Delta_n(y)$, $n \geq 0$,

$D^p_{-n}(\omega) = \tilde{D}^p_n(x) \times [0,1[$, $0 \leq n \leq p-1$. Then

(5) $$S\bar{\phi} = 2^{-q} \sum_{n=-(p-1)}^{q-1} 2^n \bar{\sigma}_n \phi_{D^p_n(\omega)} (\omega^o) + \sigma_q \bar{\phi} - \sigma_{-p+1}\, \mu(\Delta_{p,q}(\omega)).$$

Since $p_m(\omega^o, \Delta_{p,q}(\omega)) = W^m \bar{\phi}$ one has

(6) $$p_m(\omega^o, \Delta_{p,q}(\omega)) = 2^{-q} \sum_{n=-(p-1)}^{q-1} 2^n \bar{\nu}^{(m)}_{n-m} \phi_{D^p_n(\omega)} (\omega^m) +$$

$$+ \nu^{(m)}_{q-m} \phi_{D^p_q(\omega)} (\omega^m) + \mu(\Delta_{p,q}(\omega))(1 - \nu^{(m)}_{-p+1}) ,$$

where $\omega^m = B^m \omega = (x^m, y^m)$. It is easy to see from (6) that the lines $D_m(0,0) = \{\omega = (x,y) \mid x = x^m\}$, $D_m(k,n) = \{\omega = (x,y) \mid x = x^m + 2^{-n}k\}$, $n \geq 1$, $k = 1,3,\ldots 2^n - 1$, have probability $p_m(\omega^o, D_m(k,n)) \neq 0$. More precisely when $p \to \infty$ in (6) we see that

$2^{-q} \sum_{n=o}^{q-1} 2^n \bar{\nu}^{(m)}_{n-m} \phi_{D^p_n(\omega)} (\omega^m)$ gives a contribution to the line $x = x^m$ with probability density along this line $c^{(m)}_o(y^m, y) =$ $= \sum_{n=o}^{\infty} 2^n \bar{\nu}^{(m)}_{n-m} \phi_{\Delta_n(y^m)}(y)$. Taking $p \to \infty$ in $2^{-q} \sum_{n=1}^{p-1} 2^{-n} \bar{\nu}^{(m)}_{-n-m} \phi_{D^p_{-n}(\omega)} (\omega^m)$ gives a probability $c^{(m)}_1 \bar{\mu}(\Delta_q(y))$ on $x = x^m$ and $c^{(m)}_n \bar{\mu}(\Delta_q(y))$ on $x = x^m + 2^{-n}k$, $n \geq 1, k = 1,3,\ldots,2^n - 1$, $c^{(m)}_n = \sum_{\ell=n}^{\infty} 2^{-n} \bar{\nu}^{(m)q}_{-n-m}$. The term $\nu^{(m)}_{q-m} \phi_{D^p_q(\omega)} (\omega^m)$ gives, when $p,q \to \infty$, a probability $\nu^{(m)}_{\infty} = a^{-m}$ concentrated in the point ω^m.

The last term in (6) is absolutely continuous with respect to the Lebesgue measure but has a vanishing probability density since $\nu^{(m)}_{-\infty} = 1$. One can show that $p_m(\omega^o, D_m) = 1$, $D_m = D_m(0,0) \cup \bar{D}_m$, $\bar{D}_m = \bigcup_{n,k} D_m(k,n)$, then, since $p_m(\omega^o, \Delta)$ is a positive set function, all the probability is concentrated in D_m, $\mu(D_m) = 0$, and consequently is singular with respect to the Lebesgue measure. This can be summarized introducing the transition probability density $\bar{p}_m(\omega^o, \omega)$, $p_m(\omega^o, \Delta) = \int d\mu(\omega) \bar{p}_m(\omega_o, \omega) \phi_\Delta(\omega)$,

(7) $$\bar{p}_m(\omega^o, \omega) = \delta(x - x^m)[c^{(m)}_o(y^m, y) + a^{-m}\delta(y - y^m)] + \sum_{n=1}^{\infty} 2^{-n} \bar{\nu}^{(m)}_{-n-m} \sum_{k=o}^{2^n-1} \delta(x - x^m - 2^{-n}k).$$

When $m \to \infty$, $\nu_{n-m}^{(m)} \to \lambda_n$, then from (6) we conclude $(W^m - U^{-m}\Lambda)\bar{\phi} \to 0$, $m \to \infty$, which implies the same thing on the basis $\chi_{n_1} \ldots \chi_{n_k}$ of $\{1\}^\perp$, where $\chi_n = U^n \chi_o, \chi_o = 1 - 2\phi_{\Delta_o}$, and since $(W^m - U^{-m}\Lambda)$ are uniformly bounded, $m \geq 0$, we conclude $(W^m - U^{-m}\Lambda)f \underset{st.}{\to} 0$, in the strong limit, $f \in L^2(\mu)$. For initial condition $\rho_o = \tilde{\phi}(\omega) = \mu(\Delta_{p,q}(\omega^o))^{-1} \phi_{\Delta_{p,q}(\omega^o)}(\omega)$ the probability distribution at time m $\rho_m(\tilde{\phi},\omega) = W^{\star m}\tilde{\phi}$ is

$$(8) \quad \rho_m(\tilde{\phi},\omega) = 2^{-p} \sum_{n=-(p-1)}^{q-1} 2^n \bar{\nu}_n^{(m)} \phi_{D_n^p(\omega^o)}(B^{-m}\omega) + $$

$$+ \nu_q^{(m)} 2^{-(p+q)} \phi_{D_q^p(\omega^o)}(B^{-m}\omega) + 1 - \nu_{-p+1}^{(m)} .$$

The approach to equilibrium is obtained from (8) since $\nu_n^{(m)} \to 0$, $m \to \infty$, and then $\rho_m \to 1$, i.e. $W^{\star m}\tilde{\phi} \to 1$. We have $\{W^{\star m}\}$, $m \geq 0$, uniformly bounded, then, by the same argument as above, we conclude $W^{\star m} f \underset{st.}{\to} P_o f, f \in L^2(\mu)$. Using $\underline{\omega}^m(k) = \underline{\omega}^o(k+m)$ one obtains from (8) for $p \geq m+1$, that

$$(9) \quad \rho_m(\tilde{\phi},\omega) = \bar{\mu}(\Delta_{p-m}(x^m))^{-1}[\phi_{\Delta_{p-m}(x^m)}(x) \;.$$

$$\cdot \sum_{n=o}^{q-1+m} 2^n \bar{\nu}_{n-m}^{(m)} \phi_{\Delta_n(y^m)}(y) + \sum_{n=1}^{p-1-m} 2^{-n} \bar{\nu}_{-n-m}^{(m)} \sum_{k=o}^{2^n-1} \phi_{(\Delta_{p-m}(x^m)+2^{-n}k)}(x)] +$$

$$+ \nu_q^{(m)} \mu(\Delta_{p-m,q+m}(\omega^m))^{-1} \phi_{\Delta_{p-m,q+m}(\omega^m)}(\omega) + 1 - \nu_{-p+1}^{(m)} .$$

The limit $p,q \to \infty$ in (9) corresponds to $\tilde{\phi}$ going to the deterministic initial condition $\delta_{\omega^o}(\omega) = \delta(x-x^o)\delta(y-y^o)$ and we can check explicitly that $\rho_m \to \bar{\rho}_m(\omega^o,\omega)$ as given by (7). Mean values for $\rho_o = \delta_{\omega^o}$ can be computed from (7) and $\langle x\rangle^{(m)} = x^m + O(a-1)$, $\langle y\rangle^{(m)} = y^m + O(\hat{a}-1)$, i.e. the deterministic trajectory $\omega^m = B^m\omega^o$ plus correction vanishing when $a = 1$. This was to be expected since $W^{\star m}$ reduces to U^m for $a=1$). This means that for deterministic initial conditions one does not go to equilibrium as we knew since $(W^m - U^{-m}\Lambda)f \to 0$, $m \to \infty$. Nevertheless, for probabilistic initial conditions $\rho_o(\omega) = P_o\rho_o = 1$, we know that we go to equilibrium since $\langle f(\omega)\rangle^{(m)} = \int d\mu(\omega) f(\omega)\rho_m(\rho_o,\omega) = \langle f,\rho_m\rangle$ ($\langle\cdot,\cdot\rangle$ scalar product in $L^2(\mu)$), and $\langle f, \rho_m\rangle = \langle f, W^{\star m}\rho_o\rangle \to \langle f, P_o\rho_o\rangle$,

$m \to \infty$. What we want to do now is to estimate the time m for which one is near equilibrium for a given initial distribution that we take as $\tilde{\phi}$. At time m' ρ_m is given by (8) and now, as we want the limit $m \to \infty$, we take $\ell \equiv m-p \geq 0$. Putting $B^m D_n^p(\omega^o) = R^\ell_{n+p}(y^m)$ one has $R^\ell_n(y^m) = B^{\ell+j}\Delta_{y^m(\ell+j)}$, $1 \leq j \leq n$, a set depending only on y, and (8) gives

$$(10)\quad \rho_m(\tilde{\phi},\omega) = \sum_{n=1}^{p+q-1} 2^n\, \bar{\nu}^{(m)}_{n-p}\, \phi_{R^\ell_n(y^m)}(\omega) +$$

$$+\ \nu_q^{(m)} 2^{p+q} \phi_{R^\ell_{p+q}(y^m)}(\omega) + 1 - \nu^{(m)}_{-p+1}\ .$$

Using (10) one easily shows that $\langle f(x)\rangle^{(m)} = \int_0^1 dx f(x) = \langle f(x)\rangle^{eq.}$, $m \geq p$. For a function depending also in y the situation is more involved but allows us to say that at time $\bar{m}=p$ we are near the equilibrium distribution. The length ε of the initial condition along the x axis is $2^{-\bar{m}}$ and then $\bar{m} \sim \ell n(\varepsilon^{-1}), \varepsilon \to 0$, which means that $\bar{m} \to \infty$ and consequently one never arrives to equilibrium for deterministic initial conditions as stated above. Let us see in what sense the transition probability goes to equilibrium. For any $\rho_o(\omega), P_o\rho_o = 1$, one has $\int d\,\mu(\omega^o)\rho_o(\omega^o)p_m(\omega^o,\Delta) \to \mu(\Delta), m \to \infty$, since $\langle\rho_o, W^m\phi_\Delta\rangle = \langle W^{\star m}\rho_o,\phi_\Delta\rangle \to \langle 1,\phi_\Delta\rangle$.

We can conclude then that the Markov process has two different behaviors

a) For a deterministic initial condition $\rho_o(\omega)=\delta_{\omega^o}(\omega)$ equilibrium is never attained. Moreover the trajectory $(\omega^o,\omega^1,\ldots\omega^m), \omega^j = B^j\omega^o$, has probability a^{-m}, and is the only trajectory with non vanishing probability. This probability remains near one for times m such that $m(a-1) \ll 1$.

b) For a non deterministic initial condition $\rho_o(\omega)$, $\mu(\text{support } \rho_o(\mu)) \neq 0$, the probability distribution $\rho_m(\rho_o,\omega) \to 1, m \to \infty$. The time $\bar{m}$ for which one starts being near equilibrium grows as $\ell n(\varepsilon^{-1}), \varepsilon \to 0$, where ε is the length of support $\rho_o(\omega)$ along the x axis. If we take the point of view that the Markov process is a more fundamental description, then (for $a > 1$ near 1) the deterministic trajectory would be observed with probability 1 even for very large m, which means that the notion of trajectory is recuperated via the most probable trajectory (no loss of information). This situation is quite general for dynamical systems for which the canonical construction of a Markov process in the sense of [2,3] can be done and the detailed study will be presented elsewhere [5].

REFERENCES

1. I. Prigogine, C. George, F. Henin, and L. Rosenfeld: Chem. Soc.4(1973).
2. B. Misra, I. Prigogine and M. Courbage: Physica 98A,1(1979).
3. M. Courbage and B. Misra: Physica 104A,359 (1980).
4. V.I. Arnold and A. Avez: "Ergodic problems of classical mechanics", Benjamin (1968).
5. S. Martinez and E. Tirapegui, to appear.

ACKNOWLEDGEMENTS

The authors are grateful to Professor A. Brunel (University of Paris) for an interesting and stimulating discussion.

HOMOCLINIC AND HETEROCLINIC POINTS IN THE HENON MAP

G.Gómez* C.Simó**

* Secció de Matemàtiques,Universitat Autònoma de Barcelona,Bellaterra, Barcelona, Spain.

**Facultat de Matemàtiques, Universitat de Barcelona,Gran Vía 585, Barcelona 7, Spain.

1. INTRODUCTION

This communication is devoted to study the role of the homoclinic and heteroclinic points of the Hénon mapping in order to explain the capture escape boundary for the Hénon attractor previously studied numerically by Hitzl [2].

The equations of the two dimensional quadratic map T under consideration are:

$$x_{i+1} = 1 + y_i - a x_i^2 \quad , \quad y_{i+1} = b x_i \qquad i=1,2,...$$

x,y being real variables and a,b real parameters. T has a constant jacobian J = -b. If b = 0 T is a one dimensional endomorphism and if b = ± 1 T is an area reversing-preserving two dimensional difeomorphism. The implications of the imbedding of the one dimensional map into the two dimensional has been studied by Mira [3].

Under the belief that, if an unstable invariant manifold has homoclinic points and no heteroclinic points and it is bounded then its closure is a strange attractor, we have been looking for tangential homoclinic and heteroclinic points associated to the invariant manifolds of the two fixed points of T. The values a,b at which the heteroclinic tangency takes place gives us in the a,b plane the capture escape boundary for the Hénon attractor when this attractor has only one component.

It might happen, and in fact it does, that points can be attracted by sets of s-components, s>1. In this situation the attractor is associated to unstable invariant manifolds of s-periodic points. The computation of the tangential homoclinic and heteroclinic points associated to this situation will appear elsewhere.

2. COMPUTATION OF THE TANGENTIAL HOMOCLINIC AND HETEROCLINIC CURVES

The computation of the tangential homoclinic and heteroclinic curves has been carried out using Davidenko's continuation method.

For the heteroclinic tangencies, if we call p, q the two fixed points of T, we choose two points $p_1 \in W_p^u$, $q_1 \in W_q^s$ and we impose the following set of non linear equations:

$$T^n(a,b,p_1) - T^{-m}(a,b,q_1) = 0$$

$$DT^n(p_1-p) \wedge DT^{-m}(q_1-q) = 0$$

The first equation gives us the condition of having an heteroclinic point and the second one the condition of tangency. The two integer numbers n and m must be changed throughout the computations.

The computation of the homoclinic tangencies is done in a similar way.

3. RESULTS

A sample of the results, for positive values of b, for the heteroclinic tangencies is shown in Fig. 1 superposed to Hitzl's curve.

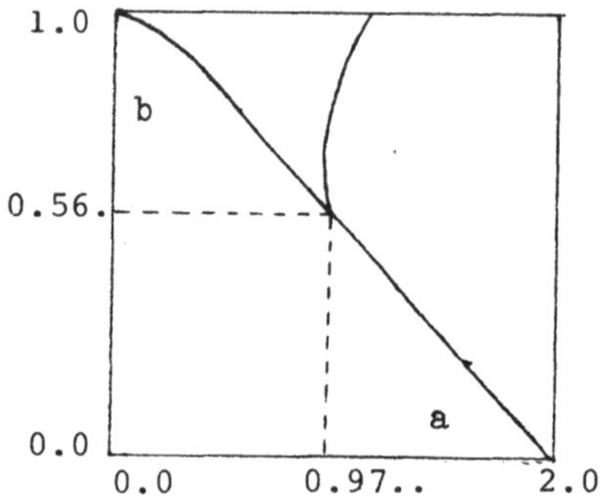

Fig. 1

Some remarkable facts must be mentioned:

i) For values of $b \in (0.0, 0.56...)$ there is good agreement between our results and those obtained by Hitzl, except that for a fixed value of b we obtain a slightly smaller value of a. This is due to the fact that Hiltz uses a finite number of iterations (5.000) in order to decide if a point is captured or not.

ii) On the almost linear segment for $b \in (0.0, 0.56...)$ several small spikes appear in Hitzl's curve at $b = 0.05, 0.21, 0.27$ and 0.49. These correspond to the beginning of secondary boundaries belonging to attracting sets of 3,6,5 and 8 components respectively. Of course we have not detected those spikes.

iii) For values of $b \in (0.56..., 1.)$ the discrepancies between both results is due to the fact that the attractor which is found before escaping to infinity has some 2^k components.

iv) For values of $b < 0$ the curve of heteroclinic tangencies is not C^1 due to the fact that the first tangency which occurs for different values of b changes from one loop to another.

v) We have proved analitically that all the 2^k tangential heteroclinic points come out from the point a=2, b=0. The homoclinic tangencies for b=0 come out for the values of $a = \mu_{2^k}$ such that $T(\mu_{2^k}, b, 0)$ belongs to an unstable $3 \cdot 2^k$-periodic orbit. See [1] for more details.

REFERENCES

1. Collet,P.Eckmann,J.P.,Iterated maps of the interval as dynamical systems. Birkhäuser, Boston 1980.
2. Hitzl,D.L.,Numerical Determination of the Capture/Escape Boundary for the Hénon Attractor. Preprint 1981.
3. Mira,Ch.,see these Proceedings.

THE SIMPLE PERIODIC ORBITS IN THE UNIMODAL MAPS

Lluís Alsedà
Departament de Teoria Econòmica
Universitat Autònoma de Barcelona
Bellaterra, Barcelona (Spain).

Rafael Serra
I.N.B. Sant Cugat
Sant Cugat del Vallès
Barcelona (Spain).

1. PRELIMINARIES

The Šarkovskiǐ's ordering in N (from now on, for definitions see [2]) is: $3 \Delta 5 \Delta 7 \Delta \ldots \Delta 2.3 \Delta 2.5 \Delta 2.7 \Delta \ldots \Delta 4.3 \Delta 4.5 \Delta 4.7 \Delta \ldots \Delta \ldots \Delta 16 \Delta 8 \Delta 4 \Delta 2 \Delta 1$.

Šarkovskiǐ's Theorem [6,7] states that if $f \in C(I)$ and $n \in P(f)$ then $m \in P(f)$ when $n \Delta m$.

Let $f \in C(I)$ and $n>1$ be the minimum of $P(f)$ in the Δ-ordering. We say that a periodic orbit of f is minimal if their period is n.

Let $P=\{P_1,\ldots,P_n\}$ be a periodic orbit of $f \in C(I)$ of period $n=2^m q>1$, where $q \geq 1$ is odd, $m>0$ and $P_1<P_2<\ldots<P_n$. For $q>1$ and any integer $m \geq 0$, we define a simple periodic orbit inductively. Suppose $m=0$ and let $t=(q+1)/2$, then we say P is simple if either (a) or (b) holds:

(a) $f(P_{t-k})=P_{t+k+1}$ for $k=0,1,\ldots,t-2$

$f(P_{t+k})=P_{t-k}$ for $k=1,2,\ldots,t-1$ and

$f(P_1)\ =P_t$

(b) $f(P_{t-k})=P_{t+k}$ for $k=1,2,\ldots,t-1$

$f(P_{t+k})=P_{t-k-1}$ for $k=0,1,\ldots,t-2$ and

$f(P_n)\ =P_t$.

Now suppose $m \geq 1$. Then we say P is simple if the two subsets $\{P_1,\ldots,P_{n/2}\}$ and $\{P_{n/2\ +1},\ldots,P_n\}$ of P are simple periodic orbits of f^2. Then we have $f(\{P_1,\ldots,P_n\})=\{P_{n/2\ +1},\ldots,P_n\}$. Finally, for $q=1$ and $m \geq 1$, we also define a simple periodic orbit inductively. If $m=1$, then P is simple. Suppose $m>1$, then we say p is simple if the two subsets $\{P_1,\ldots,P_{n/2}\}$ and $\{P_{n/2\ +1},\ldots,P_n\}$ of P are simple periodic orbits of f^2.

It is easy to find that the number of possible different behavior of the simple periodic orbits ot period $n=2^m q$ where $m \geq 0$ and $q \geq 1$ odd is: 2^{2^m-m-1} if $q=1$ and $2^{2^{m+1}-m-1}$ if $q \geq 3$.

Theorem (see [1,2] and [7]). Let $f \in C(I)$ and suppose that P is a minimal periodic orbit. Then P is simple.

The above Theorem justifies the study of the behavior of the simple periodic orbits. We shall study this subjet in the unimodal maps (from now on, for definitions see also [3]).

We say that an itinerary of length n is Quasi-min-max of period n if their first n-1 symbols coincide whith the corresponding ones of the min-max itinerary of period n.

2. RESULTS

Proposition A. Let f be a unimodal map and suppose that the itinerary of the last point of P is Quasi-min-max. Then the behavior of f on P is unique.

In the proof of proposition A we use both the lemma 11 and the statement in page 10 of [4]. In this way one can give explicitly, for each period, the only possible behavior in terms of the rotation sequence [4,5].

Theorem B. Let f be a unimodal map and P a periodic orbit. Then the itinerary of the last point of P is Quasi-min-max if and only if P is simple.

Note that Theorem B and Proposition A state that the behavior of a simple periodic orbit of period n in the unimodal maps is unique, (note the difference with the particular case). In particular we have a way to describe it.

REFERENCES

1. Ll. Alseda and J. Llibre, Minimal periodic orbits and topological entropy for one dimensional maps, preprint, Barcelona, 1981.
2. L. Blok, Simple periodic orbits of mappings of the interval. Trans. Amer. Math. Soc. 254 (1979), 391-398.
3. P. Collet and J.P. Eckmann. Iterated maps on the interval as dynamical systems, Birkhäuser, Boston, 1980.
4. J. Llibre, On the rotation sequences of the unimodal maps, preprint Barcelona, 1982.
5. Ch. Mira and I. Gumowski, Dynamique chaotique. Transformations ponctuelles. Transition ordre-désordre. Toulouse, Cépadues, 1980.
6. A. N. Šarkovskiǐ. Coexistence of cycles of a continuous map of a line into itself, Ukr. Math. Ž. 16 (1964), 61-71.
7. P. Štefan. A theorem of Šarkovskiǐ on the existence of periodic orbits of continuous endomorphisms of the real line. Comm. Math. Phys. 54 (1977), 237-248.

MODULATION PROPERTIES IN DECAYING PROCESSES OF THE CORRELATION FUNCTION IN A FAMILY OF 1-D MAPS

T. NAGASHIMA

Institut für Theoretische Physik, Universität Stuttgart
D-7000 Stuttgart-80, Fed. Rep. of Germany (*)

Since the work of Ruelle-Takens on turbulence generation [1], properties of the correlation function in dynamical systems have received a great concern [2,3]. In this report, we give an exact analysis of the correlation function for a certain family of 1-D maps.

We take a family of 1-D maps on the interval $[0,1]$ defined by

$$x_{n+1} = \varphi_\alpha(x_n) \quad ; \quad \varphi_\alpha(x) = (2x+\alpha) - [2x+\alpha] \quad , \quad (1)$$

where $0 \leqslant \alpha < 1$, $0 \leqslant x \leqslant 1$, and $[\xi]$ represents the integral part of ξ . Assuming $f(x)$ is a measurable function of x , we consider the correlation function for $f(x)$ defined as follows;

$$C_{f(x)}(m) = \left\langle \frac{\{ f(\varphi_\alpha^{(m)}(x)) - \langle f(\varphi_\alpha^{(m)}(x))\rangle\} \cdot \{ f(x) - \langle f(x)\rangle\}^*}{| f(x) - \langle f(x)\rangle |^2} \right\rangle \quad , \quad (2)$$

where $\varphi_\alpha^{(m)}(x) \equiv \overbrace{\varphi_\alpha \circ \cdots \circ \varphi_\alpha}^{m}(x)$,and $\langle F(x)\rangle$ means the phase average of $F(x)$ given by $\langle F(x)\rangle = \int_0^1 F(x)\,dx$.

Our idea in analyzing the correlation function defined by eq.(2) is, first , to expand $f(\varphi_\alpha^{(m)}(x))$ in terms of Fourier series, i.e.,

$$f(\varphi_\alpha^{(m)}(x)) = \sum_{k=-\infty}^{\infty} a_k^{(m)} e^{2\pi i k}$$

and, then, to establish a recursion relation between the Fourier coefficient $a_k^{(m)}$ and $a_j^{(n-1)}$.

(*) On leave from Institute of Precision Mechanics, Faculty of Engineering Hokkaido University, Sapporo, Japan

If we restrict our attention to the correlation function for the coordinate itself, putting $f(x) = x$, we obtain

$$C_x(m) = 2^{-m} \cdot \frac{6}{\pi^2} \sum_{k=1}^{\infty} \frac{1}{k^2} \cos\{2\pi k (2^m - 1)\alpha\}$$
$$= 2^{-m} \cdot \{6 w_m^2 - 6 w_m + 1\}_{\text{periodic funct. with period 1}}$$
$$\equiv 2^{-m} \cdot P(w_m) \qquad (3)$$

where $w_m \equiv (2^m - 1) \cdot \alpha$, and the function $P(w_m)$ is a periodically extended function with its basic period $[0,1)$.

Being based on a relation $P(w_m) = P(\varphi_\alpha^{(m)}(0))$, where $\varphi_\alpha^{(m)}(0)$ ($m=0,1,2,\cdots$) is the orbit starting at $x=0$, the temporal behaviors of the correlation function described by eq.(3) can be classified, according to the parameter α, in the following way:

1) pure exponential decay,
2) periodic oscillatory modulation of the exponential decay,
3) chaotic (non-periodic oscillatory) modulation of the exponential decay.

It is conjectured that the number of α belonging to the category 3) is uncountable in $[0,1)$.

Details of this report will be published elsewhere.

The author would like to express his gratitude to Prof. H. Haken for providing him the opportunity to stay in Stuttgart, where this manuscript has been written.

References

[1] D. Ruelle and F. Takens, Commun. Math. Phys. 20, 167 (1971).
[2] S. Grossmann and S. Thomae, Z. Naturforsch. 32a, 1353 (1977).
[3] S. Thomae and S. Grossmann, J. Stat. Phys. 26, 485 (1981).

RELAXATION TIMES AND RANDOMNESS FOR A NONLINEAR CLASSICAL SYSTEM

Mario Casartelli

Istituto di Fisica, Sezione Teorica
Università di Parma
Via M. D'Azeglio 85
43100 Parma (Italia)

We shall describe numerical experiments on a classical nonlinear system possessing a stochastic transition[1], showing that a "memory" of the ordered region survives in the stochastic region, and can be revealed through the analysis of the relaxation times.

The model consists of a chain of N particles of mass 1, interacting at the nearest neighbours with a Lennard-Jones potential $V(r) = 4\varepsilon\left[(\sigma/r)^{12}-(\sigma/r)^{6}\right]$ ($\varepsilon=27.5$, $\sigma=1$). We denote: E the total energy, $u = E/N$ the specific energy, E^h and u^h the corresponding harmonic quantities, u_k^h and ω_k the energy and the frequency of the k-th harmonic mode, $\Delta\omega = (\omega', \omega'')$ a frequency band, N_ω the number of frequencies in $\Delta\omega$.

The following facts are assumed to be known[2]: there exist two specific energies u_1 and u_2 such that for $u < u_1$ the system behaves as an integrable one, for $u > u_2$ as a stochastic one, with a continuum of intermediate dynamical situations for $u_1 < u < u_2$. Estimates from previous computations give $u_1 > 0.01$ and $u_2 < 1$ in the range $20 < N < 200$.

Let now $W_\omega = W_\omega(u,N,T,\Delta\omega)$ and $Q_\omega = Q_\omega(u,N,T,\Delta\omega)$ be defined as

$$W_\omega = \frac{1}{uN_\omega}\sum_{k\in\Delta\omega}\frac{1}{T}\int_0^T u_k^h(t)\,dt$$

$$Q_\omega = \frac{1}{E^2 N_\omega}\sum_{k\in\Delta\omega}\left(\frac{1}{T}\int_0^T (u_k^h(t)-u^h(t))\,dt\right)^2$$

W_ω and Q_ω are respectively the T-time averaged specific energy and the mean square deviation of the T-time averaged energies of the harmonic modes in $\Delta\omega$. Further divisions by u and E^2 make them dimensionless. In the stochastic region, according to classical previsions, the physical observations do not depend on initial conditions. This means that, in the limit as $T\to\infty$, $W_\omega \to W°$ (the equipartition value) and $Q_\omega \to 0$. Conversely, for finite times, the analysis of W_ω and Q_ω may show how the equipartition takes place, and the role of various parameters. Since no subset of the energy surface is to be privileged a priori, we shall use random initial conditions (RIC), namely coordinates in the phase space which are proportional to random numbers $\{r_k\}$ distributed with a fixed probability law (e.g. equidistributed in an interval). Experiments on W_ω and Q_ω are performed starting with many different sets $\{r_k\}$, $\{r'_k\}$, $\{r''_k\}$ etc., and following the corresponding trajectories.

The results are summarized in the following four points: 1) when $u-u_2$ is positive and small (e.g. u=1) the relaxation of the system is very slow in general; moreover it is not uniform with respect to various RIC. 2) the relaxation becomes faster and faster as u increases, and a sensible uniformity takes place. 3) features 1 and 2 neither depend on $\Delta\omega$ nor on the probability distribution of the set $\{r_k\}$. 4) the dependence on N is not immediately clear in the range explored (N=25,50,100). Only, property 1 seems to be lightly weakened as N increases. If such trend were confirmed by further experiments, a substantial uniformity of the relaxation times with respect to a random access in the phase space would be regained, in the thermodinamic limit, even for finite time observations.

The results above may be interpreted in the framework of the KAM theory. Invariant surfaces (n-tori filling the low energy region of the space) vanish in measure as the system approaches to ergodicity, but may survive (with measure 0) also in the stochastic region. Therefore, trajectories starting from different sectors may remain differently trapped for finite time observations, even if the energy surface is metrically transitive. In this sense there is a "memory" of the ordered region in the stochastic one. It seems also reasonable that such memory is released as u increases, so that points 1-3 do not present difficulties. As to the thermodinamic limit, further experiments are needed for a precise statement (a plausible conjecture is that the frequency spectrum may play a significant role. See[3] for a comparison with a model of radiant cavity). In any case, point 4 appears also to be compatible with the interpretation above.

REFERENCES

1. For general informations, R.Helleman in Fundamental Problems in Statistical Mechanics , E.G.D.Cohen ed. North-Holland (1980)
2. M.Casartelli, Phys.Rev. 19A, 1741 (1979)
3. G.Benettin and L.Galgani, J.Stat.Phys. 27, 153 (1982)

TOPOLOGICAL ENTROPY ON ROTATION SEQUENCES

Christian Gillot

Institut National des Sciences Appliquées
Département de Mathématiques
Av.de Rangueil 31077 Toulouse Cedex (France)

1. INTRODUCTION

Let $f:[0,1]\to[0,1]$ be a continuous unimodal map: for $u\in[0,1[$, f is strictly increasing on $[0,u]$, $f(0)>0$, $f(u)=1$, and strictly decreasing on $[u,1]$, $f(1)=0$. We assume that f is such that u is a periodic point, with period $N\geq 2$. The relative positions of the iterates of u are transcribed by some distinct integer sequence $R_N=(i_k)$, $1\leq k,i_k\leq N$, which is called the rotation sequence of the cycle. Properties of rotation sequences are given in [1] and considered known in the following. The aim of this paper is to give a direct application of these properties in the study of topological entropy. We claim that entropy can be defined on the set of rotation sequences, and, among others, a "fractal" behaviour of the entropy function is shown.

2. THEOREM

(i). *All continuous unimodal maps possessing the same rotation sequence have the same topological entropy.*

(ii). *The topological entropy of every irreducible rotation sequence with at least 3 terms has $(\log 2)/2$ as lower bound.*

(iii). *Let $B_n o B_m$ be a compound rotation sequence, with B_n, $n\geq 2$, as fundamental component. For each integer $m\geq 2$ the topological entropy is such that*

$$h(B_n o B_m)=\begin{cases} h(B_n) & \text{if } n\neq 2,\\ \frac{1}{2}\,h(B_m) & \text{if } n=2.\end{cases}$$

Proof.

(i). The map f swaps the elementary intervalls with extremities u_{i_k}, $u_{i_{k+1}}$. For $p>N$, $f^p(u_{i_k})$ is an extremal value of f^p, so we use the definition of the topological entropy [2]

$$h(f,R_N) = \lim_{p\to\infty} \frac{1}{p} \log \operatorname{Var} f^p .$$

Let $e_i^{(o)}$ be the length of the elementary intervall E_i, $1\le i\le N-1$, V_o the column vector with components $e_i^{(o)}$, and $||V_o|| = \Sigma_i\ e_i^{(o)}$. The vector $V_p = A^p V_o$ is such that $||V_p|| = \operatorname{Var} f^p$, $p\ge 1$. The non negative matrix A has a single non null diagonal term. From [3], there exists a positive latent root λ_M, such that $\lambda_M = \rho(A)$. By a standard way we have $||V_p|| = \lambda_M^p(\theta_p+\phi)$, $|\phi|<\infty$, $\lim_{p\to\infty} \theta_p = 0$. Thus

$$h(f,R_N) = \lim_{p\to\infty} \frac{1}{p} \log ||V_p|| = \log \lambda_M.$$

The latent root λ_M depends only on A, i.e on the sequence R_N. Then the entropy $h(f,R_N)$ is constant for all maps which have the same rotation sequence R_N: we can thus relate to R_N its topological entropy $h(R_N)$.

(ii). Let A be the evolution matrix of the basic sequence B_n, $n\ge 3$. The matrices A, A^2 are non negative and irreducible. From the properties P1, P2, P3, [1], A has only two non null terms in first column, and at least two non null terms in other columns. From the min-max theorem [3], we have $\rho(A^2)>2$. From the Perron-Frobenius theorem, the greatest latent root, λ_M, simple, positive, of A equals $\rho(A)$. Thus $\lambda_M>\sqrt{2}$, and $h(B_n)>(\log 2)/2$, for each $n\ge 3$. In view of studying the lower bound $(\log 2)/2$ of the entropy, let us consider the infinite set of irreducible sequences of the form

$$B_{2k+1} = (3,1,2k,2k-2,\ldots\ldots,4,5,\ldots\ldots,2k+1,2), \quad k\ge 1.$$

It is shown [4] that the secular equation of the matrix related to B_{2k+1} is

$$P_{2k}(x) = (x^{2k+1} - 2x^{2k-1} - 1) / (x+1).$$

Its greatest positive root, θ_{2k}, is such that

$$2^{1/2} - 2^{-k-1} - k\, 2^{-2k-3/2} < \theta_{2k} < 2^{1/2} - 2^{-k-1} ,$$

so that $\lim_{k\to\infty} \theta_{2k} = 2^{1/2}$. There exists thus irreducible sequences with entropy arbitrarily close to $(\log 2)/2$.

Remark. The basic sequence $B_2=(1,2)$ is an exception to the rule : its matrix is $A=[1]$, whence $h(B_2)=0$. The particular role played by B_2 can be observed in what follows.

(iii) The compound rotation sequence $B_n \circ B_m$ has a reducible matrix [1]

$$A = \begin{bmatrix} \alpha & o \\ \beta & \gamma \end{bmatrix}$$

where γ is the matrix of the fundamental component $\mathcal{B}_n$. The matrix α is weakly cyclic, α^m completely reducible [3], with diagonal blocks. From the composition of α [1], we show that these blocks are $\underline{a}$ or $\underline{e}\ \underline{a}\ \underline{e}$ matrices, $\underline{a}$ being the matrix of $\mathcal{B}_m$. These matrices have the same spectra ($\underline{e}^2$ = Identity). The same theorem[3] says that $\rho(\alpha^m)=\rho(\underline{a})$. From the reducibility of A, we have $\rho(A) = \max(\rho(\alpha),\rho(\gamma))$. Two cases are to be considered :

- The index n of $\mathcal{B}_n$ is ≥ 3. The matrix α is composed of m^2 sub-matrices[1], so that $\rho(\alpha)=(\rho(\underline{a}))^{1/m} < 2^{1/2}$, for each $m \geq 2$. As $\mathcal{B}_n$ is irreducible, we have $\rho(\gamma)>2^{1/2}$, and $\rho(A) = \rho(\gamma)$, whence $h(\mathcal{B}_n \circ \mathcal{B}_m) = h(\mathcal{B}_n)$, $n \geq 3$.

- For $n = 2$, we have $\mathcal{B}_2 = (1,2)$ and $\rho(\gamma) = 1$. The matrix α contains 2^2 sub-matrices, and $\rho(\alpha^2) = \rho(\underline{a})$. As $\underline{a}$ is irreducible, $\rho(A) = \rho(\alpha)$, whence $h(\mathcal{B}_n \circ \mathcal{B}_m) = (h(\mathcal{B}_m))/2$.

Example.

$$\mathcal{B}_2 = (1,2)\ ,\quad \mathcal{B}_3 = (3,1,2)$$

$$\mathcal{B}_3 \circ \mathcal{B}_2 = (3,6,4,1,5,2)\ ,\ \rho(A) = \rho(\underline{a}(\mathcal{B}_3)) = (1+\sqrt{5})/2\ .$$

$$\mathcal{B}_2 \circ \mathcal{B}_3 = (3,1,5,4,6,2)\ ,\ \rho(A) = \rho(\underline{a}(\mathcal{B}_3))^{1/2}$$

Remark. The distinction between different sequences with 2^k terms is to be noticed. Among these, there exists a single one, written $\mathcal{B}_2 \circ \mathcal{B}_2 \circ \dots \circ \mathcal{B}_2$, (k times). Its entropy is zero, for every $k \geq 1$. There exists another one of the form $\mathcal{B}_2 \circ \mathcal{B}_{2^{k-1}}$, $\mathcal{B}_{2^{k-1}}$ irreducible, whose entropy is $(h(\mathcal{B}_{2^{k-1}}))/2$. The entropy of the sequence $\mathcal{B}_2 \circ \mathcal{B}_2 \circ \mathcal{B}_{2^{k-2}}$ is then $(h(\mathcal{B}_{2^{k-2}}))/2^2$, etc... The same result stands for the sequences $\mathcal{B}_n$, $\mathcal{B}_2 \circ \mathcal{B}_n, \mathcal{B}_2 \circ \mathcal{B}_2 \circ \mathcal{B}_n, \dots, n \geq 3$. This illustrates the "fractal" behaviour of the topological entropy on the set of rotation sequences, and enforces the basic role played by irreducible sequences. As an application, it is easily seen that the C° map $F_s(x)=s-1-s|x|$, $1 \leq s \leq 2$, [5], whose entropy is $\log s$, admits only basic sequences for $2^{1/2}<s<2$, each of these appearing once only. Whereas the quadratic map possesses compound sequences [6], and (iii) asserts that the topological entropy is constant on all sequences of the form $\mathcal{B}_n \circ \mathcal{B}_m$, for every $m \geq 2$.

REFERENCES

1. C.Gillot, Séminaire Analyse Numérique. Univ. Toulouse III (1981).
2. M.Misiurewicz and W. Slenk, Studia Mathematica, T.LXVII (1980).
3. R.S. Varga, Matrix Iterative Analysis. Prentice Hall Series (1962).
4. C.Gillot and G.Gillot, Springer Series in Synergetics (1981).
5. L.Jonker and D.Rand, Inv.math, 62, 347-365 (1981).
6. I.Gumowski and C.Mira, Dynamique Chaotique. Cepadues Ed.Toulouse (1980).

THE TAYLOR-GREEN VORTEX :

FULLY DEVELOPED TURBULENCE AND TRANSITION TO SPATIAL CHAOS

Uriel Frisch

CNRS, Observatoire de Nice
B.P. 252
06007 Nice Cedex (France)

SUMMARY

A brief introduction to fully developed turbulence is given (cf. Refs. 1-2). We then report results obtained from numerical simulations of the Taylor-Green three dimensional vortex flow [3]. This flow is perhaps the simplest system in which one can study the generation of small scales by three-dimensional vortex stretching and the resulting turbulence. The problem is studied by both direct spectral numerical solution of the Navier-Stokes equations (with up to 256^3 modes) and by power series analysis in time.

The inviscid dynamics are strongly influenced by symmetries which confine the flow to an impermeable box with stress-free boundaries. There is an early stage during which the flow is strongly anisotropic with well-organized (laminar) small-scale excitation. The flow is smooth but has complex-space singularities within a distance $\delta(t)$ of the real space which are manifest through an exponential tail in the energy spectrum. It is found that $\delta(t)$ decreases exponentially in time to the limit of our resolution. Indirect evidence is presented that more violent vortex stretching takes place at later times, possibly leading to a real singularity ($\delta = 0$) at a finite time. These direct integration results are consistent with newly presented results extending the Morf, Orszag and Frisch [4] temporal power series analysis from order t^{40} to order t^{80}. Still, convincing evidence for or against the existence of a real singularity will require even more sophisticated analysis.

The viscous dynamics (decay) have been studied for Reynolds numbers R (based on integral scale) up to 3000 and beyond the time t_{max} at which the maximum energy dissipation is achieved. Early time, high R dynamics are essentially inviscid and laminar. Then, instabilities starting at small scales, which may be driven by viscosity, make the flow increasingly chaotic (turbulent) with extended high-vorticity patches appearing away from the impermeable walls. Near t_{max} the small scales of the flow are nearly isotropic provided $R \gtrsim 1000$. Various features characteristic of fully developed turbulence are observed near t_{max} when $R = 3000$ and $R_\lambda = 110$:

(i) A k^{-n} inertial range in the energy spectrum is obtained with $n \sim 1.6 - 2.2$ (in contrast with a much steeper spectrum at earlier times),

(ii) The energy dissipation has considerable spatial intermittency, its spectrum has a $k^{-1+\mu}$ inertial range with the codimension $\mu \sim 0.3 - 0.7$.

Details on the above material may be found in Ref. 3.

We also discuss the question of <u>chaos</u> in fully developed turbulence. For flows which are driven deterministically (by convection, shear, etc., ...) <u>temperal</u> chaos may be defined in the same way(s) as for low-order systems. Among the new features we expect to find 'partial relaminarization windows' (in the control parameter). In such a window the large scales have an almost laminar behaviour whereas the small scales retain their strongly chaotic character. This is suggested by the rather common occurence of 'coherent structures' in high Reynolds number experiments. This coexistence of order and chaos is made plausible by the observation that widely separated scales have only weak interactions (except for advection effects which are dynamically trivial).

For decaying flows, such as the Taylor-Green vortex, temporal chaos becomes meaningless. <u>Spatial</u> chaos may be defined in terms of the instantaneous topology of vortex lines: the vorticity field being divergenceless, its integral lines define a Hamiltonian dynamical system which may or may not be chaotic [5]. If the flow has no spatial chaos at $t = 0$ (as is the case for the Taylor-Green vortex), inviscid evolution during the regular phase cannot lead to chaos because the vortex lines are following the fluid motion. Viscous diffusion and reconnection of vortex lines, however, can lead to intricate topologies. For the Taylor-Green vortex we have some evidence that strong spatial chaos appears at high Reynolds numbers around $t = 7$, following the dissipation of coherent rolled-up vortex sheets which are strongly helical. It is worth emphasizing that in this 'transition to spatial chaos' the phase space is the ordinary physical space and the control parameter is the time.

REFERENCES

1. U. Frisch, P.L. Sulem and M. Nelkin, J. Fluid Mech. 87, 719 (1978).
2. U. Frisch in Chaotic Behaviour in Deterministic Systems, Les Houches 1981, North Holland (in press).
3. M. Brachet, D. Meiron, S. Orszag, B. Nickel, R. Morf and U. Frisch, Small-scale structure of the Taylor-Green vortex, J. Fluid Mech. in press (1982 or 1983).
4. R. Morf, S. Orszag and U. Frisch, Phys. Rev. Lett. 44, 572 (1980).
5. M. Hénon, C. R. Acad. Sci. 262, 312 (1966).

ANHARMONIC SYSTEMS IN EXTERNAL PERIODIC FIELDS WITH CHAOTIC BEHAVIOUR

W.-H. Steeb, A. Kunick, and W. Erig

Theoretische Physik, Universität Paderborn

D-4790 Paderborn, West-Germany

Anharmonic systems with an external periodic field appear in various physical and engineering problems. Recently, several authors (compare [1] and references therein) have investigated chaotic states of such systems. The differential equation is given by

$$\ddot{x} + \alpha\dot{x} + \frac{dV(x)}{dx} = f\cos(\Omega t) \qquad (1)$$

V is the potential and is chosen so that there is an anharmonicity in eq.(1). α is the damping coefficient ($\alpha > 0$). f denotes the amplitude of the external periodic force and Ω is the frequency of the external perturbation. In general eq.(1) cannot be solved exactly and we are forced to make numerical calculations. For this purpose it is convenient to write eq.(1) as an autonomous system, namely

$$\dot{x} = y \quad , \quad \dot{y} = -\alpha y - \frac{dV(x)}{dx} + f\cos(z) \quad , \quad \dot{z} = \Omega \quad , \qquad (2)$$

where $z(t=0) = 0$. This equation is defined on $\mathbb{R}^2 \times S^1$. We are also interested in phase portraits for the (x,y)-plane.

In the present note we investigate the chaotic behaviour of eq.(2) for the potential $V(x) = -\beta\cos x$, where $\beta > 0$. We mention that the limiting case where $f = \alpha = 0$ is the completely integrable equation of the pendulum. Whether or not chaotic behaviour occurs depends on the parameters f, Ω, α, and β. There are several possibilities to characterize the chaotic state (decay of the auto-correlation functions, one-dimensional Lyapunov exponent, Poincaré mapping, Hausdorff dimension and so on). In the following we apply the one-dimensional Lyapunov exponent. If the one-dimensional Lyapunov exponent is positive, then chaotic behaviour is indicated.

For the following parameter values we have calculated the one-dimensional Lyapunov exponent.

Case 1. We set $\alpha = 0.2$, $\beta = 1$, and $\Omega = 0.8$. The amplitude of the external force is varied within the range $0 < f < 1.5$. For $0 < f < 0.8$ we do not find chaotic behaviour. In the range $0.8 < f < 1.5$ we obtain regions with and without chaotic states (Figure 1).

Case 2. We set $\alpha = 0.2$, $\beta = 1$, $f = 1$. The external frequency Ω is varied within the range $0 < \Omega < 0.8$. Again we find regions with and without chaotic states (Figure 2).

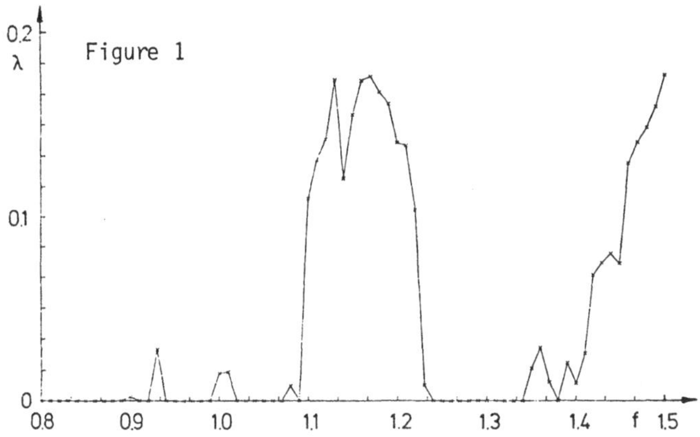

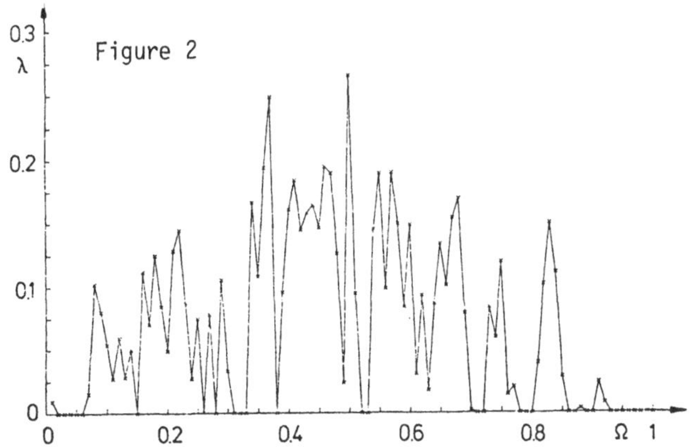

W.-H. Steeb and A. Kunick, Phys. Rev. A 25, 2889 (1982).

RENORMALIZATION OF NON-ANALYTICAL UNIMODAL MAPS

Michel Cosnard and André Eberhard

Laboratoire IMAG
B.P. 53 X
38041 Grenoble cedex (France)

1. INTRODUCTION

Let U be the set of unimodal continuous selfmaps of $[-1,+1]$. The renormalization operator R is a map from a part V of U in U defined as follows : let f be in V and $\alpha = -f(1)$

$$\forall x \in [-1,+1] \qquad Rf(x) = -\frac{1}{\alpha} f^2(-\alpha x)$$

Our purpose is to study the properties of R applied to particular subsets of U and to show the differences with Feigenbaum theory of the analytical case

2. COMPUTATION OF RENORMALIZATION CONSTANTS

First we introduce the set F of piecewise linear functions with slopes in geometric progression. This set is completely defined by two real ρ and θ. R is a map from a part $\Gamma(\rho,\theta)$ of the set $F(\rho,\theta)$. We put on $F(\rho,\theta)$ a structure of differential manifold of dimension 2.

For some values of ρ and θ, R possesses in $\Gamma(\rho,\theta)$ a unique fixed point in which R is differentiable. The eigenvalues of the derivative R' at the fixed point are real and strictly greater than 1. This is the first difference with the case of analytical maps.

In the neighbourhood of the fixed point. R admits an invariant manifold corresponding to one of the eigenvalues : λ_1. This manifold is one-dimensionnal and can be identified to a one parameter family (it corresponds to the unstable manifold of Feigenbaum theory). The parameter values μ_j associated to the bifurcation cascade converge almost geometrically to their limit and we have :

$$\lim_{j\to+\infty} \frac{\mu_j - \mu_{j-1}}{\mu_{j+1} - \mu_j} = \lambda_1$$

Another difference with the analytical case is that there exist values of ρ and θ such that R does not possess any fixed point in $\Gamma(\rho,\theta)$. We do not know the behaviour of R in this case. However, for a particular couple (ρ,θ) we show that, instead of a fixed point, R admits a cycle of order 2.

Afterwards, using a truncature process, we construct non-analytical families f_μ invariant by R such that f_{μ_∞} is a fixed point of R. We compute explicitly the associated renormalization constants. If the fixed point is Feigenbaum function, we show that the constant is different from $\delta = 4.669...$ It is another illustration of the dependance of the constants with the differentiability of the families in the neighbourhood of 0.

At last we study numerically non-analytical families, whose uniform distance to $1 - \mu x^2$ is as small as we want. The results show that the associated constants are infinite and that, when the sequence of families converges to $1 - \mu x^2$, the sequence of constants does not converge to δ.

3. CONCLUSION

These results show that the Feigenbaum theory is specific to the case of analytical functions and that in other sets, the properties of R can be very different. Moreover, this work leads us to think that it is necessary to take great care in the practical use of Feigenbaum results.

REFERENCES

1. P.COLLET and J.P.ECKMANN, Iterated maps on the interval as dynamical systems. Birkhäuser, Boston-Basel-Stuttgart (1980)
2. M.COSNARD and A.EBERHARD, Preprint IMAG (To appear).
3. M.FEIGENBAUM, J. Stat. Phys. 21, 25 (1979)
4. M.FEIGENBAUM, Commun. Math. Phys. 77, 65 (1980)
5. O.E LANFORD III Preprint IHES P/81/17 (1981)

CRITICAL FLUCTUATIONS IN A THERMO-CHEMICAL INSTABILITY

C. Van den Broeck

Departement Natuurkunde
Vrije Universiteit Brussel
Pleinlaan 2
1050 Brussel (Belgium)

1. INTRODUCTION

Thermo-chemical systems have been studied in detail in the engineering sciences [1,2], but only recently attention has been paid to the effect of fluctuations [3]. In these systems the coupling of thermal and chemical variables can give rise to instabilities leading to multiple steady states, limit cycles and chaotic behaviour. We have investigated the effect of both homogeneous and inhomogeneous fluctuations in the simplest case of an instability leading to multiple steady states [4].

2. A THERMO-CHEMICAL INSTABILITY IN A STIRRED TANK REACTOR

The most common situation involving thermo-chemical reactions are oxidation reactions in a continuously stirred tank reactor (C.S.T.R.). In this case, a constant flow of the chemical mixture passes through a reactor in which it is consumed. Typically, the macroscopic equations of evolution for the concentration x of the relevant chemical species X and for the temperature T read :

$$\partial_t x = -kx - \alpha(x - x_e) \qquad (2.1)$$

$$\partial_t T = ckx - \beta(T - T_e) \qquad (2.2)$$

Here x_e and T_e are the externally fixed reference values of concentration and temperature, α and β transfer coefficients, $k \sim \exp(-\frac{U}{k_B T})$ (U activation energy) the rate constant of a monomolecular exothermic reaction consuming X and giving rise to an increase of the temperature at a rate equal to ckx. The dependence of k on T confers to (2.1) and (2.2) a highly nonlinear character. As a consequence, one verifies that for some critical values of the control parameters $x_{e,c}$ and $T_{e,c}$, a transition occurs to a regime of multiple stationary states (cusp bifurcation).

3. CRITICAL FLUCTUATIONS

In order to introduce the effect of thermal fluctuations, one adds Gaussian white forces to the macroscopic equations (2.1) and (2.2). Following a method expounded in [5] one obtains the following result for the fluctuation in the number density X :

$$\left.\frac{\langle \partial X^2 \rangle}{\langle X \rangle^2}\right|_{\text{critical}} = \left.\frac{\langle \partial X^2 \rangle}{\langle X \rangle^2}\right|_{\substack{\text{Gaussian}\\ \text{approximation}}} \rho \left[\frac{K_{3/4}(\frac{\rho}{4})}{K_{1/4}(\frac{\rho}{4})} - 1 \right] \qquad (3.1)$$

where K_ν is the modified Bessel function and $\rho = 10^{15}\, V\varepsilon^{2/3}$. V is the volume in cm^3 of the vessel, $\varepsilon = \frac{T_{st,c} - T_{st}}{T_{st}}$ the distance from the critical point and the coefficient 10^{16} is a typical value for a reaction in gaseous phase. The function of ρ appearing in the r.h.s. of (3.1) does not depend on the details of the considered model, but is a consequence of the fact that the probability density describing the critical variable is a quartic. From (3.1) and (3.2) one concludes that strongly enhanced fluctuations will be perceived (with light scattering, $V \sim 10^{-12}$ cm^3) for $\varepsilon \sim 10^{-4}$, hence a region of 0,1 °K around the critical temperature.

The above results can be generalized to take into account the effect of inhomogeneous fluctuations. However it turns out that non classical critical behaviour can only be observed in a (experimentally unattainable) close vicinity of the critical point.

REFERENCES

1. N.V. Kondratiev and E.E. Nikitin, Gas-Phase Reactions, Springer, Berlin (1981).
2. G. Gavalas, Nonlinear Differential Equations of Chemically Reacting Systems, Springer, Berlin (1968).
3. G. Nicolis, F. Baras and M. Malek Mansour, in Nonlinear Phenomena in Chemical Dynamics, Ed. A. Pacault and C. Vidal, Springer, Berlin (1981).
4. C. Van den Broeck and E. Tirapequi, to be published.
5. C. Van den Broeck, M. Malek Mansour and F. Baras, J. Stat. Phys. 28 No 3, 557 (1982).

THE SECOND ORDER MELNIKOV INTEGRAL

APPLIED TO DETECT QUASI-RANDOMNESS

Anton Aubanell

Facultat de Matemàtiques
Universitat de Barcelona
Barcelona 7 (Spain)

1. INTRODUCTION

The crossing of invariant manifolds in a dynamical system allows us to consider the Bernoulli Shift as a subsystem and, as a consequence, to have quasi-randomness[1,4,5]. The Melnikov integral (which we will call of first order: F.O.M.I.) was used by Holmes and Marsden[3], Mallet-Paret, Chow, Hale[2], etc., in order to detect quasi-randomness. In some cases this information cannot be obtained from the F.O.M.I. In this work we introduce the second order (and higher order if necessary) Melnikov Integral (S.O.M.I.) to study these problems.

2. THE SECOND ORDER MELNIKOV INTEGRAL

Let us consider the hamiltonian dynamical system $\dot{x}=f_0(x)$, $x \in R^2$ with a hyperbolic point and a homoclinic orbit $x_0(t)$. If we perturb the system by adding a term $\varepsilon f_1(x,t)$, two invariant manifolds $(x^s_\varepsilon(t),\ x^u_\varepsilon(t))$ appear. Given a t_0 we can define as Holmes and Marsden[3]:

$$\Delta_{\varepsilon^2}(t_0,t_0) = \Omega(f_0(x_0(t-t_0)),\ x^s_\varepsilon(t,t_0)-x^u_\varepsilon(t,t_0))$$

Ω being the usual symplectic form. Δ_{ε^2} measures the distance between the invariant manifolds. For $t=t_0$ we can write:

$$\Delta_{\varepsilon^2}(t_0,t_0) = -\varepsilon M_1(t_0) - \varepsilon^2 M_2(t_0) + 0(\varepsilon^2).$$

$M_1(t_0)$ is the F.O.M.I. which is given by Holmes and Marsden[3]. If M_1 has simple or odd order zeros then there exists quasi-randomness. If $M_1 \equiv 0$ then we must study $M_2(t_0)$ which will be called "The Second Order Melnikov Integral" (S.O.M.I.). In this case $(M_1 \equiv 0)$ it has the following form:

$$M_2(t_o) = \int_{-\infty}^{\infty} \Omega(f_o(x_o(t-t_o)), \tfrac{1}{2} D^2 f_o(x_o(t-t_o)) [x_1(t,t_o)]^2 +$$

$$+ D_x f_1(x_o(t-t_o), t) x_1(t,t_o)) dt ,$$

where x_1 satisfies:

$$\dot{x}_1 = Df_o(x_o(t-t_o))x_1 + f_1(x_o(t-t_o),t)$$

with $x_1(t_o,t_o)=v$ (the first order difference between $x_\varepsilon^s(t_o,t_o) = x_\varepsilon^u(t_o,t_o)$ and $x_o(t_o)$). If M_2 has odd order zeros then there is quasi-randomness. When $M_2 \equiv 0$ we must study M_3 and so on.

Fortunately we can assert the following concerning the convergence.

Theorem:

$$M_k(t_o) \equiv 0 \Rightarrow M_{k+1}(t_o) \quad \text{convergent}$$

$$M_k(t_o) \not\equiv 0 \Rightarrow M_{k+1}(t_o) \quad \text{divergent.}$$

We have applied these results to a perturbed pendulum: $\ddot{x} + \sin x = \varepsilon \sin(t+\alpha x)$. We have proved that for almost all α, the F.O.M.I. indicates quasi-randomness and, numerically, for α such that $M_1^\alpha(t_o) \equiv 0$, then the S.O.M.I. indicates quasirandomness. We have also studied examples depending on two parameters and we observed that if $M_1 \equiv 0$ and $M_2 \equiv 0$ then $M_3 \not\equiv 0$. This fact leads us to the conjecture: "Integrable hamiltonian systems are of codimensional infinity."

REFERENCES

1. A. Aubanell, Tesina de Licenciatura. Univ. Barcelona (1981).
2. S.N. Chow, J.K. Hale and J. Mallet-Paret, J. Diff. Eq. 37, 351 (1980).
3. P.J. Holmes and J.E. Marsden, Arc. for Rat. Mech. and Anal. 76, 135 (1981).
4. J. Llibre and C. Simó, Math. Ann. 248, 153 (1980).
5. J. Moser in Stable and Random motions in dynamical Systems, Princeton Univ. Press (1973).

THE FOKKER-PLANCK EQUATION AS A DYNAMICAL SYSTEM

A. Muñoz Sudupe and R.F. Alvarez-Estrada

Departamento Física Teórica
Universidad Complutense
Madrid-3 (SPAIN)

In dealing with non-equilibrium Statistical Mechanics it is useful to consider, as starting point, the phenomenological stochastic equations of the Langevin or Fokker-Planck types. In order to understand the properties of such dynamical equations, it is quite natural to use, in analogy with statics, the path integral formalism and field-theory techniques. Up to now the most extended formalism is the one introduced by Martin, Siggia and Rose (MSR) which has been applied with good results by various authors |1|. We present here an alternative formulation which seems more powerful not only in perturbative calculations but also regarding rigorous results. We use as basic theoretical starting point the Fokker-Planck equation for the probability distribution f(q,t) for continous modes [2], where q(x,t) is the order parameter. The stationary distribution $\partial f_\circ/\partial t=0$ is $f_\circ=\exp-\frac{2}{Q}\int d^dx\left[\frac{\alpha}{2}q^2+\frac{\gamma}{2}(\nabla q)^2+\frac{\beta}{4}q^4\right]$. Upon introducing $\phi=f_\circ^{-1/2}f(q,t)$ one reduces the above F-P equation to a "parabolic " type Hamiltonian (which is completely equivalent to the Liouvillian operator used by other authors [3]).

$$H\phi=-\partial\phi/\partial t\ ;\ H=\int d^dx\left[-\frac{Q}{2}\ \frac{\delta^2}{\delta q^2}+\frac{1}{2Q}(\alpha q-\gamma\Delta q+\beta q^3)^2-\frac{\alpha}{2}\delta(o)+\frac{\gamma}{2}(\Delta\delta)_{x=0}-\frac{3\beta}{2}q^2\,\delta(o)\right] \quad (1)$$

Let $q(x,t)=\frac{1}{(2\pi)^d}\int\frac{d^dk}{\omega}\left[a(k)e^{ikx-\omega t}-a^+(k)e^{-ikx+\omega t}\right]$, where a, a^+ can be interpreted as creation or anihilation operators for "thermal modes". We have proved: i) that the volume divergences $\delta(o)$, $(\Delta\delta)_{x=0}$ in (1) are responsible for the Wick ordering of the rest of the terms in (1) and, ii) that they cancel exactly so that if H is expressed in terms of a, a^+ all coeficients are finite. These divergences are not characteristic of our model because they are also present in the MSR formalism where they assure some kind of causality [1]. Perturbative checks have also been carried out successfully. We will concentrate hereafter in the functional formalism. The new partition function (functional generator) associated to (1) writes:

$$Z(J)=N\int[dq]\ \exp-\int d^dxdt\left[\frac{1}{2Q}\left(\frac{dq}{dt}\right)^2+\frac{1}{2Q}(\alpha q-\gamma\Delta q+\beta q^3)^2-\frac{3\beta}{2}q^2\delta(o)+Jq\right] \quad (2)$$

This generating functional which only depends on q(x,t) is not identical with the one commonly used in literature, which also involve additional auxiliary fields (compare with [1]). The volume divergences in (2) cancel in the correlation function $\delta^2 Z/\delta J(x,t)\delta J(x',t')\big|_{J=0}$, which still have the usual ultraviolet divergences. The Feynman rules in this formalism are somewhat simpler than in the usual one|1|, once we have cancelled the volume divergence. Just one propagator $G(p,\omega)=|\omega^2+\Delta^2(p)|^{-1}$ with $\Delta(p)=\alpha+\gamma p^2$, and two kinds of vertices: $\frac{\beta}{Q}\,\Delta(p)$  and $\frac{\beta^2}{2Q}$  . There are no closed loops attached to the vertex by , due to volume divergence cancellation. Neither response propagator nor response functions are necessary here.

Renormalization is perhaps the most sensitive test in which the present formalism can be checked. In our case it has been shown that the theory defined by Eqs. (1) or (2) is indeed a renormalizable one if $d\leq 4$. The Fluctuation-Dissipation Theorem (FDT) has played a fundamental role in solving it. Relating statics and dynamics we have been able to translate the static renormalization conditions to the dynamical problem. A similar philosophy using the standard rules and auxiliary fields, response propagators, etc... was advocated in [1]. We shall discuss briefly the divergences which appear and in the 1PI vertex functions $\Gamma^{(2)}$, $\Gamma^{(4)}$ and how they are substracted in space-dimension d=3 and d=4.

In dimension d=3, $\Gamma^{(2)}$ is quadratically divergent and $\Gamma^{(4)}$ is linearly divergent. Then a second order mass renormalization (static) is necessary to render $\Gamma^{(n)}$ finite. In dimension d=4 (critical dimension) $\Gamma^{(2)}$ is quartically divergent and $\Gamma^{(4)}$ is quadratically divergent. In this dimension, three static renormalization conditions are required to start with, namely: "mass" $\alpha_o=Z_\alpha\alpha_1$, "coupling constant" $\beta_o=Z_\beta\beta_1$ and field strength $q=Z_q 1/2 q_R$. After these have been accomplished, $\Gamma^{(4)}$ is finite but we are still left with a logarithmic divergence in $\Gamma^{(2)}$ to order two loops. One extra dynamical renormalization $Q_o= Z_Q Q_1$ and a new redefinition of the whole parameters $\alpha = \frac{\alpha_1}{Q} Z_Q$, $\beta= \frac{\beta_1}{Q} Z_Q$, $\gamma = \frac{\gamma_1}{Q} Z_Q$, is still required After all these renormalizations, all $\Gamma^{(n)}$ are finite and their static limit coincide with the renormalized static correlation functions.

REFERENCES

[1] Martin, P.C., Siggia, E.D., Rose, H.A.:Phys. Rev. A8, 423 (1973)
C. De Dominicis, L. Peliti. Phys. Rev. B18, 353 (1978)
H.K. Janssen: Z. Physik B23 , 377 (1976)

[2] H. Haken "Synergetics" Springer-Verlag (1977)

[3] L. Garrido, M. San Miguel Prog. of Theor. Phys. 59, 1, 40-63 (1978)

ON INTEGRABILITY OF QUADRATIC AREA PRESERVING MAPPINGS IN THE PLANE

Ernest Fontich

ETSEIB
Universitat Politècnica de Barcelona
Diagonal, 647
Barcelona 28 (Spain)

In this work we study quadratic area preserving mappings (Q.A.P.M.) in the plane with a fixed point[2,4].

After a translation and a linear change of variables we can write the mapping

$$T_D(x,y) = (y,\ -x+2y^2+2Dy) \quad D \in R. \tag{1.1}$$

For $D \in (1,3)$ T_D has an elliptic fixed point at $(1-D,1-D)$ and has an extra fixed point $(0,0)$ which is hyperbolic. For $D=1$ one gets a double fixed parabolic point; if $D > 3$ the two fixed points are hyperbolic. Finally, $D < 1$ is equivalent to $D > 1$ through a suitable change of variables. Other Q.A.P.M. not contained in the family (1.1) are trivial[2].

We observe that $T_D = G'_D \circ G_D$ where

$$G_D(x,y) = (y,x) \quad \text{and} \quad G'_D(x,y) = (x, y-2(y-x^2-Dx))$$

and this shows us certain useful symmetries.

Our purpose is to study the invariant manifolds W_D^s, W_D^u of the hyperbolic point in order to obtain information about homoclinic points and transversality. As is well known[1], if the angle between the invariant manifolds in a homoclinic point is different from zero, the diffeomorphism cannot be integrable.

For each D we define two sequences of functions (f_k), (g_k) $k \geq 1$ by

$$f_1(x) = 2x^2 + 2Dx \qquad g_1(x) = f_1^{-1}(x)$$
$$f_k(x) = 2x^2 + 2Dx - g_{k-1}(x) \qquad g_k(x) = f_k^{-1}(x)$$

in suitable intervals and we restrict f_k in a way so as to have g_k well defined. These sequences converge to two functions f and g, whose graphs are W_D^u and W_D^s until the first fold. We call $[\alpha_D, 0]$ the interval between the fold and the hyperbolic point.

Using monotony properties of the functions of the sequence and their derivatives we can prove that the Taylor expansion $f_D(x) = \sum_1^\infty A_n(D)x^n$ of the invariant manifolds has radius of convergence α_D.

Also, the monotony properties of $A_n(D)$ allow us to prove that the function $f: H \longrightarrow R$ defined by $f(D,x) = f_D(x)$ where $H = \{(D,x) \in R^2 : \alpha_D < x \leq 0\}$ is analytic. We can also prove that there exists a point P, different from (0,0), on the y=x line such that $P \in W_D^s \cap W_D^u$. Then using the implicit function theorem for $F(D,x) = f(D,x)-x$ we can prove the following.

Theorem

The angle between W^s and W^u at P is an analytic function of D for $D > 1$.

We have obtained the following positive lower bound for the angle for values of $D > 1.8$:

$$2 \operatorname{arctg} \left(\frac{s+1}{s-1}\right) \qquad \text{where} \qquad s = 1 - \left(16 \frac{(D-1)^2}{2D-1} + 1\right)^{\frac{1}{2}}$$

so the angle is not identically zero and at most a sequence of values of D tending to 1 with the angle equal to zero may exist. We can also prove that if these values exist they must be less than 1.5.

We have computed the angle numerically with enough accuracy and we found that the angle is positive and it behaves as $\exp(-A/(D-1)^\alpha)$, $A, \alpha > 0$ as D-1 is small[2,3]. However the computations don't give reliable information for $1 < D < 1.2$.

Combining the theorem obtained with some extensions and numerical computations, it seems that the family (1.1) is only integrable for D=1.

REFERENCES

1. J. Moser, Stable and Random Motions in Dynamical Systems, Princeton Univ. Press (1973).
2. C. Simó in Actas del V Congreso de la Agrupación de Matemáticos de Expresión Latina, 361-369 (1978).
3. V.I. Arnold and A. Avez, Ergodic Problems of Classical Mechanics, Benjamin (1967).
4. M. Hénon, Quart. of App. Math. 27, 291-312 (1969).

RESONANCES: KEY ELEMENTS TO THE UNDERSTANDING OF NON LINEAR OSCILLATIONS

I. Gumowski

Dynamic Systems Research Group
University of Toulouse 3

1. INTRODUCTION

Systems of real valued non autonomous differential equations

(1) $\dot{x} = g(x,y,t)$, $\dot{y} = f(x,y,t)$, $' = d/dt$

where t = time, x,y = scalar variables, describing physical evolution processes are considered, where f,g are sufficiently smooth functions of x,y,t and in addition periodic in t with a fixed period $T > 0$. In order to fall within the scope of most informative existence, uniqueness and continuity theorems it is assumed that f,g are analytic with respect to x,y inside a given finite (but not necessarily small) domain $A(x,y)$ and possessing a sufficient number of continuous derivatives with respect to t inside a given finite interval $B(t)$: $t_{min} \leq t \leq t_{max}$. Let $x(t_0) = x_0$, $y(t_0) = y_0$ be an initial condition, subject to $T \leq t_{max} - t_0$, t_0 inside B, (x_0,y_0) inside A. In order to avoid certain distracting pathologies it is also assumed that f,g and A are such that the solutions $x(t),y(t)$ do not reach the boundary of A too soon, for example for $t - t_0 < T$. It follows then that a solution $x(t),y(t)$ of (1) satisfying $x(t_0) = x_0, y(t_0) = y_0$ exists, is unique and quite smooth (but not necessarily analytic) for $t_0 \leq t \leq t_1 = t_0 + T$. The smoothness of f,g implies then that $x_1 = x(t_1), y_1 = y(t_1)$ depend smoothly on x_0,y_0. If (t_0,x_0,y_0) is replaced by (t_1,x_1,y_1), and no runoff to the boundary of A is encountered for $t_1 \leq t \leq t_2 = t_1 + T < t_{max}$, then $x(t_2) = x_2, y(t_2) = y_2$ also exists, is unique, quite smooth, and depends smoothly on x_1,y_1 and thus also on x_0,y_0. More generally, if $t_n = t_0 + nT$, $n = 0,1,2,\ldots$ and no runoff is encountered for $n \leq N$, then the intrinsically local existence, uniqueness and smoothness theorems valid in a sequence of finite intervals $t_{n-1} \leq t \leq t_n$ allow to represent (1) in the fully equivalent global form

(2) $x_{n+1} = \bar{g}(x_n,y_n)$, $y_{n+1} = \bar{f}(x_n,y_n)$,

where $\bar{f},\bar{g}$ are smooth fuctions of their arguments. The reduction of (1) to (2) by what is usually called the Poincaré method of sections has the main advantage that in many cases it is easier to establish the validity of (2) for $N \to \infty$ by discrete induction than to establish an extended existence, uniqueness and smoothness theorem applying to (1) for $t_{max} \to \infty$. A similar reasoning applies when n is replaced by $-n$. The expression (2) is called an autonomous second order recurrence. With some sacrifice of physical content it can also be considered as a point-mapping of A onto itself (see Chapt. 2 and 6 of [1] for more details).

Another advantage of (2) is that its suitably defined phase portrait (see [1,2] for a definition and several relevant properties) is two-dimensional. Like in the case of autonomous differential equations

$$(3) \qquad \dot{x} = \tilde{g}(x,y) \ , \quad \dot{y} = \tilde{f}(x,y)$$

with sufficiently smooth $\tilde{f},\tilde{g}$ [3], the phase portrait of (2) can be subdivided into cells, each defined essentially by resonances and filled by continuous trajectories, i.e. suitably defined invariant curves (see Chapt. 2 and 6 of [1]) of the same qualitative type. In contrast to (3), however, where the smoothness and bounded variation of $\tilde{f},\tilde{g}$ (characterized by a finite number of real roots of $\tilde{f}(x,y) = 0$, $\tilde{g}(x,y) = 0$) implies an "orderly" phase portrait, i.e. in general a finite number of phase cells [3], the phase portrait of (2) contains as a rule some "chaotic" elements even if $\bar{f},\bar{g}$ are globally analytic. The chaos in the phase portrait of (2) manifests itself by the existence of an infinity of phase cells, some of which contain trajectories which cannot occur in (3).

Let the general solutions of (1),(2),(3) be

$$(1a) \quad x = G(x_0,y_0,t) \ , \quad y = F(x_0,y_0,t) \ ,$$
$$(2a) \quad x = \bar{G}(x_0,y_0,n), \quad y = \bar{F}(x_0,y_0,n) \ ,$$
$$(3a) \quad x = \tilde{G}(x_0,y_0,t), \quad y = \tilde{F}(x_0,y_0,t) \ ,$$

respectively, with n extended to real values in (2a), cf [1]. In some cases these solutions may turn out to be only local, implicit functions being required for a global representation, but this peculiarity will be disregarded when it plays no role. Accumulated knowledge about (2a) and (3a) indicates that for $\bar{f},\bar{g}$ and $\tilde{f},\tilde{g}$ of comparable smoothness the functions $\bar{F},\bar{G}$ have a structural complexity which is considerably greater than that of $\tilde{F},\tilde{G}$, i.e. $\bar{F},\bar{G}$ possess not only more singularities, but some of these singularities are of types which do not occur in F,G. From this rather obvious fact it follows that, although there is an equivalence between (1a) and (2a), as a rule no equivalence can exist between (1a) and (3a). In other words, except in very special cases, there exists no regular transformation linking F,G and $\tilde{F},\tilde{G}$, or what is the same, f,g and $\tilde{f},\tilde{g}$. This property implies in particular that none of the widely used reduction methods of (1) to (3) (perturbations [4], convergent or asymptotic expansions [5,6], successive iterations [7], various averaging schemes [8]) can be completely successful. More specifically, Siegel's assertion of the non-convergence of sequences of canonical transformations linking (1) to (3) when f,g derive from a Hamiltonian becomes quite obvious [9], no matter which other variant of reduction is also attempted. What can be attained at most is a local approximation by (3),(3a) of some features of the phase portrait of (2),(2a). A similar conclusion holds when (1) is weakly damped. For a sufficiently strong damping the phase portrait of (2) becomes orderly, and a global reduction of (1) to (3) becomes possible. Such cases are not considered here.

The objective of this paper is a critical discussion of problems related to the study of dynamically characteristic properties of solutions of specific equations of form (1) occuring in mechanics, like for example

(4a) $\dot{x} = y \ , \ \dot{y} = -\alpha y - \omega_0^2(1 + h\cos\omega t)x + \varepsilon x^n$,

(4b) $\dot{x} = y \ , \ \dot{y} = -\alpha(1 + h\cos\omega t)y + \omega_0^2 x + \varepsilon x^n$,

(4c) $\dot{x} = y \ , \ \dot{y} = -\alpha y - \omega_0^2 x + \varepsilon x^n + h\cos\omega t$,

(4d) $\dot{x} = y \ , \ \dot{y} = -\alpha y - \omega_0^2(1 + h\cos\omega t)\sin x$,

(4e) $\dot{x} = y \ , \ \dot{y} = -\alpha y - \omega_0^2\sin x + h\cos\omega t$,

....

when the damping coefficient α vanishes (Hamiltonian case) or is not too large. Although equations (4) admit for $\alpha = 0$ quasiperiodic solutions describable by KAM tori [10,11], i.e. solutions located "sufficiently far" from resonances, accumulated experience in the study of specific equations has shown (see for example [1,2,12] and the references therein) that it is more efficient to study the phase plane structure of the associated recurrence (2) by identifying resonances, described by periodic solutions of (1), i.e. by fixed points and cycles of (2). With the additional help of some invariant curves more characteristic features of the phase portrait are obtained in this manner.

From a more analytical point of view it is, however, remarkable that (an accuracy controlled) numerical integration of (4), coupled with the Newton method of root finding, discloses readily a considerable number of periodic solutions (from a dozen to a few hundreds), whereas so far only very few of these solutions (usually not more than four or five) can be described successfully by explicit expressions involving known functions, no matter which analytical method is tried. Some of the "standard" methods fail in principle, while others present technical difficulties which could not yet be overcome (mainly because not enough effort was brought to bear on them, cf [19, 20]). Another objective of this paper is to determine some reasons for such an unexpected failure.

2. NATURAL DEPENDENCE ON PARAMETERS AND A THEOREM OF POINCARÉ

Consider first an algebraic equation described by a smooth function $f(x,y,\alpha) = 0$, where x,y are two scalar variables and α is a scalar parameter. Assume that at least one of the implicitly defined roots is known explicitly: $y = F(x,\beta)$, $\beta = \beta(\alpha)$. Like in the case of (1) assume also that all variables, functions and parameters are real valued. For conciseness α is called a primary parameter, β a secondary parameter, and the function $\beta(\alpha)$ the natural dependence of the secondary parameter on the primary one [13]. The notion of natural dependence plays a key role in the study of equations (4). In fact, it is of major importance to know how at least three secondary parameters of periodic solutions (amplitude, phase and resonance width) depend on primary parameters. The determination of at least approximate explicit expressions of secondary parameters constitutes the main body of most treatises on oscillations.

Concrete cases of algebraic equations $f = 0$, for example polynomials in x,y or in y alone, suggest that the dependence between primary and secondary parameters can be quite varied: the functions f and F,β may have different types of smoothness. In particular, when f is analytic (admits a convergent McLaurin expansion) in x,y <u>and</u> α, the functions F,β need not be analytic in α. An elementary example is: $f = y^2 - \alpha$, $\alpha > 0$, $F = \beta = \sqrt{\alpha}$! The same kind of restricted "smoothness transfer" between primary and secondary parameters carries over into functions defined implicitly by differential equations and recurrences. For example, when all parameters are positive in (4a) and $n = 3$, it is well known (p. 220 of [6]), and it can be readily guessed on physical grounds, that the amplitude of the $\frac{1}{2}$-subharmonic solution is a function of $1/\sqrt{c}$ and not of c. This amplitude is thus definitely not an analytic function of c.

An elementary differential equation, illustrating in a transparent manner some relations between natural dependence and analyticity, is analogous to the preceding algebraic one:

$$\begin{aligned} &\dot{x} = g(x,y) = y \ , \ \dot{y} = f(x,y) = \alpha x \ ; \quad \alpha > 0 \ , \ \beta = \sqrt{\alpha} \\ &x = G(t,\beta) = C(\alpha)e^{-\beta t} + \bar{C}(\alpha)e^{\beta t} \quad , \ y = \frac{d}{dt}G(t,\beta) \quad . \end{aligned} \tag{5}$$

The functions f,g in (5) are analytic in t,x,y,α, whereas the solution functions $F = G', G$ are analytic in t, but not necessarily in α. In fact, the dependence on α is a function of both the natural parameter dependence $\beta(\alpha) = \sqrt{\alpha}$ and the dependence of the integration constants $C(\alpha), \bar{C}(\alpha)$ on α. By definition the parameter dependence of the integration constants can be fixed arbitrarily and is not determined by the form of the differential equation, i.e. by the functions f,g.

Smoothness and analyticity theorems concerning parametric dependence can thus apply only to particular solutions, specified by initial conditions of a special form. Such initial conditions are not necessarily consistent with the existence of periodic solutions, because the amplitudes and phases of the latter have usually to be roots of certain algebraic equations. Hence, neither their "smallness" nor their natural dependence can be imposed a priori. In this context it is useful to recall a theorem of Poincaré (often misinterpreted), which is stated here only for two dimensions (for a more general version see for example [14] vol. 3, No. 461, p. 17 and No. 463, p. 20): Let the differential equations depending on a scalar parameter ε be of the form

(6) $\dot{z} = h(z,t,\varepsilon)$, $t \geq t_0$, $z = \binom{x}{y}$, $h = \binom{g}{f}$, t_0 = fixed.

The function h is assumed to be analytic in z,ε at least for $|z|$ and $|\varepsilon|$ sufficiently small, and smooth (at least continuous) in t. In other words, the series expansion

(7) $h(z,t,\varepsilon) = a_0(t) + \sum_{i+k=1}^{\infty} a_{ik}(t)\, z^i \varepsilon^{k-i}$

is assumed to be convergent for all relevant t (i.e. for $t_0 \leq t \leq t_{MAX}$) and at least for $|z| < \bar{z}$, $|\varepsilon| < \bar{\varepsilon}$, where $\bar{z},\bar{\varepsilon}$ are fixed strictly positive constants. The Poincaré theorem affirms then that the solution of (6) verifying the initial conditions

(8) $z(t_0) = z_0$, $\partial z_0/\partial\varepsilon \equiv 0$, $|z_0| < \bar{z}$,

can be expressed in the form of the series

(9) $z = \sum_{m=0}^{\infty} z_m(t,z_0)\, \varepsilon^m$, $z_m = \sum_{n=0}^{\infty} b_n(t)\, z_0^n$, $z_m(0,z_0) = \begin{cases} z_0, & m = 0 \\ 0, & m > 0 \end{cases}$,

which is convergent for $|z_0| \leq \bar{z}_1$, $|\varepsilon| < |\bar{\varepsilon}_1|$, where $0 < \bar{z}_1 \leq \bar{z}$, $0 < \bar{\varepsilon}_1 \leq \bar{\varepsilon}$ and

(10) $0 < t - t_0 < \bar{t} = \bar{t}(t_0,z_0,\varepsilon)$.

It should be stressed that the constants $\bar{z}_1, \bar{\varepsilon}_1, \bar{t}$ defining the convergence of (9) are strictly positive, but their values are not known in advance. Moreover, the bounding constant $\bar{t}$ in (10) is not relatively "free", like in the assumption about (7), but depends strongly on t_0, z_0 and ε. This inherent limitation of $\bar{t}$ cannot be removed without some very special constraints on the shape of the functions f,g in h. Widespread contrary opinion notwithstanding [15,16], the legitimacy of the extension $\bar{t} \to \infty$ is rather exceptional and needs in every case a justification. One example of the failure of (9) is the previously mentioned $\frac{1}{2}$-subharmonic resonance described by (4a) with n = 3. Since the Poincaré theorem obviously applies when $\varepsilon = c$, the convergence bound $\bar{t}$ is necessarily less than the period $T = 4\pi/\omega$ for the z_0 corresponding to the periodic solution. A more transparent scalar example is

(11) $\dot{z} = (z + \varepsilon)^2$, $t \geq t_0$, $z(t_0) = z_0$ = fixed constant.

The function $h = (z + \varepsilon)^2$ admits a series expansion of form (7) for all z, ε and t, and is thus maximally smooth. The explicit solution of (11) being

$$(12) \qquad z = -\varepsilon + (z_0 + \varepsilon)/[1 - (z_0 + \varepsilon)(t - t_0)] \, ,$$

its series expansion of form (9) involves obviously an inherently finite $\bar{t}$, defined by $|(z_0 + \varepsilon)(\bar{t} - t_0)| < 1$. The limitation of $\bar{t}$ by the properties of h other than smoothness (again widespread contrary opinion notwithstanding, especially for Hamiltonian equations) has a profound influence on any smoothness-based method of determining periodic solutions of (1), and in particular on various variants of the method of von Zeipel [4], and of convergent or asymptotic "small parameter" expansions, like those described in [5,6]. If the convergence bound $\bar{t}$ (as was already pointed out, unknown in advance) is less than the period T of the corresponding periodic solution, then the results are dubious, in spite of any apparent "methodological harmony". In other words, when a sequence of canonic transformations, or several successive terms of a formal expansion appear to "work" without any indications of internal inconsistencies, this property by itself does not guarantee the correctness of the results obtained. An independent verification is needed, in the most unfavourable case by resorting to numerical computations.

Particular cases of (1) with unlimited convergence properties of (9) do no doubt exist, best known examples being some linear equations and certain periodic solutions of the Van der Pol or the Duffing equations. In the presence of unlimited convergence properties almost any method is crowned by success, even imbedding of non conservative equations into higher-dimensional Hamiltonian ones (cf [17]). Such isolated successes constitute of course no basis for any generalizations, because examples of failure are widespread (most unpublished). In this context a conceptually important phenomenon in mechanics is covered by the so-called (n/k)-resonances with $n > k$. Consider in fact a special (Hamiltonian) case of equation (4c):

$$(13) \qquad \ddot{x} + \omega_0^2 x = \varepsilon(\omega_0^2 \cos \omega t + c x^3) \, , \quad \omega_0^2 - \tfrac{9}{4}\omega^2 = \Delta\omega \, , \quad |\Delta\omega| \ll \omega^2 ,$$

for which the following periodic solution has been obtained by using an unjustified smoothness argument [18]:

$$(13a) \qquad x = A\cos(\tfrac{3}{2}\omega t + \theta) + \frac{\varepsilon\omega_0^2 A}{4\omega^2}\cos(\tfrac{\omega}{2}t + \theta) - \frac{\varepsilon\omega_0^2 A}{8\omega^2}\cos(\tfrac{5}{2}\omega t + \theta) + \frac{\varepsilon c A^2}{72\omega^2}\cos(\tfrac{9}{2}\omega t + 3\theta) + \cdots ,$$

where A is a real root of

$$(13b) \qquad 5\varepsilon^2 c^3 A^4 - 72\varepsilon c\omega^2 A^2 + 16(\varepsilon c A^2 - 6\omega^2 + \Delta\omega) = 0 \, .$$

Direct numerical integration of (13) failed to confirm the validity of (13a),(13b). On conceptual grounds the existence of (n/k)-harmonic resonances with $n > k$ would imply the existence of cycles of (2) of fractional order $k = k/n < 1$. No such cycles have ever been found and none have been postulated on mathematical grounds. The second term in (13a) suggests that the cycle giving rise to the (3/2)-resonance might be a cycle of order $k = 2$, i.e. (13a) might describe in principle a resonance of the third harmonic of the basic subharmonic $\omega/2$. No numerical confirmation could be found for this alternate possibility.

Another example of the formal failure of (9) is equation (4c) with $\alpha = 0$, $n = 2$, $\omega_0 = \omega$, $\varepsilon = h$. No periodic solution of period $T = 2\pi/\omega$ can be found with the help of expansion (9), a contradiction is encountered. By guessing correctly the natural dependence of the oscillation amplitude on h it is possible to deduce a modified convergent expansion: (9) with ε^m replaced by $\varepsilon^{m/3}$. In other words, the periodic solution in question is described by

$$(14) \qquad x = \sum_{m=1}^{\infty} x_m(t).\, \varepsilon^{\frac{m}{3}} \quad , \quad x_m(t + T) = x_m(t) \quad ,$$

in which there are no small denominators in spite of the resonance. From the periodicity of $x_3(t)$ it follows that

$$(14a) \qquad x = A\cos\omega t + O(h^{\frac{2}{3}}) \quad , \quad A = -(4h/3c)^{\frac{1}{3}} \quad .$$

Other examples of non analytic expansions of periodic solutions can be found in literature (some typical cases are discussed in [20]). It should be recalled that an analytic expansion in fractional powers of a primary parameter can be found already in the works of Poincaré, in connection with what is now known as the Poincaré-Hopf bifurcation of a centre, giving rise to a limit cycle.

3. CONCLUSION

The study of several equations of form (1) with analytic right-hand sides occuring in mechanics has shown that their periodic solutions have amplitudes and phases which are generally non analytic functions of primary parameters. Different resonances require thus qualitatively different mathematical representations for their efficient quantitative description. In this context a knowledge of resonant periodic solutions appears to have a higher information content than that of KAM-tori.

REFERENCES

1. I. Gumowski, C. Mira, "Recurrences and discrete dynamic systems", LN in Math. No. 809, Springer Verlag, (1980)
2. I. Gumowski, C. Mira, "Dynamique chaotique", Cépadues Editions, Toulouse, (1980)
3. A. A. Andronov, E. A. Leontovich, I. I. Gordon, A. G. Maier, "Bifurcation theory of dynamic systems in the phase plane", Nauka, Moscow, (1967), and references therein

4. H. von Zeipel, "Recherches sur le mouvement des petites planètes", Arkiv Math. Astron. Fysik, 11(1916), No. 1,7
5. I. G. Malkin, "Some problems of the theory of non linear oscillations", GITTL, Moscow, (1956)
6. N. N. Bogoliubov, Iu. A. Mitropolsky, "Asymptotic methods in the theory of non linear oscillations", Fizmatgiz, Moscow, (1958)
7. G. Schmidt, "Parametererregte Schwingungen", DVW, Berlin, (1975)
8. Iu. A. Mitropolsky, "Averaging method in non linear mechanics", Naukova Dumka, Kiev, (1971), and the references therein (till 1682)
9. G. L. Siegel, "Über die Normalform analytischer Differntialgleichungen in der Nähe einer Gleichgewichtslösung", Nachr. Akad. Wiss. Göttingen, Math.-Phys. Kl., No. 5, (1952), p. 21-30
10. A. N. Kolmogorov, "On the conservation of quasi periodic motions for small changes of the Hamiltonian function", Dokl. Akad. Nauk SSSR, 98(1954), p. 527-530, and V. I. Arnold, "Proof of the theorem of Kolmogorov on the conservation of quasi periodic motions for small changes of the Hamiltonian function", Uspekhi Mat. Nauk, 18(1963), No. 5(113), p. 13-40
11. J. Moser, "On invariant curves of area-preserving mappings of an annulus", Nachr. Akad. Wiss. Göttingen, Math.-Phys. Kl. IIa, (1962), p. 1-20
12. F. M. Izrailev, B. V. Chirikov, "Stochasticity of a simple dynamic model with separated phase space", Preprint No. 191, Inst. Iad. Fiziki, Novosibirsk, (1968), and B. V. Chirikov, "Research concerning the theory of non linear resonance and stochasticity", CERN Trans. No. 71-40, Geneva, (1971)
13. I. Gumowski, "Some relations between differential equations, recurrences and functional iterates", 9th ICNO, Kiev, (1981)
14. E. Goursat, "Cours d'analyse mathématique", vol. 1,2,3, Gauthier-Villars, Paris, (1943)
15. V. K. Melnikov, "On the stability of the center for time-periodic perturbations", Trans. Moscow Math. Soc., 12,1,(1963), p. 1-56
16. P. Holmes, "A non linear oscillator with a strange attractor", Phil. Trans. Roy. Soc. London, 292,1394,(1979), p. 419-448
17. A. A. Kamel, "Perturbation method in the theory of non linear oscillations", Celestial Mechanics, 3(1970), p. 90-106
18. L. F. Shulezhko, "Non linear vibration of a plate", Approximate Methods of solving differential equations", Izd. Akad. Nauk Ukr. SSR, Kiev, (1963), p. 141-154
19. I. Gumowski, R. Thibault, "Dynamic systems with a singular parametric resonance", EQUADIFF-78, Firenze, (1978), p. 91-98
20. I. Gumowski, "Periodic steady states of dynamic systems and their smooth dependence on parameters", Colloque Intern. du CNRS No. 332 sur la théorie de l'itération et ses applications, Toulouse, (1982)

ON SYSTEMS PASSING THROUGH RESONANCES

Amadeo Delshams

Facultat de Matemàtiques
Universitat de Barcelona
Barcelona 7 (Spain)

1. INTRODUCTION

Let us consider the system:

$$\dot{I} = \varepsilon F(I,\phi,\varepsilon), \qquad \dot{\phi} = \omega(I,\varepsilon)+\varepsilon f(I,\phi,\varepsilon) \tag{1.1}$$

with $I=(I_1,\ldots,I_1)$, $\phi=(\phi_1,\ldots,\phi_s)$ (mod.2π); ω,F,f analytic for $I \in G$, $|\mathrm{Im}\ \phi| < \rho$, $|\varepsilon| < \varepsilon_o$, where $G \subset C^1$ is compact, and ε_o, ρ are positive real numbers.

In order to study the system (1.1), it is usual to consider the <u>averaged system</u>:

$$\dot{J} = \varepsilon\bar{F}(J), \qquad \bar{F}(J) = (2\pi)^{-s}\int_{T^s} F(J,\phi,0)\, d\phi \quad , \tag{1.2}$$

and to expect that the solutions of the two systems will be close for a long time, i.e.

$$\|I(t)-J(t)\| \ll 1, \quad \text{if } |t| < 1/\varepsilon,\ I(0)=J(0).$$

This is true[4] for s=1, but not, in general, for $s \geq 2$, due to the passage through resonances[3].

2. RESONANCES

We say that $I \in G$ is in a <u>resonance</u> for the system (1.1) if there exists $m \in Z^s$, $m \neq 0$, orthogonal to $\omega(I)$: $(m,\omega(I))=0$ (from now on, we will not write the dependence on ε). Given $m \in Z^s$, $m \neq 0$, <u>the resonant manifold associated to m</u> is $Vr(m) = \{I \in G: (m,\omega(I))=0\}$ (generally, a hypersurface).

Let us comment on the effect of the resonances. If a solution $(I(t),\phi(t))$ of (1.1) encounters a resonant manifold Vr(m) and remains on it, i.e., the solution is "locked" in a resonance, then the system (1.2) will be a bad approximation of the system (1.1), due to the fact that the frequencies $\omega_1,\ldots,\omega_s$ are commensurable on Vr(m), and $\bar{F}$, the space average of F, is, in general, different from the time average of F[2]. To avoid this, we impose some resonance unlocking:

Definition. We say that the system (1.1) satisfies the first order resonance unlocking (f.o.r.u.) if for each $I \in G$, $\omega(I) \neq 0$, and

$$\text{if } (m,\omega(I)) = 0 \quad \text{for} \quad m \neq 0, \text{ then } (m, D\omega(I)F(I,\phi,\varepsilon)) \neq 0.$$

This relation asserts that if I is in a resonant manifold $Vr(m)$, then its transversal velocity to $Vr(m)$ is different from zero (in the same way, we could also define the kth-order resonance unlocking [7]). We can now announce the following theorem (whose proof will be given elsewhere):

Theorem. If the system (1.1) satisfies the f.o.r.u., then there exist positive constants C_1, C_2, ε_1, such that for each ε with $|\varepsilon| < \varepsilon_1$, we have:

$$\|I(t)-J(t)\| < C_1 \, \varepsilon^{3/2-r} \exp(C_2 \varepsilon^{1-r}), \text{ if } |t| < \varepsilon^{-r}, \; I(0)=J(0). \quad (2.1)$$

3. REMARKS

a. If $0 < r \leq 1$, $|t| < \varepsilon^{-r}$, then $\| I(t)-J(t)\| \ll 1$, and this result generalizes the previous work of Arnold[1] and Neishtadt[5], valid only for $s=2$, $r=1$.

b. We cannot avoid the exponential term in (2.1), because $\| J_1(t)-J_2(t) \| \sim \|J_1(0)-J_2(0)\| \exp(ct)$. In this sense, (2.1) is the best possible result that can be obtained.

c. If (1.1) is hamiltonian, then $s=1$ and $\bar{F}=0$, and the result of Nekhoroshev[6]: $\|I(t)-I(0)\| < \varepsilon^b$, if $|t| < \frac{1}{\varepsilon} \exp(\frac{1}{\varepsilon^a})$ is valid if the unperturbed hamiltonian $H_o(I)$ $(\omega(I) = (\partial H_o(I))/(\partial I))$ is sufficiently "steep". For these systems, in general, we obtain worse estimates for the time of stability, but better estimates for $\| I(t)-I(0)\|$.

4. REFERENCES

1. Arnold, V.I. Sov. Math. Dokl. 6,331-334 (1965).
2. Arnold, V.I. Chapitres supplémentaires de la théorie des équations différentielles ordinaires. Mir, Moscou (1980).
3. Arnold, V.I., Avez, A. Problèmes ergodiques de la mécanique classique. Gauthier-Villars,Paris (1967).
4. Bogolubov, N.N., Mitropolski, Yu. A. Les méthodes asymptotiques en théorie des oscillations non lineaires. Gauthier-Villars, Paris (1962).
5. Neishtadt, A.I. Sov. Phys. Dokl. 20, 189-191 (1975).
6. Nekhoroshev, N.N. Russ. Math. Surveys 32, 1-65 (1977).
7. Simó, C. La variedad de órbitas Keplerianas y la teoría general de perturbaciones. Tesis, Universidad de Barcelona, (1974).

THE LYAPUNOV CHARACTERISTIC NUMBERS AND THE NUMBER OF ISOLATING INTEGRALS IN GALACTIC MODELS

Pierre Magnenat

Observatoire de Genève
1290 Sauverny/Switzerland

1. INTRODUCTION

Observations show that the dynamical studies of galaxies must be undertaken in the frame of dynamical systems with 3 degrees of freedom. A first step consists of understanding the main properties of orbits in the simplest case of a 3-D conservative potential taking into account the most important non-linear effects. Particularly the conditions of existence of 1 or 2 isolating integrals besides the Hamiltonian ("non-classical" integrals) are one of the main problems to be examined. For some years the Geneva Observatory stellar dynamics group has undertaken numerical investigations of the stellar orbital behaviour in the potential

$$V = \tfrac{1}{2}\,(Ax^2 + By^2 + Cz^2) - \varepsilon xz^2 - \eta yz^2 \qquad (1.1)$$

Particularly the resonance ratio 6:4:3 (A = .9, B = .4, C = .225) has been extensively studied[1,2,3]. The observation, by means of the surface of section method, of the progressive dissolution of invariant tori when one of the perturbation terms is increased, showed, qualitatively, cases where only one isolating integral seems to exist besides the Hamiltonian[1]. For his part C. Froeschlé[4,6], in the case of the 3-body 3-D restricted problem, observed a quasi-simultaneous disappearance of both non-classical integrals.

The Lyapunov Characteristic Numbers (LCN) constitute a quantitative tool for measuring the stochasticity of an orbit. Each orbit of a Hamiltonian system with n degrees of freedom can be associated with 2n LCN χ_p, p = 1,...,2n. In our case, these numbers are known to be of the form $\chi_1 \geq \chi_2 \geq \chi_3 = 0$, and $\chi_p = -\chi_{6+1-p}$, $p = 1,2,3$. (For details see Benettin et al[5]). A useful property of these numbers is that if 1, 2 or 3 isolating integrals exist in the neighbourhood of the orbit, 2, 4 or 6 LCN vanish.

2. RESULTS

Calculations were performed, in the cases examined in [1] by the method described in [5], of the functions $\chi_p(t)$ whose values are approximations of χ_p for large t. In the case of $\eta = .465$ (right part of Fig. 1), only the energy integral subsists: only $\chi_3(t)$ tends towards zero. When $\eta = .4$ (left) χ_2 and χ_3 tend towards 0, whereas χ_1 tends towards a finite limit, which implies the existence of one non-classical integral. These results confirm the observations made in [1]: cases are found where 1 integral exists besides the energy. This fact could be of importance in the construction of realistic galactic models.

REFERENCES

1. L. Martinet and P. Magnenat, Astron. Astrophys. 96, 68 (1981).
2. P. Magnenat, Celes. Mech. (1982) in press.
3. G. Contopoulos, P. Magnenat and L. Martinet, Physica D (1982) in press.
4. C. Froeschlé, Astron. Astrophys. 4, 115 (1970).
5. G. Benettin, L. Galgani, L. Giorgilli and J.M. Strelcyn, Meccanica 15, 21 (1980).
6. R. Gonczi and C. Froeschlé, Celes. Mech. 25, 271 (1981).

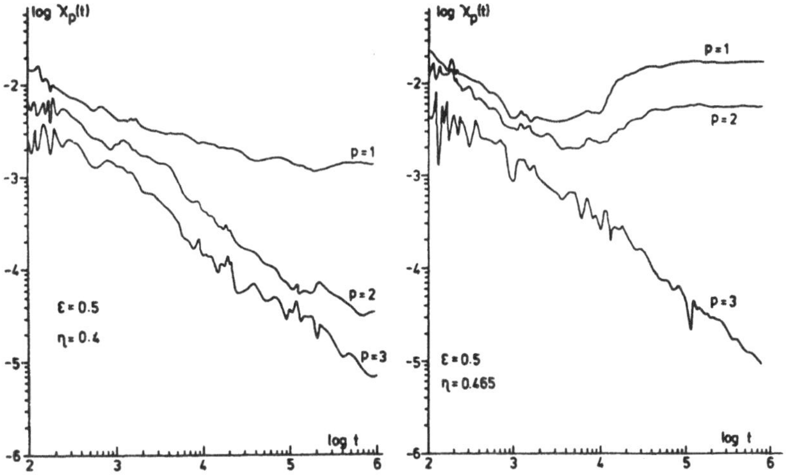

Figure 1. Estimation of the LCN in a case where 1 integral subsists besides the energy (left) and in a case where only the energy is an integral (right).

ON THE PERIODIC ORBITS OF THE CONTOPOULOS HAMILTONIAN

Miquel Grau, Facultat de Matemàtiques, Universitat de Barcelona, Gran Vía 585, Barcelona 7 (Spain)

Jaume Llibre, Secció de Matemàtiques, Universitat Autònoma de Barcelona, Bellaterra (Spain)

Rosa M. Ros, Depart. de la Física de la Tierra y el Cosmos, Universidad de Barcelona, Diagonal 647, Barcelona 28 (Spain)

1. INTRODUCTION

Looking for a third integral in the motion of a particle in a galactic potential with cylindrical symmetry, Contopoulos[3,4] was lead to the Hamiltonian

$$H(x,y) = \frac{1}{2}(y_1^2 + y_2^2) + \frac{1}{2}(x_1^2 + x_2^2) - x_1 x_2^2 ,$$

where the position is given by $x = (x_1, x_2)$ and the momentum by $y = (y_1, y_2)$. The Hamiltonian system X_H is

$$\dot{x}_1 = y_1 , \quad \dot{x}_2 = y_2 , \quad \dot{y}_1 = -x_1 + x_2^2 , \quad \dot{y}_2 = -x_2 + 2x_1x_2 .$$

This system has the origin $Q_1 = (0,0,0,0)$ as a critical point at energy level $h = 0$ with repeated eigenvalues $\pm i$, and two additional critical points $Q_2 = (2^{-1}, -2^{-1/2}, 0, 0)$ and $Q_3 = (2^{-1}, 2^{-1/2}, 0, 0)$ at energy level $h=1/8$ with eigenvalues ± 1 and $\pm 2^{1/2} i$.

This communication surveys and states results on the simple periodic orbits of X_H which start in their critical points.

2. THE FAMILIES OF SIMPLE PERIODIC ORBITS

We restrict to some energy level $H = h$. Cutting it by $x_2 = 0$ we get a 2-dimensional manifold S. Then consider the Poincaré map T which maps S into itself in the following way. Take a point $p \in S$ with coordinates (x_1, y_1), $x_2 = 0$ and $y_2 \geq 0$ obtained from $H(p) = h$. Then T(p) is the point given by the next cut of S by the orbit through p, if it exists. Fixed points under T are associated with the so called simple periodic orbits of X_H.

Table I summarizes the results about the existence of the simple periodic orbits of X_H which start in the critical points Q_i, for $i = 1,2,3$.

In a similar way to [6] we have proved that Figure 1 (resp. Figure 2) gives us the projections on the position plane of the periodic orbits π_i for $i = 1,2,3,4,5,6$ (resp. $i = 1,3,5,6,7,8$) when the energy $h \in (0,1/8)$ (resp. $h \in (1/8,+\infty)$).

<u>Proposition</u>: For all h the orbits π_i for $i = 1,2,3,4,5,6$ are mutually linked in the energy level h, if they exist.

The proof of the Proposition follows as in the Hénon-Heiles problem[10].

The simple periodic orbit	starts in the critical point	and exists for the following values of h.	Different proofs of the existence of the periodic orbits π_i according to h. $0 \quad \epsilon \quad \frac{1}{8} \quad \frac{1}{8}+\delta \quad h \rightarrow$
π_1	Q_1	$(0,+\infty)$	****oooooooooooooooooooooooo
π_2		$(0,1/8)$	****oooo
π_3			****oooo
π_4		$(0,+\infty)$	****++++++++++-------------
π_5			****-----------------------
π_6			****-----------------------
π_7	Q_2	$(\frac{1}{8},+\infty)$	.....---------------
π_8	Q_3		.....---------------

Table I. $\epsilon > 0$ and $\delta > 0$ are sufficiently small.

**** For these values of h the existence of π_i for $i=1,\dots,6$ was proved by Braun[1] in 1973 using the Gustavson normal form. Also Kummer[9] proved it in 1976 using a reformulation of Braun's method. Again Broucke[2] showed it in 1981 using the Lindstedt's method.

oooo The existence of π_i for $i=1,2,3$ follows from the invariance by the flow of the three planes $P_1=\{x_2=y_2=0\}$, $P_3=\{x_2=2^{1/2}x_1,\ y_2=2^{1/2}y_1\}$ and $P_2=\{x_2=-2^{1/2}x_1,\ y_2=-2^{1/2}y_1\}$ respectively.

++++ For these values of h the existence follows using the fact that X_H is R-reversible[8] with respect to the reversing involution $R(x_1,x_2,y_1,y_2) = (x_1,-x_2,-y_1,y_2)$. The proof is the same than the proof for the periodic orbit π_4 of the Hénon-Heiles Hamiltonian[6].

.... Here the existence follows easily from the Liapounov-Moser's Theorem[5,11].

---- For these values of h we have computed numerically the periodic orbits until values of their stability parameter of order 10^5.

In order to study the stability of the families π_i we compute numerically the stability parameter $k=\mathrm{Trace}(A)$, where A is the monodromy matrix associated to the periodic orbit[6]. A periodic orbit is elliptic, parabolic or hyperbolic according to $|k|<2$, $|k|=2$ or $|k|>2$, respectively. A qualitative picture of the stability parameter against the energy is given in Figures 3,4,5,6 and 7, for π_i with $i=1,2,\ldots,8$. It is known that the periodic orbit π_1 (resp. π_2 or π_3) alternates between ellipticity and hyperbolicity infinitely often as $h \to +\infty$ (resp. $h \to \frac{1}{8}$)[6,7].

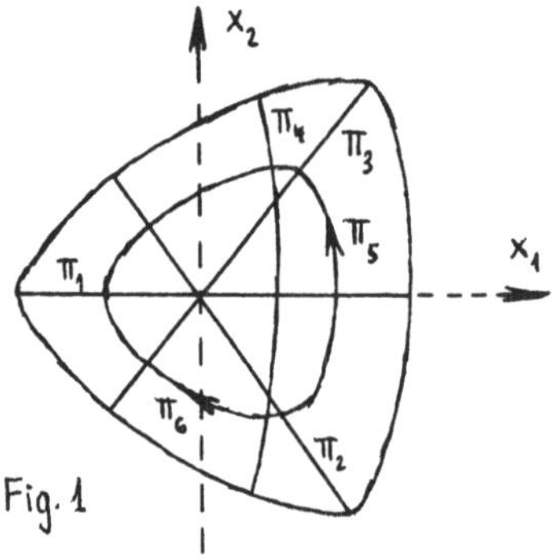

Fig. 1

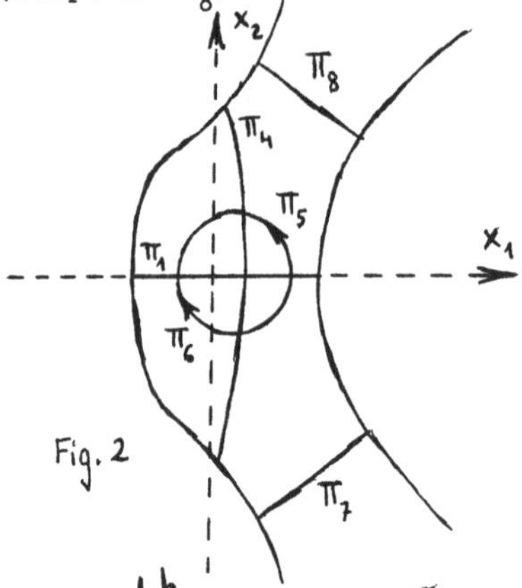

Fig. 2

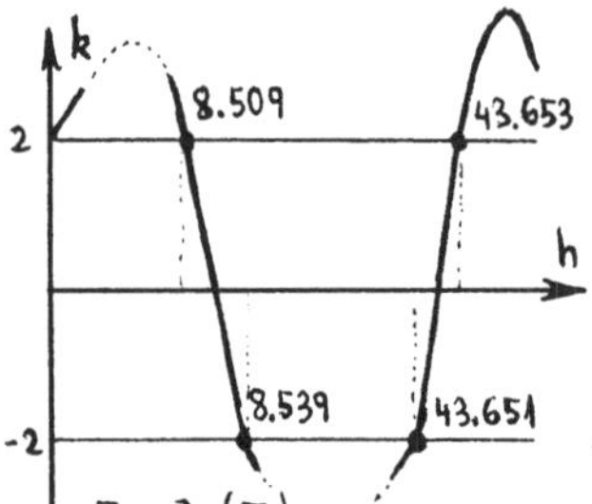

Fig. 3 (π_1)

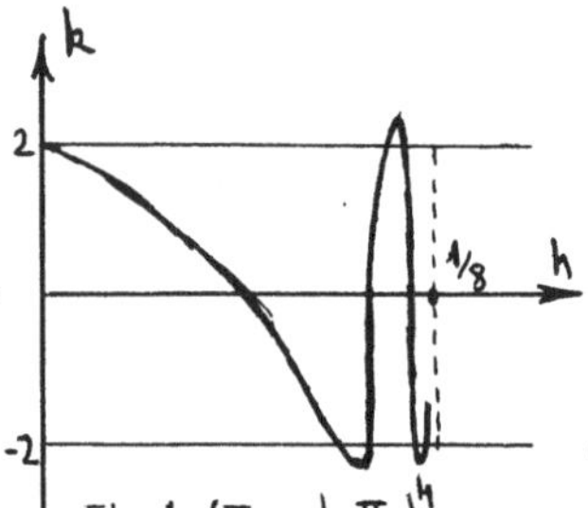

Fig. 4 (π_2 and π_3)

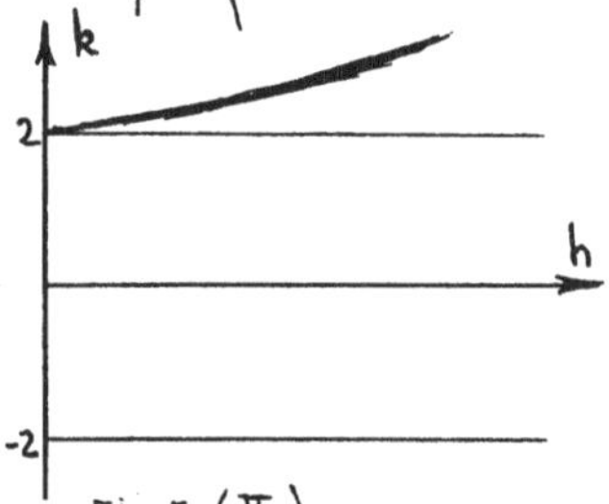

Fig. 5 (π_4)

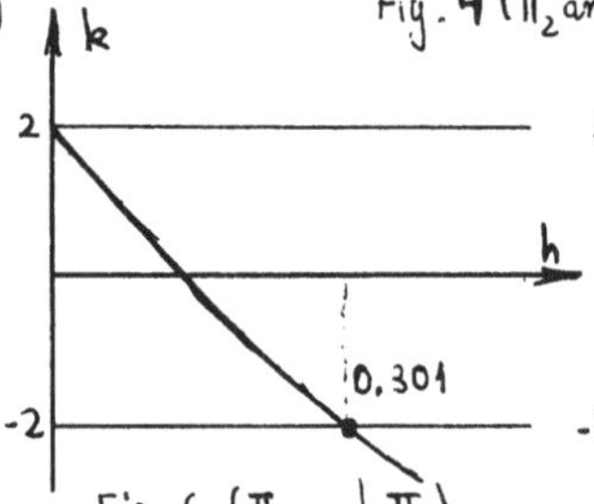

Fig. 6 (π_5 and π_6)

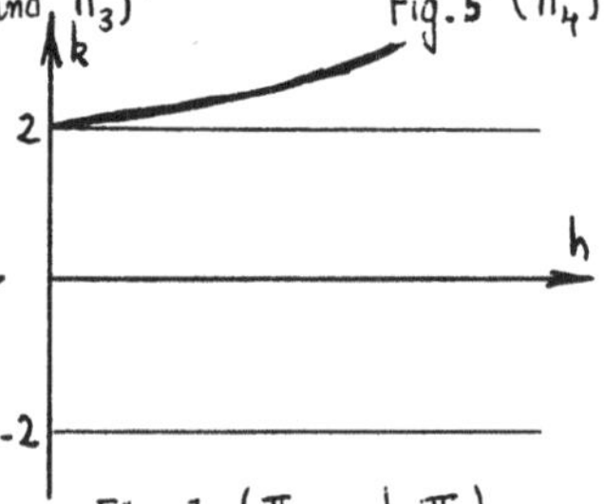

Fig. 7 (π_7 and π_8)

REFERENCES

1. M. Braun, J. Differential Equations 13,300(1973).
2. R. Broucke, Comp. Math. with Appls. 7,451(1981).
3. G. Contopoulos, Astrom. J. 75,96(1970).
4. G. Contopoulos and Zikides, Astron. Astrophys 90,198(1970).
5. C. Conley, J. Differential Equations 5,136(1969).
6. R.C. Churchill, G. Pecelli and D.L. Rod, Lectures Notes in Physics 93, 76(1979).
7. R.C. Churchill, G. Pecelli and D.L. Rod, Archive for Rat. Mech. Anal. 73,313(1980).
8. R.L. Devaney, Trans. Amer. Math. Soc. 218,89(1976).
9. M. Kummer, Commum. Math. Phys. 48,53(1976).
10. J. Llibre and C. Simó, Actas III Congreso de Ecs. Diferenciales y Aplicaciones, Universidad de Santiago,183(1980).
11. J. Moser, Commun. Pure Appl. Math. 11,257(1958).

FEASIBILITY OF CALCULATING DIMENSION AND TOPOLOGICAL ENTROPY

Thea Pignataro
Joseph Henry Laboratories of Physics
Princeton University
Princeton, NJ 08544 U.S.A.

1. INTRODUCTION

Presented here is (i) a summary of some previous work[1] on the application, to several numerical models displaying chaos, of an algorithm suggested by Takens[2] to calculate the capacity of attractors from the time series of a single observable, (ii) the description of some new preliminary results on using the algorithm to calculate topological entropy.

2. THE ALGORITHM

Let the measured observable be expressed as y, a smooth function from the state space of a smooth dynamical system (with flow ϕ) to the real numbers. Assume an attractor exists for the system and is contained in an invariant manifold, M, which is also attractive. Ways[2,3] for reconstructing M from the experimental time series $\{a_i = y(\phi_{i\alpha}(x))\}_{i=0}^{\infty}$ have been suggested, and Takens has proven that if ϕ, y, x and α possess certain generic properties, then the set of n+1-dimensional vectors

$$\{(a_i, a_{i+1}, \ldots, a_{i+n})\}_{i=0}^{\infty} \tag{2.1}$$

form an embedding of M into $\mathbb{R}^{n+1}$, as long as $n \geq 2 \times \dim M$. And, there is a one-to-one correspondence between the positive limit set $L^+(x)$ and the limit points of the sequence defined by (2.1) (i.e. between the asymptotic behavior of the real and reconstructed systems). For an experimental system, one chooses x on the attractor (i.e. all transients are gone), and the set $\{a_i\}$ is finite, so (2.1) gives an approximation of M. Takens then gives a specific algorithm for choosing a subset $J_{n,\varepsilon}$ of the vectors in (2.1) which differ by at least an amount ε. The capacity then equals

$$D(L^+(x)) = \lim_{n\to\infty} \liminf_{\varepsilon\to 0} (-\log C_{n,\varepsilon}/\log \varepsilon) \tag{2.2}$$

and the topological entropy of $\phi_\alpha|L^+(x)$ is

$$h(L^+(x)) = \lim_{\varepsilon\to 0} \limsup_{n\to\infty} (\log C_{n,\varepsilon}/n) \quad (2.3)$$

where $C_{n,\varepsilon}$ = cardinality of $J_{n,\varepsilon}$ and one has the inequalities Hausdorff dimension ≤ capacity, measure theoretic entropy ≤ h.

The practicality of this method for analyzing experimental data can be tested by applying it to well-known models to see if and when these limits exist numerically. Plotting log $C_{n,\varepsilon}$ vs. -log ε or n gives the desired quantities (as the asymptotic slopes) more readily than formula 2.2 or 2.3.

3. NUMERICAL TESTING OF THE ALGORITHM

a. Previous results on dimension

Our studies[1] of the 2/3 Cantor set, the map $x \to 1-2x^2$, and the Hénon[4], Lorenz[3] and Curry[3] models showed that the capacity could be easily obtained for the sets lying in a 1 or 2-dimensional phase space (the first three examples), but, for the latter two models, an impractical number of data points would be necessary. (We saw no convergence, but obtained lower bounds on the capacity of 2.06 and 2.4, respectively, by using Lyapunov exponents.) In general, we found n ≥ dim M to suffice.

b. Current results on dimension and entropy

The question arises whether topological entropy is even more complicated (in the sense of speed of convergence) to calculate than dimension. First, the map $x \to 1-2x^2$ on [-1,1], with initial condition $x_0 = 0.2$, was studied. It has known dimension = 1 and topological entropy h = log 2. For 50,000 data points, n = 2 and $10^{-1} \leq \varepsilon \leq 10^{-3}$ the capacity was 0.99 ± 0.02 (verifying the previous results). For the same data set, $2 \leq n \leq 4$, and $\varepsilon = 10^{-1}$ and 10^{-2}, exp(h) = 1.9 ± 0.2.

For the x variable of the Hénon map[4] (with parameter values a = 1.4, b = 0.3; initial conditions $x_0 = 0.63135448$ and $y_0 = 0.18940634$; and discarding the first 10,000 points as transients) the capacity for 100,000 points, n = 3,4,and 5, $10^{-1} \leq \varepsilon \leq 10^{-2}$ was 1.25 ± 0.02 (also in good agreement with previous results). The entropy for $\varepsilon = 10^{-1.5}$ and 10^{-2} was found to be 0.44 ± 0.02. For the same parameter values, Curry obtains[5] approximately 0.4 for the *measure theoretic* entropy.

Since, in these preliminary tests, the entropy calculations were performed only in a range where convergence in the capacity was first seen, the stated values of the entropy give only an estimate of the ones which would be obtained if a more thorough search for its convergence were made. They do, however, already indicate that topological entropy seems to be as attainable as capacity, with approximately the same number and accuracy of data points necessary for calculation. (ε must be greater than the errors in measurement, therefore dictating the accuracy needed in the experiment.)

4. CONCLUSIONS

It must be emphasized that the purpose of this investigation was not to calculate these quantities in the most efficient or clever way (i.e. using knowledge of the maps or their symmetries), but rather to apply the algorithm "blindly" (although numerically efficiently) as one would in the case of real experimental data, when little is known of the underlying dynamics. It seems that systems of $\lesssim 2$ dimensions can be analyzed by these methods. Higher dimensional systems would require the use of a large (on the order of a million) number of data points, the possible use of a Poincaré section to reduce the dimension by one, and extreme efficiency in the computational methods, in order to have any hope of being studied by the algorithm.

I would like to thank Henry Greenside for first introducing me to the work of Takens, and I would like to acknowledge his collaboration, as well as that of J. Swift and A. Wolf in performing the work described in Section 3a.

REFERENCES

1. Greenside, Wolf, Swift and Pignataro, Phys. Rev. A 25, 3453 (1982).
2. F. Takens in Dynamical Systems and Turbulence, Warwick 1980, Springer-Verlag (1981).
3. See references cited in 1.
4. M. Hénon, Comm. Math. Phys. 50, 69 (1976).
5. J. Curry, preprint IHES/P/81/18.

DIFFUSIONS GENERATED FROM DYNAMICAL SYSTEMS

Michael Williams

Department of Mathematics
Virginia Polytechnic Institute and State University
Blacksburg, Virginia 24061 (USA)

1. INTRODUCTION

In recent years much attention has been focused on the problem of determining the statistical characteristics of dynamical systems subject to small random perturbations.[3,9,11,12,14,16,17,19,20] To show the nature of the type of questions of interest here, a simple demonstrative example is first discussed.

Consider a continuously stirred tank reactor for which specification of the temperature of the tank and the concentration of one of the chemical species determines the state. Suppose, as is typical in the simplest models, that there are three critical points for this system of two ordinary differential equations and that at least two of them, say point one and point two, are stable. Because of imperfections in the tank system, measurement errors, random events, etc., the actual performance of the reactor is only approximately described by the deterministic ODE's and could more accurately be modeled by a stochastic system. Suppose now that it is desired that the reactor should operate at the steady state represented by point one; the "random" affects will cause the path of the state to jump about and, perhaps after possibly a long time, these fluctuations might be sufficient to cause the state to leave the domain of attraction of point one, resulting in the reactor system being swept into point two. This may be a very undesirable circumstance. It is of interest to be able to compute the various statistical properties of this stochastic system relative to the exit problem. This problem has been considered in many other contexts, particularly in biological models (genetic extinction, population extinction, etc.).[11,12,16]

In this work the probability distribution of the random variable defined to be the length of time the process remains within some fixed domain before leaving is determined in the asymptotic limit as the strength of the noise in the system tends to zero. The method used here is an interplay between probabilistic tools and the techniques of differential equations and singular perturbations.

2. FORMULATION AND DISCUSSION OF THE PROBLEM

Let $x(t) \in D \subset R^n$, D bounded, be a state variable for some system evolving according to the deterministic equation

$$\dot{x} = b(x) \tag{2.1}$$

where b is a vector function from R^n to R^n and $\cdot$ stands for differentiation with respect to time. There are many ways in which noise or random perturbations might enter into the system; the form of the randomness treated here is that of a small, non-degenerate additive white noise which modifies (2.1) to produce

$$dx^{\varepsilon} = b(x^{\varepsilon})\ dt + \varepsilon^{\frac{1}{2}}\sigma(x^{\varepsilon})\ dw \tag{2.2}$$

where $\varepsilon << 1$, $\sigma(x)$ is an $n \times n$ uniformly positive-definite matrix function and w is a standard n-dimensional Brownian motion. As usual, owing to the almost sure non-differentiability of the Brownian paths, (2.2) is written in differential form, rather than in terms of total derivatives, and must be interpreted as a stochastic integral equation. It is of no consequence to the later analysis which of the many stochastic integral definitions is adopted; for its simplicity and widespread use, the Ito stochastic calculus is used throughout to define (2.2).

As is well known, the process, $x^{\varepsilon}(t)$, defined by (2.2) is a diffusion, and therefore, with probability one the process will exit D in finite time. Thus, natural questions are: What is the conditional distribution of the exit point on ∂D? What are the statistics of τ_x^{ε}, the length of time until exit defined by

$$\tau_x^{\varepsilon} = \inf\{t : x^{\varepsilon}(t) \notin D,\ x^{\varepsilon}(0) = x\}\ ? \tag{2.3}$$

For a general survey of results on these and related questions the reader should see the review article of Schuss.[16]

We restrict our attention to the distribution of τ_x^{ε} problem. More specifically, we wish to consider processes for which the underlying deterministic process is inward at the boundary (i.e. $n \cdot b < 0$, n the outward normal on ∂D) and for which b has a simple unique critical point in D. Two basic results on the distribution of τ_x^{ε} in this case are available in the literature. First, rigorous exponential decay rates for $P[\tau_x^{\varepsilon} < T]$ for fixed T as $\varepsilon \to 0$ were obtained by Ventcel and Freidlin[17], Friedman[3,4] and Fleming[1,2]. Secondly, when the generator of the process,

L^ε, is self-adjoint with principal eigenvalue $-\lambda^\varepsilon$, the formal result, $E(\tau_x^\varepsilon) \sim 1/\lambda^\varepsilon$ $(\varepsilon \to 0)$ was observed by Ludwig[11] and Matkowsky and Schuss[14]. Also, at a rigorous level the relation between the principal eigenvalue and the exit time was observed by Friedman[4] using techniques based on the Fredholm alternative and the maximum principle. It was shown that

$$\lambda^\varepsilon = \sup\,\{\lambda > 0;\ \sup_{x\varepsilon D} E(e^{\lambda\tau_x^\varepsilon}) < \infty\}\ . \tag{2.4}$$

The result is recovered in our formulation.

One of two approaches is usually taken in attacking such problems (though Fleming[2] uses a trick to get a result via stochastic control theory). One approach is to study the process directly and derive pathwise estimates generally based on the Cameron-Martin Girsanov formula[3,17]. Alternatively, certain probabilistic quantities (e.g. $E(\tau_x^\varepsilon)$) can be shown by use of Ito's formula to satisfy differential equations involving the generator of the process and then analytic methods of studying differential equations may be employed to obtain probabilisitic information.[11,14,16,20]

The second approach is used in this paper.

One can easily show[4] that $u^\varepsilon(x,t) = P[\tau_x^\varepsilon > t]$ satisfies

$$\begin{gathered} L^\varepsilon u^\varepsilon = u_t^\varepsilon \\ u^\varepsilon(x,0) = 1\ ,\ x\,\varepsilon\, D \qquad u^\varepsilon(x,t) = 0\ ,\ x\,\varepsilon\,\partial D\ , \end{gathered} \tag{2.5}$$

so that the distribution of the exit time could be obtained as the solution of this problem. If L^ε can be made self-adjoint by a scaling (the so-called potential case), a separation of variables leads to an eigenfunction expansion of u^ε given by

$$u^\varepsilon = \sum_{n=0}^{\infty} c_n^\varepsilon\, v_n^\varepsilon(x) e^{-\lambda_n^\varepsilon t} \tag{2.6}$$

where λ_n^ε, v_n^ε are the eigenvalues and scaled eigenfunctions respectively of $-L^\varepsilon$. The problem is then to determine the behavior of this series as $\varepsilon \to 0$; this is a non-trivial problem in singular perturbation theory. Formally, based on the asymptotic character of the eigenfunctions, one expects that the first term will dominate the series as $\varepsilon \to 0$ suggesting the result we seek. In fact, Ludwig[11] using similar argument conjectured the exponential character of the exit time that we show. This heuristic argument provided the motivation for this work.

To facilitate the exposition we restrict attention to the one-dimensional case and extend the results later. The generator corresponding to (2.2) in this case is given by

$$L^\varepsilon \equiv \varepsilon \frac{a(x)}{2} \frac{d^2}{dx^2} + b(x) \frac{d}{dx} \tag{2.7}$$

where $0 < a(x) = \sigma^2(x)$ and a and b are scalar functions on an interval $[-A,B]$, $A, B > 0$. It is assumed that $b(x)$ has a simple unique zero at $x = 0$ and $x \cdot b(x) < 0$ except at $x = 0$. The prototype example satisfying these requirements is the Ornstein-Uhlenbeck process with generator $\frac{\varepsilon}{2} \frac{d^2}{dx^2} - x \frac{d}{dx}$. An explicit calculation of this example verifies that $E(\tau_x^\varepsilon) \sim 1/\lambda^\varepsilon$, and moreover, that $\lambda^\varepsilon \tau_x^\varepsilon$ converges in distribution to a unit exponential distribution as $\varepsilon \to 0$.

In order to study this limit in general, a direct application of Ito's formula (with $f(x,t) = u_s(x)e^{-st}$) shows that

$$u_s^\varepsilon(x) = E_x(e^{-s\lambda^\varepsilon \tau_x^\varepsilon}) \tag{2.8}$$

satisfies the boundary value problem,

$$L^\varepsilon u_s^\varepsilon = s\lambda^\varepsilon u_s^\varepsilon \qquad u_s^\varepsilon(-A) = 1 = u_s^\varepsilon(B) \ . \tag{2.9}$$

u_s^ε is the generating function of the scaled exit time and s is the transform variable.

The attack is straight-forward; if we show that $u_s^\varepsilon(x)$ converges pointwise in s and x to $1/(1 + s)$ for $x \in (-A,B)$, $s > 0$ then by (2.8) $\lambda^\varepsilon \tau_x^\varepsilon \Rightarrow \tau$ where τ has a unit exponential distribution since there is corresponding convergence of the generating functions. This result is stated in the following theorem.

THEOREM. Suppose a, b are Lipschitz continuous in $[-A,B]$ with $a > 0$, $\gamma \equiv -2b/(x\,a) > 0$ and γ continuous on the interval. If L^ε is defined as in (2.7) and with λ^ε and τ_x^ε defined as in the above discussion, then $\lambda^\varepsilon \tau_x^\varepsilon$ converges in distribution to a unit exponential.

3. THE ANALYSIS

As the preceding discussion indicates the theorem relies on explicit computations of the problem (2.9) and the principal eigenvalue problem for $-L^\varepsilon$ given by

$$-L^\varepsilon v^\varepsilon = \lambda^\varepsilon v^\varepsilon \qquad v^\varepsilon(-A) = 0 = v^\varepsilon(B) \ . \tag{3.1}$$

The behavior of λ^ε as $\varepsilon \to 0$ is a well studied problem and the reader is referred to [6,8,18] for the results quoted here. Depending on the sign of $I \equiv \int_{-A}^{B} t\gamma(t)dt$, the leading order behavior of λ^ε is given by

$$\lambda^\varepsilon \sim \frac{|b(\xi)|}{a(\xi)} \sqrt{\frac{a(0)\,|b'(0)|}{\pi}}\; \varepsilon^{-\frac{1}{2}}\, e^{(2/\varepsilon)\int_0^\xi \frac{b(t)dt}{a(t)}} \tag{3.2}$$

where $\xi = A$ or B as $I > 0$ or $I < 0$. If $I = 0$, λ^ε is asymptotic to the sum of $I > 0$ and $I < 0$ asymptotic behaviors.

It turns out that both (2.9) and (3.1) as singularly perturbed boundary value problems exhibit a phenomenon called Ackerberg-O'Malley resonance[13]. There have been several nice treatments of this problem in the literature,[6,10,13,15] but they fall short of describing the problems of interest here. This leads to the treatment of the resonance problem by the author[18] by variational methods used here [see also[5]].

To establish the theorem, it must be shown that the solution of (2.9) is asymptotic to $1/(1 + s)$ as $\varepsilon \to 0$. Rewriting (2.9) in the current setting gives

$$\varepsilon u_s^{\varepsilon\,''} - x\gamma(x)u_s^{\varepsilon\,'} - \frac{2s\lambda^\varepsilon}{a(x)}\, u_s^\varepsilon = 0 \qquad u_s^\varepsilon(-A) = 1 = u_s^\varepsilon(B) \tag{3.3}$$

where λ^ε satisfies (3.2) depending on the value of $I \equiv \int_{-A}^{B} \frac{b(t)}{a(t)}dt$.

It is well known that u_s^ε has a uniform asymptotic expansion of the form

$$v_s^\varepsilon = C_0(s) + w^\varepsilon \tag{3.4}$$

where w^ε is the boundary layer expansion which satisfies

$$w^\varepsilon \sim (1 - C_0)e^{-A\gamma(A)\frac{A+x}{\varepsilon}} + (1 - C_0)e^{-B\gamma(B)\frac{B-x}{\varepsilon}} \tag{3.5}$$

$C_0(s)$ is to be determined as the stationary point of the functional,[18]

$$J[y] = \tfrac{1}{2}\int_{-A}^{B}\left[\varepsilon y'^2 + \frac{2s\lambda^\varepsilon}{a(x)}\, y^2\right] e^{-\frac{1}{\varepsilon}\int_0^x t\gamma(t)dt}\, dx \tag{3.6}$$

which upon introduction of (3.4) gives

$$J(C_0) \sim \tfrac{1}{2}\Big\{B\gamma(B)(1 - C_0)^2\, e^{-\frac{1}{\varepsilon}\int_0^B t\gamma(t)dt} \tag{3.7}$$
$$+ A\gamma(A)(1 - C_0)^2\, e^{-\frac{1}{\varepsilon}\int_0^A t\gamma(t)dt} + 2sC_0^2\lambda^\varepsilon\, \frac{\varepsilon^{\frac{1}{2}}}{a(0)}\sqrt{\frac{2\pi}{\gamma(0)}}\Big\}.$$

Note that the above procedure could also be used to determine the asymptotics of the principal eigenvalue, λ^{ε}. Setting s = -1 in (3.7), the leading order of λ^{ε} will be the value required to make $J(C_0)$ singular, i.e., drop from quadratic in C_0 to linear in C_0.

If I < 0, (3.7) reduces to

$$J(C_0) \sim \tfrac{1}{2}B\gamma(B)\, e^{-\frac{1}{\varepsilon}\int_0^B t\gamma(t)dt} ((1 - C_0)^2 + sC_0^2) \qquad (3.8)$$

which is easily seen to have its stationary point at $C_0(s) \sim 1/(1 + s)$ which establishes the required behavior of the solution of (3.3) in this case. The other two cases follow easily in the same fashion.

It is clear that if additional terms in the expansions of the boundary layers and of λ^{ε} were computed, higher order terms in the expansion of $C_0(s)$ could be computed. These additional terms give higher order corrections for the distribution of $\lambda^{\varepsilon}\tau_x^{\varepsilon}$ (or τ_x^{ε}).

In the case where the spatial dimension is greater than one, the preceding method breaks down in general due to the lack of an appropriate variational principle. However, in the potential case the basic problem may be rescaled to be self-adjoint and a variational principle is available. In this event the analysis above may be carried through with few changes to produce the same result.[19]

An indication that the stated result is true more generally (i.e. in the non-potential case) may be seen by making use of the techniques initiated by Matkowsky and Schuss[14] and studied rigorously by Kamin.[7,8] Since these techniques are so well-documented in the cited papers, details will not be presented here. The exponential limit result may be obtained by successively applying the second Green's identity to a solution of an associated adjoint problem and the principal eigenfunction of L^{ε} and then again to the adjoint solution and the solution of the n-dimensional version of (2.9) and finally then eliminating the integrals involving the adjoint solution between them by making use of the known asymptotic forms of the respective functions. This result, easily obtained, is as before.

The difficulty with this procedure is that the validity of the asymptotics of the associated adjoint solution is not settled. Kamin[7,8] has shown that given a hypothesis ("H*") on the solutions of a certain nonlinear first order partial differential equation, then all the formulas are correct. It appears, however, that H* is not generally verifiable in the non-potential case, thus leaving open the questions of whether the result holds in this case.

4. ACKNOWLEDGEMENT

The author wishes to thank the Department of Energy (grant number DE-AS05-80ER10711) and the Laboratory for Transport Theory and Mathematical Physics for their support during this project.

5. REFERENCES

1. W. H. Fleming, Rocky Mountain J. Math., 4, 407 (1974).
2. W. H. Fleming, Appl. Math. Optim., 4, 329 (1978).
3. A. Friedman, Indiana Univ. Math. J., 24, 533 (1974).
4. A. Friedman, Stochastic Differential Equations and Applications, Volume 2, Academic Press (1976).
5. J. Grasman and B. J. Matkowsky, SIAM J. Appl. Math., 32, 588 (1977).
6. P. P. N. deGroen, SIAM J. Math. Anal., 11, 1 (1980).
7. S. Kamin, Indiana Univ. Math. J., 27, 935 (1978).
8. S. Kamin, Comm. in PDE, 4, 573 (1979).
9. R. Z. Khasminskii, Isv. Akad. Nauk SSSR Ser. Mat., 27, 1281 (1963).
10. N. Kopell, SIAM J. Appl. Math., 37, 436 (1979).
11. D. Ludwig, SIAM Rev., 17, 605 (1975).
12. M. Mangel, SIAM J. Appl. Math., 36, 544 (1979).
13. B. J. Matkowsky, SIAM Rev., 17, 82 (1975).
14. B. J. Matkowsky and Z. Schuss, SIAM J. Appl. Math., 33, 365 (1977).
15. F. W. J. Olver, SIAM J. Math. Anal., 9, 328 (1978).
16. Z. Schuss, SIAM Rev., 22, 119 (1980).
17. A. D. Ventcel and M. Freidlin, Uspehi Mat. Nauk., 25, 3 (1970). Russ. Math. Surveys, 25, 1 (1970).
18. M. Williams, SIAM J. Appl. Math., 41, 288 (1981).
19. M. Williams, SIAM J. Appl. Math., 42, 149 (1982).
20. R. G. Williams, Thesis, California Institute of Technology, (1977).

REPORT ON THE DRIVEN JOSEPHSON EQUATION

Jan A. Sanders

Subfaculteit Wiskunde en Informatica
Vrije Universiteit
De Boelelaan 1081
1081 HV Amsterdam (The Netherlands)

In this paper I would like to report on some current research on the Josephson equation. A review of the asymptotic theory of the (driven) Josephson equation

$$\beta\ddot{\phi} + (1 + \gamma\cos\phi)\dot{\phi} + \sin\phi = \alpha + k\omega\sin\omega t$$

will appear in Sanders (1982 A) with a list of references, while a more technical question has been treated in Sanders (1982 B).

In collaboration with Prof. M. van Veldhuizen, an attempt has been made to write a program to construct an asymptotic approximation to the solution of this equation for small β and large ω.

When $|\gamma| > 1$, this is quite complicated, due to the jumps in the solution, and some more subtle phenomena, dips and slices, the approximation of which put some tough requirements on the whole procedure. The first difficulty one encounters, is that the reduced equation ($\beta = 0$) cannot be integrated analytically. So one has to use numerical calculations from the very start.
The calculation of the jumping time proceeds along classical lines, using the first zero of the Airy function, except when the solution is nearly extremal (tangent to the boundary layer $1 + \gamma\cos\phi = 0$). In that case the analysis needs the computation of the modulus ν and the first zero of D_ν, the parabolic cylinder function of order ν.
This poses no problems, unless ν is very small or very large; in both cases one can use asymptotic results to improve the calculations.

A viable alternative is to integrate the Riccati equation numerically in the boundary layer.

When $\nu > 0$, the solution jumps, and when $\nu < 0$ it dips, i.e. it leaves the boundary layer along the slow stable manifold.
But for small ν, the solution tends to follow the slow unstable manifold, until it falls off. This leads to a 'slice' (Littlewood's terminology).
To approximate slices uniformly is more than one can hope for: relative differences of order 10^{-6} give rise to variations of order 1 in the solution. This is illustrated in Fig. 1.

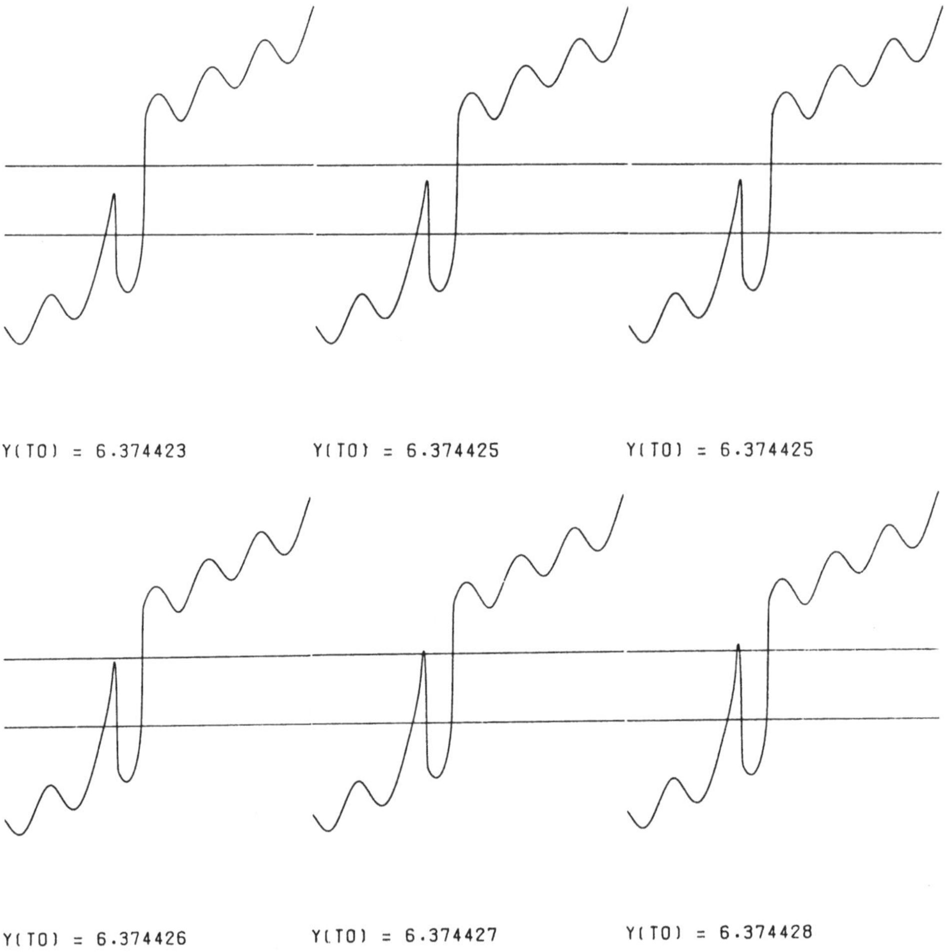

Fig. 1: *Some numerical results illustrating a 'slice', for six different initial conditions. The horizontal lines indicate the boundary layers where* $1 + \gamma \cos \phi = 0$.
Values of the parameters:

$\beta = 0.001$ $\qquad \dot{y}(t0) = 10$
$\gamma = 1.5$ $\qquad t0 = 0.505$
$\omega = 45.68128$ $\qquad k = 1$
$\qquad \alpha = 11$

Not to lose too much accuracy outside the boundary layer, we use a time dependent time-shift in the solution of the reduced equation, obtained by formal calculations, which, by the implicit nature of its definition avoids the blow-up of higher order approximations near the boundary layer (cf. Sanders 1982 B).

REFERENCES

Sanders, J.A., 1982 A, to appear in: Asymptotic Analysis II, ed. F. Verhulst, Springer Lecture Notes in Mathematics.

Sanders, J.A., 1982 B, to appear in: Proc. Equadiff '82, Springer Lecture Notes in Mathematics.

Springer Series in

Synergetics

Series Editor: **H. Haken**

Volume 1: **H. Haken**
Synergetics
An Introduction. Nonequilibrium Phase Transitions and Self-Organization in Physics, Chemistry and Biology. 2nd enlarged edition. ISBN 3-540-08866-0

Volume 2:
Synergetics
A Workshop. Proceedings of the International Workshop on Synergetics at Schloß Elmau, Bavaria, May 2-7, 1977
Editor: **H. Haken**
ISBN 3-540-08483-5

Volume 3:
Synergetics
Far from Equilibrium. Proceedings of the Conference Far From Equilibrium: Instabilities and Structures, Bordeaux, France, September 27-29, 1978
Editors: **A. Pacault, C. Vidal**
ISBN 3-540-09304-4

Volume 4:
Structural Stability in Physics
Proceedings of Two International Symposia on Applications of Catastrophe Theory and Topological Concepts in Physics, Tübingen, Federal Republic of Germany, May 2-6 and December 11-14, 1978
Editors: **W. Güttinger, H. Eikemeier**
ISBN 3-540-09463-6

Volume 5:
Pattern Formation by Dynamic Systems and Pattern Recognition
Proceedings of the International Symposium on Synergetics at Schloß Elmau, Bavaria, April 30-May 5, 1979
Editor: **H. Haken**
ISBN 3-540-09770-8

Volume 6:
Dynamics of Synergetic Systems
Proceedings of the International Symposium on Synergetics, Bielefeld, Federal Republic of Germany, September 24-29, 1979
Editor: **H. Haken**
ISBN 3-540-09918-2

Volume 7: **L. A. Blumenfeld**
Problems of Biological Physics
ISBN 3-540-10401-1

Volume 8:
Stochastic Nonlinear Systems
in Physics, Chemistry and Biology. Proceedings of the Workshop, Bielefeld, Federal Republic of Germany, October 5-11, 1980
Editors: **L. Arnold, R. Lefever**
ISBN 3-540-10713-4

Volume 9:
Numerical Methods in the Study of Critical Phenomena
Proceedings of a Colloquium, Carry-le-Rouet, France June 2-4, 1980
Editors: **J. Della Dora, J. Demongeot, B. Lacolle**
ISBN 3-540-11009-7

Volume 10: **Y. L. Klimontovich**
The Kinetic Theory of Electromagnetic Processes
ISBN 3-540-11458-0

Volume 11:
Chaos and Order in Nature
Proceedings of the International Symposium on Synergetics at Schloß Elmau, Bavaria, April 27 - May 2, 1981
Editor: **H. Haken**
ISBN 3-540-11101-8

Volume 12:
Nonlinear Phenomena in Chemical Dynamics
Proceedings of an International Conference, Bordeaux, France, September 7-11, 1981
Editors: **C. Vidal, A. Pacault**
ISBN 3-540-11294-4

Volume 13: **C. W. Gardiner**
Handbook of Stochastic Methods
for Physics, Chemistry and the Natural Sciences
ISBN 3-540-11357-6

Volume 14: **W. Weidlich, G. Haag**
Concepts and Models of a Quantitative Sociology
The Dynamics of Interacting Populations
ISBN 3-540-11358-4

Volume 15: **W. Horsthemke, R. Lefever**
Nonequilibrium Transitions Induced by External Noise
ISBN 3-540-11359-2

Volume 16: **L. A. Blumenfeld**
Physics of Bioenergetic Processes
ISBN 3-540-11417-3

Volume 17:
Evolution of Order and Chaos
in Physics, Chemistry, and Biology
Proceedings of the International Symposium on Synergetics at Schloß Elmau, Bavaria, April 26 - May 1, 1982
Editor: **H. Haken**
ISBN 3-540-11904-3

Springer-Verlag
Berlin
Heidelberg
New York

Lecture Notes in Physics

Vol. 144: Topics in Nuclear Physics I. A Comprehensive Review of Recent Developments. Edited by T.T.S. Kuo and S.S.M. Wong. XX, 567 pages. 1981.

Vol. 145: Topics in Nuclear Physics II. A Comprehensive Review of Recent Developments. Proceedings 1980/81. Edited by T. T. S. Kuo and S. S. M. Wong. VIII, 571-1.082 pages. 1981.

Vol. 146: B. J. West, On the Simpler Aspects of Nonlinear Fluctuating. Deep Gravity Waves. VI, 341 pages. 1981.

Vol. 147: J. Messer, Temperature Dependent Thomas-Fermi Theory. IX, 131 pages. 1981.

Vol. 148: Advances in Fluid Mechanics. Proceedings, 1980. Edited by E. Krause. VII, 361 pages. 1981.

Vol. 149: Disordered Systems and Localization. Proceedings, 1981. Edited by C. Castellani, C. Castro, and L. Peliti. XII, 308 pages. 1981.

Vol. 150: N. Straumann, Allgemeine Relativitätstheorie und relativistische Astrophysik. VII, 418 Seiten. 1981.

Vol. 151: Integrable Quantum Field Theory. Proceedings, 1981. Edited by J. Hietarinta and C. Montonen. V, 251 pages. 1982.

Vol. 152: Physics of Narrow Gap Semiconductors. Proceedings, 1981. Edited by E. Gornik, H. Heinrich and L. Palmetshofer. XIII, 485 pages. 1982.

Vol. 153: Mathematical Problems in Theoretical Physics. Proceedings, 1981. Edited by R. Schrader, R. Seiler, and D.A. Uhlenbrock. XII, 429 pages. 1982.

Vol. 154: Macroscopic Properties of Disordered Media. Proceedings, 1981. Edited by R. Burridge, S. Childress, and G. Papanicolaou. VII, 307 pages. 1982.

Vol. 155: Quantum Optics. Proceedings, 1981. Edited by C.A. Engelbrecht. VIII, 329 pages. 1982.

Vol. 156: Resonances in Heavy Ion Reactions. Proceedings, 1981. Edited by K.A. Eberhard. XII, 448 pages. 1982.

Vol. 157: P. Niyogi, Integral Equation Method in Transonic Flow. XI, 189 pages. 1982.

Vol. 158: Dynamics of Nuclear Fission and Related Collective Phenomena. Proceedings, 1981. Edited by P. David, T. Mayer-Kuckuk, and A. van der Woude. X, 462 pages. 1982.

Vol. 159: E. Seiler, Gauge Theories as a Problem of Constructive Quantum Field Theory and Statistical Mechanics. V, 192 pages. 1982.

Vol. 160: Unified Theories of Elementary Particles. Critical Assessment and Prospects. Proceedings, 1981. Edited by P. Breitenlohner and H.P. Dürr. VI, 217 pages. 1982.

Vol. 161: Interacting Bosons in Nuclei. Proceedings, 1981. Edited by J.S. Dehesa, J.M.G. Gomez, and J. Ros. V, 209 pages. 1982.

Vol. 162: Relativistic Action at a Distance: Classical and Quantum Aspects. Proceedings, 1981. Edited by J. Llosa. X, 263 pages. 1982.

Vol. 163: J. S. Darrozes, C. Francois, Mécanique des Fluides Incompressibles. XIX, 459 pages. 1982.

Vol. 164: Stability of Thermodynamic Systems. Proceedings, 1981. Edited by J. Casas-Vázquez and G. Lebon. VII, 321 pages. 1982.

Vol. 165: N. Mukunda, H. van Dam, L.C. Biedenharn, Relativistic Models of Extended Hadrons Obeying a Mass-Spin Trajectory Constraint. Edited by A. Böhm and J.D. Dollard. VI, 163 pages. 1982.

Vol. 166: Computer Simulation of Solids. Edited by C.R.A. Catlow and W.C. Mackrodt. XII, 320 pages. 1982.

Vol. 167: G. Fieck, Symmetry of Polycentric Systems. VI, 137 pages, 1982.

Vol. 168: Heavy-Ion Collisions. Proceedings, 1982. Edited by G. Madurga and M. Lozano. VI, 429 pages. 1982.

Vol. 169: K. Sundermeyer, Constrained Dynamics. IV, 318 pages. 1982.

Vol. 170: Eighth International Conference on Numerical Methods in Fluid Dynamics. Proceedings, 1982. Edited by E. Krause. X, 569 pages. 1982.

Vol. 171: Time-Dependent Hartree-Fock and Beyond. Proceedings, 1982. Edited by K. Goeke and P.-G. Reinhard. VIII, 426 pages. 1982.

Vol. 172: Ionic Liquids, Molten Salts and Polyelectrolytes. Proceedings, 1982. Edited by K.-H. Bennemann, F. Brouers, and D. Quitmann. VII, 253 pages. 1982.

Vol. 173: Stochastic Processes in Quantum Theory and Statistical Physics. Proceedings, 1981. Edited by S. Albeverio, Ph. Combe, and M. Sirugue-Collin. VIII, 337 pages. 1982.

Vol. 174: A. Kadić, D.G.B. Edelen, A Gauge Theory of Dislocations and Disclinations. VII, 290 pages. 1983.

Vol. 175: Defect Complexes in Semiconductor Structures. Proceedings, 1982. Edited by J. Giber, F. Beleznay, J. C. Szép, and J. László. VI, 308 pages. 1983.

Vol. 176: Gauge Theory and Gravitation. Proceedings, 1982. Edited by K. Kikkawa, N. Nakanishi, and H. Nariai. X, 316 pages. 1983.

Vol. 177: Application of High Magnetic Fields in Semiconductor Physics. Proceedings, 1982. Edited by G. Landwehr. XII, 552 pages. 1983.

Vol. 178: Detectors in Heavy-Ion Reactions. Proceedings, 1982. Edited by W. von Oertzen. VIII, 258 pages. 1983.

Vol. 179: Dynamical Systems and Chaos. Proceedings, 1982. Edited by L. Garrido. XIV, 298 pages. 1983.

Selected Issues from

Lecture Notes in Mathematics

Vol. 809: I. Gumowski and Ch. Mira, Recurrences and Discrete Dynamic Systems. VI, 272 pages. 1980.

Vol. 810: Geometrical Approaches to Differential Equations. Proceedings 1979. Edited by R. Martini. VII, 339 pages. 1980.

Vol. 816: L. Stoica, Local Operators and Markov Processes. VIII, 104 pages. 1980.

Vol. 819: Global Theory of Dynamical Systems. Proceedings, 1979. Edited by Z. Nitecki and C. Robinson. IX, 499 pages. 1980.

Vol. 821: Statistique non Paramétrique Asymptotique. Proceedings, 1979. Edited by J.-P. Raoult. VII, 175 pages. 1980.

Vol. 826: Ph. G. Ciarlet et P. Rabier, Les Equations de von Kármán. VI, 181 pages. 1980.

Vol. 827: Ordinary and Partial Differential Equations. Proceedings, 1978. Edited by W. N. Everitt. XVI, 271 pages. 1980.

Vol. 831: Representation Theory I. Proceedings, 1979. Edited by V. Dlab and P. Gabriel. XIV, 373 pages. 1980.

Vol. 832: Representation Theory II. Proceedings, 1979. Edited by V. Dlab and P. Gabriel. XIV, 673 pages. 1980.

Vol. 836: Differential Geometrical Methods in Mathematical Physics. Proceedings, 1979. Edited by P. L. García, A. Pérez-Rendón, and J. M. Souriau. XII, 538 pages. 1980.

Vol. 837: J. Meixner, F. W. Schäfke and G. Wolf, Mathieu Functions and Spheroidal Functions and their Mathematical Foundations Further Studies. VII, 126 pages. 1980.

Vol. 838: Global Differential Geometry and Global Analysis. Proceedings 1979. Edited by D. Ferus et al. XI, 299 pages. 1981.

Vol. 840: D. Henry, Geometric Theory of Semilinear Parabolic Equations. IV, 348 pages. 1981.

Vol. 841: A. Haraux, Nonlinear Evolution Equations- Global Behaviour of Solutions. XII, 313 pages. 1981.

Vol. 842: Séminaire Bourbaki vol. 1979/80. Exposés 543–560. IV, 317 pages. 1981.

Vol. 843: Functional Analysis, Holomorphy, and Approximation Theory. Proceedings. Edited by S. Machado. VI, 636 pages. 1981.

Vol. 845: A. Tannenbaum, Invariance and System Theory: Algebraic and Geometric Aspects. X, 161 pages. 1981.

Vol. 849: P. Major, Multiple Wiener-Itô Integrals. VII, 127 pages. 1981.

Vol. 851: Stochastic Integrals. Proceedings, 1980. Edited by D. Williams. IX, 540 pages. 1981.

Vol. 852: L. Schwartz, Geometry and Probability in Banach Spaces. X, 101 pages. 1981.

Vol. 856: R. Lascar, Propagation des Singularités des Solutions d'Equations Pseudo-Différentielles à Caractéristiques de Multiplicités Variables. VIII, 237 pages. 1981.

Vol. 858: E. A. Coddington, H. S. V. de Snoo: Regular Boundary Value Problems Associated with Pairs of Ordinary Differential Expressions. V, 225 pages. 1981.

Vol. 861: Analytical Methods in Probability Theory. Proceedings 1980. Edited by D. Dugué, E. Lukacs, V. K. Rohatgi. X, 183 pages. 1981.

Vol. 866: J.-M. Bismut, Mécanique Aléatoire. XVI, 563 pages. 1981.

Vol. 878: Numerical Solution of Nonlinear Equations. Proceedings, 1980. Edited by E. L. Allgower, K. Glashoff, and H.-O. Peitgen. XIV, 440 pages. 1981.

Vol. 881: R. Lutz, M. Goze, Nonstandard Analysis. XIV, 261 pages. 1981.

Vol. 888: Padé Approximation and its Applications. Proceedings, 1980. Edited by M. G. de Bruin and H. van Rossum. VI, 383 pages. 1981.

Vol. 898: Dynamical Systems and Turbulence, Warwick, 1980. Proceedings. Edited by D. Rand and L.-S. Young. VI, 390 pages. 1981.

Vol. 901: Séminaire Bourbaki vol. 1980/81 Exposés 561–578. III, 299 pages. 1981.

Vol. 904: K. Donner, Extension of Positive Operators and Korovkin Theorems. XII, 182 pages. 1982.

Vol. 905: Differential Geometric Methods in Mathematical Physics. Proceedings, 1980. Edited by S.J. Andersson, H.-D. Doebner, and. H.R. Petry. VI, 309 pages. 1982.

Vol. 909: Numerical Analysis. Proceedings, 1981. Edited by J.P. Hennart. VII, 247 pages. 1982.

Vol. 912: Numerical Analysis. Proceedings, 1981. Edited by G. A. Watson. XIII, 245 pages. 1982.

Vol. 920: Séminaire de Probabilités XVI, 1980/81. Proceedings. Edité par J. Azéma et M. Yor. V, 622 pages. 1982.

Vol. 921: Séminaire de Probabilités XVI, 1980–81 Supplément: Géométrie Différentielle Stochastique. Proceedings. Edité par J. Azéma et M. Yor. III, 285 pages. 1982.

Vol. 922: B. Dacorogna, Weak Continuity and Weak Lower Semicontinuity of Non-Linear Functionals. V, 120 pages. 1982.

Vol. 923: Functional Analysis in Markov Processes. Proceedings, 1981. Edited by M. Fukushima. V, 307 pages. 1982.

Vol. 926: Geometric Techniques in Gauge Theories. Proceedings, 1981. Edited by R. Martini and E.M.de Jager. IX 219 pages. 1982.

Vol. 927: Y. Z. Flicker, The Trace Formula and Base Change for GL (3). XII, 204 pages. 1982.

Vol. 928: Probability Measures on Groups. Proceedings 1981. Edited by H. Heyer. X, 477 pages. 1982.

Vol. 929: Ecole d'Eté de Probabilités de Saint-Flour X – 1980. Proceedings, 1980. Edited by P.L. Hennequin. X, 313 pages. 1982.

Vol. 930: P. Berthelot, L. Breen, et W. Messing, Théorie de Dieudonné Cristalline II. XI, 261 pages. 1982.

Vol. 931: D.M. Arnold, Finite Rank Torsion Free Abelian Groups and Rings. VII, 191 pages. 1982.

Vol. 932: Analytic Theory of Continued Fractions. Proceedings, 1981. Edited by W.B. Jones, W.J. Thron, and H. Waadeland. VI, 240 pages. 1982.

Vol. 934: M. Sakai, Quadrature Domains. IV, 133 pages. 1982.

Vol. 935: R. Sot, Simple Morphisms in Algebraic Geometry. IV, 146 pages. 1982.

Vol. 936: S.M. Khaleelulla, Counterexamples in Topological Vector Spaces. XXI, 179 pages. 1982.

Vol. 937: E. Combet, Intégrales Exponentielles. VIII, 114 pages. 1982.

Vol. 938: Number Theory. Proceedings, 1981. Edited by K. Alladi. IX, 177 pages. 1982.

Vol. 942: Theory and Applications of Singular Perturbations. Proceedings, 1981. Edited by W. Eckhaus and E.M. de Jager. V, 363 pages. 1982.

Vol. 953: Iterative Solution of Nonlinear Systems of Equations. Proceedings, 1982. Edited by R. Ansorge, Th. Meis, and W. Törnig. VII, 202 pages. 1982.